Indigo Plantations and Science in Colonial India

Prakash Kumar documents the history of agricultural indigo, exploring the effects of global processes on a colonial industry in South Asia. Kumar discusses how the knowledge of indigo culture thrived among peasant traditions on the Indian subcontinent in the early modern period. Caribbean planters and French naturalists then developed and codified this knowledge in widely disseminated texts. European planters who began to settle in Bengal with the establishment of British rule in the third quarter of the eighteenth century drew on this network of information. Through the nineteenth century, indigo culture in Bengal became more modern, science based, and expert driven. When a cheaper and purer synthetic indigo was created in 1897, the planters and the colonial state established laboratories to find ways to cheapen the cost of the agricultural dye and improve its purity. This indigo science crossed paths with the colonial state's effort to develop a science for agricultural development and the effort by the Indian intelligentsia to develop a science for the nation. For two decades, natural indigo survived the competition of the industrial substitute. The indigo industry's optimism faded only at the end of the First World War, when the use of German synthetic indigo for textile dyeing and printing became almost universal.

Prakash Kumar is Assistant Professor of South Asian History at Colorado State University.

Indigo Plantations and Science in Colonial India

PRAKASH KUMAR

Colorado State University

CAMBRIDGE
UNIVERSITY PRESS

CAMBRIDGE
UNIVERSITY PRESS

University Printing House, Cambridge CB2 8BS, United Kingdom

One Liberty Plaza, 20th Floor, New York, NY 10006, USA

477 Williamstown Road, Port Melbourne, VIC 3207, Australia

314-321, 3rd Floor, Plot 3, Splendor Forum, Jasola District Centre, New Delhi - 110025, India

79 Anson Road, #06-04/06, Singapore 079906

Cambridge University Press is part of the University of Cambridge.

It furthers the University's mission by disseminating knowledge in the pursuit of
education, learning and research at the highest international levels of excellence.

www.cambridge.org
Information on this title: www.cambridge.org/9781107023253

© Prakash Kumar 2012

First published 2012

A catalogue record for this publication is available from the British Library

Library of Congress Cataloging in Publication data
Kumar, Prakash, 1967–
Indigo plantations and science in colonial India / Prakash Kumar.
p. cm.
Includes bibliographical references and index.
ISBN 978-1-107-02325-3 (hbk. : alk. paper)
1. Indigo industry – India – Bengal – History. 2. Plantations – India –
Bengal – History. 3. Indigo – India – History. 4. Crop science – India –
History. 5. Agriculture – India – History. 6. India – History – British
occupation, 1765–1947. I.Title.
HD9019.I321445   2012
338.1´7337–dc23   2012016494

ISBN 978-1-107-02325-3 Hardback

Dedicated to the memory of my father,
Akhileshwar Prasad Sinha

Contents

Figures

Maps

Tables

Preface

This book explores the construction of marginality for agricultural indigo via colonial conditions and global processes. The history of the dye's development on the Indian subcontinent exemplifies colonial power and conditions. The life cycle of indigo on the subcontinent also illustrates participation of colonial India in a global order of uneven structures. The Indian peasantry's labor for the production of indigo was appropriated for a world market. European planters were the agents of this appropriation. This cheap Bengal indigo produced with the labor of Indian peasantry pushed out Spanish and French indigo from the world markets, making dyers across the world beneficiaries of the colonial enterprise in Bengal. The subsequent marginalization of agricultural indigo by synthetic dyes was itself the product of new global rankings among major industrial powers when Germany emerged to be the leading industrial producer of hydrocarbon-derived dyes. The "global" in this study, to deploy the Africanist James Ferguson's use of the phrase, is "globe-hopping, not globe covering." The global emerges at certain points as processes, multiscalar in dimension, are connected in historically contingent ways. It does not efface the other forces characterized as local, colonial, national, metropolitan, and imperial but is integrated with them.

Acknowledgments

This book developed over more than a decade and has accumulated debts to a wide range of scholars and institutions on multiple continents.

My academic career started in India. Sunil Kumar and Sumit Sarkar at University of Delhi first taught me to appreciate the values and pleasures of historical research during the M.Phil. stage of my training. Deepak Kumar of Jawaharlal Nehru University introduced me to history of science, while Majid Siddiqui and Neeladri Bhattacharya provided key input at important stages in my developing research. In my early years in the teaching profession, I was privileged to be in the company of very committed fellow historians at Ramjas College at the University of Delhi. My senior colleagues, Mukul Manglik, Hari Sen, Sudhakar Singh, and G. B. Upreti, reinforced in me the importance of remaining a historian tuned in to important issues of the times. In Delhi through the years, friends like Shahana Bhattacharya, Pankaj Jha, Dipu Saran, and Ravikant Sharma shared their own insights over cups of tea at the Nehru Memorial Library of Teen Murti Bhavan.

The project on the history of indigo took concrete shape as part of my doctoral research at the Georgia Institute of Technology in Atlanta. In 2000, I first found papers on indigo experiments in colonial India among agricultural files during a summer visit to the Bihar State Archives in Patna. These papers ignited my initial interest in the subject and became the core of a developing project on colonial history of indigo for my Ph.D. thesis. As my principal supervisor John Krige at Georgia Tech helped define this project, and his exceptional support was critical to the launching of this project and its development through to the very end. John was an ideal supervisor and remains a very important influence on me. Among other members on the Ph.D. committee, Steve Usselman critically influenced the writing of the dissertation. He served as principal investigator for a dissertation improvement grant from the National Science Foundation that enabled a research trip to the United Kingdom to consult archives. Hanchao Lu on the committee encouraged me to explore the agricultural and social dimensions of the history of laboratory science, a suggestion that I took to heart while developing the book manuscript. Anthony Travis as the external member on the Ph.D. committee prodded me to contextualize indigo history within a broader history of the dye industry and responded to countless queries while I was revising the dissertation. Two independent studies on British imperialism and South Asian historiography with Jonathan Schneer and James Heitzman (now deceased) and a separate course on the comparative history of labor, industry, and technology jointly taught by Mike Allen of Georgia Tech, Matt Payne of Emory University, and Michelle Brattain of Georgia State were helpful to the task of broadly defining my doctoral research. To many faculty at Georgia Tech, I owe an intellectual and personal debt, particularly to Mike Allen, Doug Flamming, Gus Giebelhaus, and Carol Moore. The company and friendship of many fellow graduate students, Ashok, Chris, Devorah, Hannes, Haven, Hyungsub, Jahnavi, Jay, Josh, LeeAnn, Leslie, Patrick, Phil, Raul, Suzanne, Tim, and Vin, made life as a graduate student less stressful. Final chapters of the dissertation were written while I had the sole responsibility to look after my son, Yukt, then a toddler. Many sessions of writing were completed with Yukt in my lap.

Many institutions have generously funded this project. At Chemical Heritage Foundation (CHF) in Philadelphia, I was the Glenn E. and Barbara Hodsdon Ullyot Scholar in 2001–2 and returned for an annual sojourn as Edelstein International Student in 2002–3. The Hagley Museum and Library in Wilmington awarded me a research grant followed by a Henry Belin du Pont Dissertation Fellowship in 2003. Numerous Fellows and scholars at CHF and Hagley provided a stimulating environment. I particularly cherished my several discussions with David Brock, John Ceccatti, Arthur Daemmrich, Gerry Fitzgerald, Tom Lassman, Gabriella Petrick, Erik Rau, Leo Slater, and Kathy Steen at CHF and with Roger Horowitz and Phil Scranton at Hagley. In Philadelphia, I also benefited from conversations with Eve Buckley, Henrika Kuklick, Susan Lindee, and Ian Petrie. Georgia Tech awarded a Raymond Riddle Fellowship that supported a research trip to the UK in 2003. The National Science Foundation's dissertation improvement grant in 2004

(NSF – 0350040/2004) helped wind up archival consultations in England for the dissertation. In the United Kingdom, David Edgerton, David Jeremy, and Phillip Sykas provided valuable support. Peter Morris very helpfully responded to my numerous queries each time I asked. I have enjoyed the friendship and benefited from the advice of Sabine Clarke and Rebekah Higgitt in the UK over the years. Ernst Homburg and Harro Maat in the Netherlands supplied key information on dyes and on Dutch colonial research laboratories in Java.

The task of revising the dissertation started during postdoctoral years at Yale University, and many individuals provided critical comments that formed the basis of my effort. Daniel Kevles read the entire dissertation and offered key suggestions that set the parameters for revision. Dan has remained a key inspiration for this book. His mentoring and influence are critical to what I am as an academic. John Krige once again stepped in to provide key suggestions as I started envisioning the structure of the future book. Daniel Headrick read the entire dissertation and provided key comments for revision. At Yale I became closely acquainted with Jim Scott and regularly attended the talks of the Agrarian Studies Colloquium for two years, which have left an indelible mark on this work. Jim Scott himself has inspired a more intense interrogation of the agrarian context of indigo history. I also benefited from attending the History of Science and Medicine colloquium at Yale University. I learned from Bettyann Kevles, Naomi Rogers, and John Warner and cherished friendly discussions with Lloyd Ackert, Eva Ahren, Ziv Eisenberg, Elizabeth Hanson, Julia Irwin, Beth Linker, Brendan Matz, Ole Molvig, and Sage Ross. Radhika Singha, Jayeeta Sharma, and Ravi Vasudevan were at Yale during my postdoctoral tenure, and we got together in many fruitful discussions. A South Asia workshop on textile history at Yale organized by Tirthankar Roy and Douglas Haynes began a new series of communication with colleagues in Indian economic history and the history of consumption. Pedro Machado, Abigail McGowan, and Ian Wendt during the workshop, and hence, have provided useful input. Tirthankar Roy read the entire dissertation and offered suggestions toward revision. Douglas Haynes in particular provided critical feedback for developing the book. Looking back, I can appreciate the enormous influence of Doug's advice at each step of the writing of this book; I owe Doug a huge intellectual debt.

When I joined the History Department at Colorado State University, I decided to expand the chronological width of the manuscript. Suzanne Moon invited me to a stimulating workshop at Harvey Mudd College in

2007 that afforded me an opportunity to discuss the subject of colony and technology with leading specialists in the field. Michael Adas in particular provided concrete feedback on my work at this point. A series of four Professional Development Fund grants and a Faculty Development Fund grant from the College of Liberal Arts and Department of History of Colorado State University offered financial support to visit archives in Britain and India. The National Science Foundation's (NSF's) Scholar's Award (0824468) in 2009 and a supplementary grant in 2011 provided the most critical support for the final phase of research and writing. The NSF funding allowed me to take a leave of absence from teaching during 2009–10, when I completed much of the writing.

At conferences and in personal meetings, and many times by e-mail, many scholars have contributed a lot to the final shape in which this work finds itself. I mention their names as a small token of gratitude and to highlight how collaborative this project has been: Antoinette Burton, Joyce Chaplin, Prachi Deshpande, David Gilmartin, Richard Grove, Sumit Guha, David Ludden, Sudhir Mahadevan, Clapperton Mavhunga, Dilip Menon, Thomas Metcalf, Projit Mukharji, Abena Dove Osseo-Assare, Ishita Pande, Kavita Philip, Gyan Prakash, Mridu Rai, Peter Robb, Willem van Schendel, Alison Shah, Mrinalini Sinha, K. Sivaramakrishnan, and Anand Yang.

Many have lately spared their valuable time to read my chapters. Michael Fisher, Jacques Pouchepadass, and Sucheta Mazumdar read the entire revised manuscript and pointed out some glaring omissions that saved me a few embarrassments. Douglas Haynes read the entire manuscript in its revised form and offered some comments that shaped my Conclusion. Mark Finlay read three chapters; David Arnold read two chapters; and John Krige, in continuing engagement with my work, once again read all of the chapters in their final form.

My colleagues at Colorado State University provided a stimulating environment and read my work. Mark Fiege, Fredrik Jonsson, and Jared Orsi were part of a reading group where we discussed the crossover themes between environmental history and the history of science. Thaddeus Sunseri very generously read several chapters, and discussions with him over the years have made me rework my arguments. I thank him for his input as a specialist and for his valuable advice on numerous occasions. Doug Yarrington answered multiple queries on the chapter dealing with the Caribbean and Spanish America. Fred Knight read the chapter on early indigo plantations. Jared Orsi read almost the entire manuscript in its final stages and helped me above all in thinking clearly through some of the dense materials. I thank him for his support and friendship. Ruth Alexander read the manuscript and gave perhaps the most comprehensive written feedback this project ever received. Her comments were important to the final shape of this work as a whole. Doug Yarrington and Diane Margolf as Chairs of the History Department always ensured a positive academic environment conducive to research. They deserve credit for that, because in the end I do believe that the intellectual environment at Colorado State offered me the best possible opportunity to develop this work.

My association with two groups in the final phases of revision of the manuscript in 2011–12 proved critical to its framing. As a Fellow of the Framing the Global group of Indiana University I have gained immensely from the discussions by participants. The interaction with fourteen other colleagues in multiple disciplines in the group has shaped my thinking about the nature of the "global." Invited speakers to the group like Saskia Sassen and Carolyn Nordstrom and interventions of extremely capable Fellows through online discussions influenced my thinking as I was finalizing the structure of the book. Early last year Jon Curry-Machado took the initiative in forming the Indigo Academic Network, an informal association of scholars with interest in the history of indigo. I received specific help from the members of this group. Ghulam Nadri and Alexander Engel deserve special mention. Ghulam shared his M.Phil. dissertation, which I have immensely benefited from, and separately helped clarify matters of detail on the early modern history of indigo. Alexander shared his forthcoming articles and clarified many specific points of note. It is gratifying to note the vast number of active, ongoing works on indigo drawing on multilingual archives and focusing on distinct regions and periods, which is only a reflection of its multifaceted history waiting to be explored. My honest hope is that my own work will pave the way for future works on this subject by the extremely capable fellow historians of this group. The students of a capstone seminar on Global History that I taught in the fall of 2011 at Colorado State deserve special mention. This group of very motivated students helped me think more deeply about the issues of global flows and networks in a crucial period of revision of the manuscript.

Several libraries and archives across continents have helped me with acquiring relevant materials for my research. The Inter Library Loan department of Morgan Library at CSU has been indispensable to my work. Maggie Cummings, Cristi MacWaters, and Theresa Spangler have gone beyond the call of duty to procure books and other research materials for me over the past six years. The Morgan Library's regional consortium for borrowing books has been a blessing whenever required material was not available locally. Naomi Lederer in the reference section of the library has been exceptional in terms of meeting my numerous requests for books. I would like to thank the staff of the libraries at Georgia Tech, the Hagley Museum, Yale University, Chemical Heritage Foundation, Philadelphia University, South Asia in Special Collections, University of Chicago, and the staff of the National Agricultural Library in the United States. The staff at the National Archives of India in New Delhi and Bihar State Archives in Patna, where I completed a major part of the work on this project, deserve special mention. The Asian and African Studies staff at the British Library has my highest praise for the diligence and efficiency with which they have supported my work during numerous research trips to the United Kingdom. I have also benefited from the help afforded by the staff at numerous other centers in the UK: Imperial College; Science Museum; Royal Society; Public Record Office, London; University of Leeds; Manchester Archives and Local Studies; and Gloucestershire Archives.

Over the years, this work has been presented at numerous meetings and at universities. I thank the audiences at the meetings of the American Historical Association, Association for Asian Studies, South Asia Conference at Madison and Berkeley, History of Science Society, and the Society for the History of Technology particularly. I also presented parts of this work at Clemson University, University of Delhi, University of Maryland, Johns Hopkins University, and Yale University, and I benefited from the comments of participants. The three referees for Cambridge University Press offered detailed feedback toward reshaping arguments and including fuller documentation in some portions. The Conclusion of the book was specifically shaped by feedback from the referees. Raphael Ruiz and Jason Chambers, graduate students in the Anthropology Department at CSU, helped me with scans of figures and maps. I thank them all. Eric Crahan and Lewis Bateman at Cambridge University Press deserve special mention for their critical suggestions in the final round of editing. Abby Zorbaugh at Cambridge ensured that the entire project moved forward through the Press in a timely manner.

Family and friends have provided utmost support to me during the writing of this book, and I continue to look to them for motivation. I always sought inspiration from my parents, Kanti Sinha and (the late) Akhileshwar Prasad Sinha; from my five elder sisters, Mira Sinha, Bina Shahi, Nita Sharma, Namrata Thakur, and Arti Sinha; and from my elder brother, Prabhat Kumar. My brothers-in-law have stood by me. One of them, N. K. P. Sinha, is a retired professor of modern Indian history himself. My in-laws, Nirmala Sinha and Dineshwar Prasad Sinha, most graciously hosted me during my early research at the Bihar State Archives in Patna. I thank them both for their warmth and affection over the years. As part of a very large family, we often share laughs with a slew of nephews and nieces and now their children. My wife, Vidushi's, grit and determination to keep us going through nine years (and counting) of living in separate cities must rank among the strongest examples of personal support to an academic husband. This book could not have been written without Vidushi's patience and commitment and I truly owe this book to Vidushi. It is her affection that keeps me going. Through difficult times, our son Yukt's passion for excellence and our toddler daughter Manasi's laughter provided example and pleasant support. Vidushi supports by sharing all my goals. And Yukt and Manasi together always remind me what is beautiful and important in life and ensure that I do not lose perspective. Thanks to Skype, we are still able to live together as a family. My everlasting friendship with Divyesh, Shantenu, and Sumit is an example of a bond that has survived the fact that we often lived on three separate continents. The last few months of copyediting and proofreading were marked by deep personal loss and grief. I lost my sister-in-law, and then my father passed away. My father's death was the third major loss in the family after the death of my fourth brother-in-law two years ago. All these losses bore heavily on me. In some ways they made me think of the sense of loss that my historical actors experienced as they witnessed an era of agricultural indigo pass them by. It is a matter of coincidence that this book touches on the history of the Imperial Institute at Pusa. The latter developed into modern Rajendra Agricultural University, which my father, Prof. A. P. Sinha, headed as Vice Chancellor. As a token of appreciation for his leadership in academia and family, I dedicate this book to him. He almost lived long enough to hold this book in his hands.

Introduction

The Odyssey of Indigo

Indigo was the quintessential blue dye in the era when dyes were extracted from plants and minerals. The world, it seemed, had been extracting dye from the indigo plant forever. Knowledge of indigo as a source of blue dye would have been widespread wherever the plant grew. After all, the leaves of indigo yielded the color in small quantities on mere pressing or squeezing. In those times deep in history when the knowledge of indigo culture had not become specialized or attached to large-scale production for commerce, indigo dye's prevalence to a large extent was determined by climate. Indigo was fundamentally a plant of the Tropics that could not be grown in temperate climates. Europeans mostly drew their supply of blue dye from another plant of a related family, woad. A good amount of indigo was also obtained from the Orient even though the difficulty of transportation over long, land-based routes drove up its price and curbed its full-scale use in the West. It is known that Oriental indigo was available in ancient Egypt and in the Greco-Roman world going back to the second millennium B.C. The oldest global networks of indigo production had significant ties to the Indian subcontinent even though it is hard to specify those connections precisely.[1] The use of indigo rose in

---

[1] The Indian subcontinent was central to the early history of indigo. Scholars have commonly inferred that indigo was in use in India in the protohistoric city of Mohenjodaro in the second millennium B.C. Etymological evidence connects indigo's deep history with India. Jenny Balfour-Paul has given an account of indigo's history in the ancient and medieval periods. She wrote, "The last two millennia of indigo's economic history are neatly encapsulated in its names." The word "indigo" derives from the Greek *indikon* or Latinized *indicum*, which meant a substance from India. In addition, the Sanskrit word *nila*, or deep blue, spread from India to both Southeast Asia and the Near East and from the latter through Arab Muslim merchants to northern Africa and Spain and Portugal and lies at

the medieval period though monarchical states in Europe enforced a ban on its import in an attempt to shore up the local woad industry, and the centrality of Asia and the Indian subcontinent as a source of the blue dye continued.[2] The European blockade of indigo ended at the cusp of the modern era. The new Europe in the age of commercial revolution began to procure indigo from Asia in vast quantities as the emerging trading companies improved the connectivity of Europe with Asia and brought down the price of imported indigo. As a consequence, plant indigo was able to defeat woad on its own terrain in the West and emerged as the universal blue colorant of the modern world.[3]

The history of indigo entered a new phase with the emergence of plantations in the seventeenth century. The plantations involved large-scale cultivation of indigo, managerial supervision by European planters, and the use of servile labor in various forms. The plantation life of indigo started in the Caribbean, South and Central America, and the American South, where modernizing trends in its production were apace before their appearance in South Asia. From the mid- to the late seventeenth century, the West Indian colonies, both English and French, and the Spanish controlled parts of Central America came to be the major suppliers of indigo to Europe. As English colonies in the Caribbean moved to the cultivation of other products, British imperial trade interests ensured the onset of indigo cultivation elsewhere within the empire. Indigo plantations first arose in South Carolina in the last three decades of the seventeenth century. But Carolina indigo could not meet the needs of the home market in Britain. It was also of an inferior quality compared to varieties emerging from Spanish Guatemala or the French Saint Domingue. A century later, the English East India Company officials based at Surat and Cambay were still trying to introduce indigo manufacturing in the presidencies of Madras and Bombay. But such efforts proved to be nonstarters.[4] Similar efforts, however, to shepherd the birth of indigo plantations in Bengal

---

the root of the Iberian word *anil* for indigo. Jenny Balfour-Paul, *Indigo*, London: British Museum Press, 1998; see chapters 2 and 3, pp. 11–88, quote on p. 11.

[2] Jenny Balfour-Paul, *Indigo*, London: British Museum Press, 1998.

[3] For the contest between woad and indigo in seventeenth-century Europe, see, Gosta Sandberg, *Indigo Textiles: Technique and History,* Asheville, N.C.: Lark Books, 1989, pp. 24–43.

[4] As Ghulam Nadri has recently pointed out, English officials made concerted efforts to poach on the surviving techniques of indigo culture in Gujarat, still an important manufacturer region, in order to start indigo production elsewhere on the subcontinent. Ghulam Nadri, *Eighteenth-Century Gujarat: The Dynamics of Its Political Economy, 1750–1800,* Leiden and Boston: Brill, 2009, p. 133; also see, notes 23 and 24.

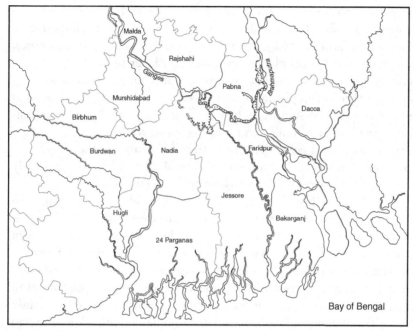

MAP 1. Indigo manufacturing districts in Bengal in the early nineteenth century

by company officials in the last quarter of the eighteenth century proved magnificently successful. European planters, speculators, bankers, and traders responded positively to the encouragement given by the colonial state. After initial fits and starts, the plantations in Bengal began to expand at a feverish pace. At the beginning of the nineteenth century, Bengal had emerged as the predominant supplier of indigo to the world. The indigo plantations were mainly concentrated in the Lower Bengal districts as shown in Map 1.

This book is a story of indigo based on a case study of plantations in colonial Bengal.[5] The Bengal plantations had a lineage extending back to

---

[5] The dominant period of indigo plantations lasted from roughly the mid-seventeenth century to the end of the nineteenth century. In this age, plant indigo ruled the world of dyes. The use of indigo for dyeing and printing in blue was practically universal in this period. While there were a few minor blue dyes made from other sources such as minerals the currency of such blues was minor. Augusti Nieto-Galan, *Colouring Textiles: A History of Natural Dyestuffs in Industrial Europe,* Boston: Kluwer Academic Publishers, 2001, pp. 17–19. Also, the color shellfish purple was obtained from shellfish in the Roman and Byzantine Empires. But no reference is found to its use after the mid-fifteenth century. *Cf.* Jenny Balfour-Paul, *Indigo,* London: British Museum Press, 1998, pp. 14–15.

the period of the early rise of indigo plantations in the Caribbean. For a century and a half Bengal indigo was the object of major efforts to give it a modern form driven by changes in the worlds of knowledge, science, and trade. The Bengal plantations also turned out to be the last major holdout against the expanding sway of synthetic indigo. They lasted long after all other indigo plantations had decayed in the face of competition from the synthetic substitute. In short, the indigo of Bengal was central to the history of indigo from the beginning to the end. Thus Bengal plantations offer a compelling case study for analyzing the early genealogy, nineteenth-century consolidation, and late nineteenth- and early twentieth-century crisis and demise of agricultural indigo.

This is also a history of Bengal indigo in a global dimension. The history of indigo was not constrained by developments in the colonial locality alone. Rather, at each stage, this history was imbricated with genealogies extending to the prior period on the subcontinent and, in parallel, to developments beyond the subcontinent. This bifurcated genealogy was apparent even at the moment of the launch of plantations in Bengal in the last quarter of the eighteenth century and has previously drawn the attention of historians. For example, the historian of indigo enterprise in Bengal, Blair B. Kling, tried to capture Bengal indigo's antecedents on the subcontinent, saying, "From the seventeenth to the twentieth centuries indigo was a fugitive among industries, wandering from Gujarat in western India to the West Indies and then back to Bengal in eastern India."[6] Indigo production evidently had a long history in India. The chroniclers of the sixteenth- and seventeenth-century Mughal court in India and travelers have attested to the high level of output of indigo on the Indian subcontinent. The early records of European trading companies also attest that a good part of India's indigo was shipped at first by Armenian merchants and then increasingly by the Portuguese, Dutch, and English traders to the West.[7] As Kling rightly implied, any history of Indian indigo that singularly focuses on the novelty of the indigo enterprise in colonial Bengal runs the risk of minimizing indigo's local lineages

[6] Blair B. Kling, *The Blue Mutiny: The Indigo Disturbances in Bengal, 1859–1862*, Philadelphia: University of Pennsylvania Press, 1966, p. 15.

[7] W. H. Moreland and P. Geyl (trans. and ed.), *Jahangir's India: The Remonstrantie of Francisco Pelsaert*, Delhi: Idarah-i-Adabiyat-i-Delli, 2009, pp. 10–18; Jean Baptiste Tavernier, *Travels in India by Jean-Baptiste Tavernier, Baron of Aubonne*, translated from the original French edition of 1676 by V. Ball, second edition, edited by William Crooke, 2 vols., London: Oxford University Press, 1925; see vol. 2, *Concerning Indigo*, on pp. 8–12.

on the subcontinent. Indigo was prevalent in many regions of India and may even have been cultivated, used, and imported into Bengal before Europeans launched their plantations.[8] This earlier system of indigo manufacturing on the subcontinent already contained the basic elements involved in the culture of indigo.[9] It is also possible to show lines of influence between the prior culture of indigo in India and the emergent modern plantations in the Western Hemisphere as well as in Bengal. But despite such continuity, the indigo in Bengal in the last quarter of the eighteenth century had clearly entered a new career. For one, it departed from all prior Indian indigo production systems in that colonial capital was its primary driver. Even more importantly, however – and this is critical in terms of the existing gaps in indigo historiography – any claims of continuity from premodern local roots on the Indian subcontinent overlook the major changes in techniques and knowledge of indigo culture that transpired in a transnational dimension over the seventeenth and the eighteenth century. It is in the knowledge dimension that the Bengal plantations make their best claim for novelty from the previous regimes of production.[10]

From the beginning of the nineteenth century colonial South Asia was the largest exporter of indigo and remained so for the rest of the century alongside Java in Southeast Asia, Guatemala in Central America, and a few other minor areas of production. One turning point occurred in the middle decades of the nineteenth century. A movement against

---

[8] See a discussion of this aspect in Indrajit Ray, *Bengal Industries and the British Industrial Revolution (1757–1857)*, London and New York: Routledge, 2010, p. 210.

[9] Iqtidar Alam Khan, "Pre-modern Indigo Vats of Bayana," *Journal of Islamic Environmental Design Research Center* (1986): 92–8; K. K. Tivedi, "Innovation and Change in Indigo Production in Bayana, Eastern Rajasthan," *Studies in History* 10 No. 1 n.s. (1994): 53–79, see, p. 68.

[10] The historiography of indigo production in colonial South Asia is rich. The violence of indigo manufacturing as a colonial enterprise and the exploitation of Bengal peasantry engaged in the cultivation of indigo have been well documented. But these studies typically do not examine the technical context of indigo plantations even as they participate in the critique that the colonial industry was based on "antiquated" technique and primarily geared to exploit the cheap labor of natives. Benoy Bhushan Chowdhury, *Growth of Commercial Agriculture in Bengal, 1757–1900*, Calcutta: India Studies, 1964, 80–124; especially see his summation on 123–4. A major critique of indigo manufacturing by Chowdhury is articulated around the fact that profits from the colonial enterprise were not plowed back into improving the industry and its workings, but rather repatriated to Britain. Jacques Pouchepadass, *Champaran and Gandhi: Planters, Peasants and Gandhian Politics*, Delhi: Oxford University Press, 1999; see, 49–58 for a description of manufacturing labor, 127–36 for his analysis of appropriation of surplus from the peasantry, and 65–6 for a summation of his critique of the primitive characteristic of cultivation and manufacturing.

planters between 1859 and 1862 wiped out the indigo industry from Lower Bengal, which was the geographical pivot of indigo manufacturing on the subcontinent. This movement, called "blue mutiny," reflected the anger of the indigo peasantry against the excesses of European planters. But the popular movement failed to banish the industry from Bengal. The center of gravity of the colonial indigo industry simply shifted elsewhere to north Bihar within the Bengal Presidency.

Meanwhile rumblings of a deeper change in the world of dyes that were slowly coming to the surface would determine the future of Bengal indigo. The plant-derived dye began to face the new reality of market competition from synthetic dyes that were extracted industrially from coal tar–based hydrocarbons. The synthetic dye industry had its birth in England and France in the middle decades of the nineteenth century. But as the century progressed, Germany took the lead in the industrial production of synthetic dyes. Most critical advances in dye science based on organic chemistry and innovations in dye manufacturing took place in Germany, which cornered much of the world's trade in synthetic dyes in the last quarter of the nineteenth century. Synthetic dyes were pure and generally cheaper than natural dyes. Many variants of synthetic dyes, the anilines and alizarins in particular, were made available to the users of natural indigo. But by and large these alternates only supplemented the supply of blue and in the context of increasing use of indigo worldwide never really displaced natural indigo from the market. The plant-derived indigo held its position against the early synthetic blues.[11]

The challenge from synthetics intensified, however, with the launching of synthetic indigo by the German company, Badische Anilin and Soda Fabrik (BASF), in 1897. Synthetic indigo progressively ate into natural indigo's erstwhile markets in the West as additional German companies and their subsidiaries also started producing the synthetic substitute. The planters in Bengal were now called upon to take measures to produce cheaper and purer agricultural dye in order to compete with synthetic

---

[11] There exist many histories of the rise of the synthetic dye industry based on the archives of synthetic dye companies in the West. These studies have predictably not focused on the longevity of natural dyes and the resistance offered by them. A singular focus on synthetic dyes in the histories of transition tends to "naturalize" the death of agricultural dyes and furnishes a somewhat triumphalist understanding of the rise of synthetic dyes. John J. Beer, *The Emergence of the German Dye Industry*, Urbana: University of Illinois Press, 1959; Anthony S. Travis, *The Rainbow Makers: The Origins of the Synthetic Dyestuffs Industry in Western Europe*, Bethlehem, Pa.: Lehigh University Press, 1993; Carsten Reinhardt and Anthony Travis, *Heinrich Caro and the Creation of Modern Chemical Industry*, Dordrecht: Kluwer, 2000.

indigo in the international market. The effort to improve natural indigo after the introduction of synthetic indigo was largely determined by the nature of competition from a consistent and cheaper industrial product. The planters embraced modern laboratory science in an effort to make the natural dye purer and to lower its cost of production. This program of improvement in the colony continued for more than two decades. It was finally stopped when trade losses produced a sense of hopelessness with regard to the prospect of revival of indigo plantations. By the closing years of the First World War synthetic indigo seemed to have won unequivocally in the market, and, as a result, scientific efforts to improve natural indigo in the colony ceased in 1920.

Analyzing the Knowledge of Indigo Culture: Historiographical Constraints

In following the odyssey of plant indigo this book's fundamental project lies in uncovering the various knowledge forms surrounding indigo. This knowledge existed in multiple forms, such as textual knowledge describing ideal methods of cultivation and processing, information passed along continents and circulating among indigo manufacturers, the awareness of optimal agricultural and environmental conditions, the epistemic component of techniques of indigo manufacturing in use, and the knowledge attached to the practice of indigo culture. At its functional end, this knowledge was geared toward improvement[12] of the indigo plant and the processes involved in indigo culture. Improvement from the planter's perspective meant increasing the yield of the crop in the field and growing better plants with a higher content of color. On the manufacturing side, improvement meant reducing the price of the commodity either by

---

[12] Slightly different studies of "improvement" as a societal goal or imperial project have appeared elsewhere. Joyce Chaplin, *An Anxious Pursuit: Agricultural Innovation and Modernity in the Lower South, 1730–1815*, Chapel Hill and London: University of North Carolina Press, 1993; see particularly pp. 23–65; David Arnold, *The Tropics and the Traveling Gaze*, Delhi: Permanent Black, 2005; see the stating of his position on improvement on p. 6; Richard Drayton, *Nature's Government: Science, Imperial Britain, and the "Improvement" of the World*, New Haven, Conn., and London: Yale University Press, 2000. The construct of improvement as a colonial project for changing Indian society appears frequently in South Asian historiography: "[Quite early on,] by the end of Lord Cornwallis's years as governor-general (1786–93), the British had put together a fundamental set of governing principles. For the most part these were drawn from their own society, and included the security of private property, the rule of law, and the idea of 'improvement.'" Thomas Metcalf, *The New Cambridge History of India: Ideologies of the Raj*, Cambridge: Cambridge University Press, 1995, p. 17.

cutting labor costs on factory processes or by making extraction efficient. The claims regarding knowledge of improvement had to be validated in the marketplace. The market players placed value on the dye produced in certain ways. The users accepted or rejected the claims of improvement. The knowledge of indigo culture was evidently created in different locations and spaces and validated in laboratories and far-off markets.

The methods of knowledge production evolved during the long history of indigo plantations. Flow and exchange of information across the planters' diasporas in the Atlantic system of the seventeenth and eighteenth centuries and the honing of skills and craft practices at the hands of planters and peasants laid the foundation of the indigo knowledge system. Useful information on indigo cultivation and manufacturing was committed to and codified in a few foundational texts, which were then translated and disseminated widely. Some of these important indigo texts found their way to Bengal on the Indian subcontinent and influenced planters' craft even as they were modified in a local context of application. Indigo itself had to adjust to local landscapes in Bengal – both physical and social. Local climate put a limit as to what type of knowledge could be actually put into practice in the cultivation and processing of indigo.

The rise of empiricist trends in modern science over the nineteenth century altered how the knowledge of indigo culture was generated. These changes undergirded the transformation of indigo knowledge from its earlier moorings in natural history and craft practices into the shape of a more formal, discipline-based, and laboratory-edified science in the second half of the nineteenth century. Toward the close of the nineteenth century these methods underwent further transformation as planters and the colonial state established laboratories and agricultural stations to alter plant indigo. The establishment of these institutions transpired in the shadow of the rise of agricultural stations as the predominant institution for applying principles of agricultural science to the practical task of improving productivity.[13]

---

[13] The more specific form of "experiment stations" rose in mid-nineteenth-century Germany and then later spread to other nations and continents. Margaret W. Rossiter, *The Emergence of Agricultural Science: Justus Liebig and the Americans, 1840–1880*, New Haven, Conn.: Yale University Press, 1975; it has been argued that England's premier agricultural research station at Rothamsted itself underwent reorganization in the model of continental experiment stations in the early twentieth century. For England's longer-term tradition of agricultural research, see, E. J. Russell, *A History of Agricultural Science in Great Britain, 1620–1954*, London: George Allen & Unwin, 1946.

This is a social history that focuses on the nature and circumstances of production of indigo knowledge. Seen one way, indigo knowledge was a body of information associated with a technical process. Seen another way, indigo knowledge was also "colonial knowledge" as variously inter-preted in the existing social histories of South Asia, that is, a social form embedded within colonial relations. The tracks followed by the history of science and South Asian history are both indispensable for generat-ing a complete understanding of the knowledge of indigo culture. The former helps uncover the constructed nature of the technical aspects of indigo culture. But a full recovery of the social dimension of indigo cul-ture requires analysis of the larger context of colonial relations as studied in area study approaches in South Asian history. Indeed the book con-tends that the study of indigo knowledge requires a simultaneous consid-eration of textual knowledge, natural history, modern scientific practice, institutional dynamics, colonial relations, and the political economy of colonialism.

But the study of science in South Asian historiography has so far evolved along two parallel tracks – works that cover colonial science and works that cover the social history of science in colonial South Asia. Their respective philosophical orientations and theoretical borrowings have led them in different directions, and they have built their own respective momentums in isolation from one another. Thus South Asia historians who study "science" fall into one group or the other. The partiality in favor of analysis in one or the other framework also accounts for the apparent chasm that separates the study of science so far. This mutual obliviousness is unfortunate because each field has much to contribute to the other and even more so because, as this study argues, the gap between the two fields is not unbridgeable.

The field of colonial science in South Asia, while invested in analyzing broader questions of colonialism and modernity, has maintained a sep-arate identity because of the agenda that the field has defined for itself. With its broad moorings within the classical history of science in the West and imperialism studies, the field has dwelled in a collection of issues around transfer and movement of knowledge, science as a tool of empire, and the nature of engagement between Western and native knowledge systems.[14] It is within this larger agenda that the field of colonial science approaches the "social."

---

[14] The history of science field delved into sociology of knowledge with a correspond-ing focus on the social context in a separate trajectory of its own. As Jan Golinski has

On the other hand, the social study of science in South Asian historiography has emerged out of traditions of investigating knowledge as a social form, evolving on two parallel tracks. Among the earlier generation of historians, Bernard Cohn used the trope of knowledge to investigate the social history of colonial India. Cohn argued that the initiatives ranging from revenue measures to the creation of Orientalist scholarship on India or the collection of historical artifacts from the Indian past were not benign but rather represented an effort to understand, codify, and rule India. Thus, in this rendering, India's castes and tribes or its museums or brahmanical religion, all were alibis for or stood for colonial knowledge.[15] Another important moment in the expanding use of knowledge frameworks was reached with the publication of Edward Said's *Orientalism* in 1978. Said addressed the knowledge implicit in Western texts as European representations of non-Western peoples. South Asianists combined Said's analysis with the new understandings of knowledge from the fields of postcolonial theory and literary theory to focus on discursive formations and on language as sites of knowledge formation. In particular, Michel Foucault's notion of discursive formations made a major impact with ideology, science, and social science theory – all of the latter, as "domains of objectivity" – becoming open

illustrated, postwar accounts of history of science opened up to inputs from the growing field of sociology of knowledge. The time was opportune because history of science was gradually moving out of the older tracks of documenting progress and the discovery of a preordained "nature" toward a more historicist project of contextualizing. In this new intellectual environment historians were ready to embrace an understanding of historical development of knowledge marked by "discontinuities and transformations" rather than mere "forward movements." But Golinski also pointed out a subsequent countermovement in the consideration of the social in the research program of the history of science. He argued that the rise of the constructivist program from the 1970s has caused a reduction in the scale of the "social" that is analyzed by historians of science. The targeted focus on specific episodes, controversies, and ethnographic look-in at the laboratory has caused a "trend away from macrosocial explanations." Jan Golinski, *Making Natural Knowledge: Constructivism and the History of Science*, Cambridge: Cambridge University Press, 1998, pp. 2–4, 10–11.

[15] Bernard Cohn's writings appeared in various publications from the 1950s to the 1980s and are published as two major collections: Bernard Cohn, *An Anthropologist among the Historians and Other Essays*, Delhi: Oxford University Press, 1990; *Colonialism and Its Forms of Knowledge: The British in India*, Princeton, N.J.: Princeton University Press, 1996. These understandings of social forms and things as knowledge were the result of Cohn's ties with the historicist traditions within American anthropology in which culture was the "common frontier" of anthropology and historiography. See "Introduction" by Ranajit Guha in Bernard Cohn, *An Anthropologist among the Historians and Other Essays*, pp. vii–xxvi, and "Foreword" by Ronald Inden in Bernard Cohn, *Colonialism and Its Forms of Knowledge*, pp. ix–xvii.

to critique as knowledge or discourse.[16] Other South Asia historians dissent from such culturist interpretations of South Asian social history. These historians work in traditions that do not entirely depend upon the explanatory templates of discourse and representation but nonetheless cover subjects that would pass for knowledge in the postcolonial and cultural formulations.[17]

The different renderings of knowledge and the associated difference in perspective on colonial power, the resilience of the local society, and the place of native traditions have left their imprint on the studies of science, medicine, and disease within the historiography of South Asia. The existing studies of science and medicine reflect those broad differences. On the one end, Gyan Prakash has used postcolonial theories within a broad cultural history approach to study modern science *as* knowledge in the colony. Prakash's study focused on the cultural authority of science launched by the colonialists, which was subsequently embraced by elite

---

[16] "Whenever one can describe between a number of statements, such a system of dispersion, whenever, between objects, types of statement, concepts, or thematic choices, one can define a regularity (an order, correlations, positions and functionings, transformations), we will say, for the sake of convenience, that we are dealing with a *discursive formation* – thus avoiding words that are already overladen with conditions and consequence, and in any case inadequate to the task of designating such a dispersion, such as 'science', 'ideology', 'theory', or 'domain of objectivity.'" Michel Foucault, *The Archaeology of Knowledge and the Discourse on Language*, French edition, 1969, trans. A. M. Sheridan Smith, New York: Pantheon Books, 1972, p. 38; Homi Bhabha (ed.), *Nation and Narration*, London: Routledge, 1990; *The Location of Culture*, London: Routledge, 1994.

[17] Christopher Bayly's study of social communication in colonial India is an excellent representative example of the study of "information" as against knowledge in a social history tradition. C. A. Bayly, *Empire and Information: Intelligence Gathering and Social Communication in India, 1780–1870*, Cambridge: Cambridge University Press, 1996. There has been a raging critique of the analytical path of seeking and explaining discursive knowledge in South Asian history by those who follow the tracks of social history. Aside from several specific notes that have pointed out gaps in the postcolonial program built around Orientalism, representation, and discourse, others have spoken of the merit of social history. Sumit Sarkar has strongly argued the benefits of the tradition of evidence-based social history that attends to social, political, economic, and intellectual contexts in greater detail. In particular, he has spoken of the values of social history traditions exemplified in the works of E. P. Thompson. Sarkar's critique represented a broader response to the "cultural turn" represented by the subaltern school in Indian historiography. Sumit Sarkar, "Orientalism Revisited: Saidian Frameworks in the Writings of Modern Indian History," *Oxford Literary Review* 16 (1994): 205–24; *Writing Social History*, New York and Delhi: Oxford University Press, 1997. A number of other scholars have alluded to what is left out in the generalizations built around the study of discursive knowledge formations. See, for instance, Benita Parry, "Problems in Current Theories of Colonial Discourse," *Oxford Literary Review* 9 Issue 1–2 (1987): 27–58.

nationalism and still later by the postcolonial elites. The latter inscribed "difference" on this science in their counterhegemonic drive even as they pressed their own claim to modernity through science.[18] Other historians have applied the notion of discursive knowledge to varying extents. David Arnold has taken a somewhat middle-of-the-road approach between cultural history and social history. Arnold emphasized the power of colonial discourse but in addition to considering discursive formations, he focused on the social and political process with which colonial power was complicit. Thus his study of exclusions and pathologizing of natives in the colony was pivoted on the examination of institutions like the army, jails, and hospitals, and of elements of social control evident in state measures to fight back epidemic diseases.[19] On the other end of the spectrum, the social history approach was best epitomized in Christopher Bayly's study of astral and medical sciences.[20]

A case can be made that the separate fields of history of science and South Asian history share some common goals in the analysis of knowledge despite their distinct analytical paths. All of these approaches are seemingly engaged with the question of exchange of knowledge, either between the "West" and the "East" or between the colonizers and the colonized. Historians of colonial science in South Asia have drifted away from simplistic assumptions that modern science was of Western origins and that it gradually diffused to the rest of the world. They have increasingly focused on the local development of science in the colony.[21] Historians of South Asia have similarly engaged with the question of knowledge and its transformation in the colonial context. The subaltern

[18] Gyan Prakash, "Science Gone 'Native' in Colonial India," *Representations* 40 (Autumn 1992): 154–78; *Another Reason: Science and the Imagination of Modern India*, Princeton, N.J.: Princeton University Press, 1999.

[19] David Arnold, *Colonizing the Body: State Medicine and Epidemic Disease in Nineteenth Century India*, Berkeley: University of California Press, 1986; a somewhat similar position is taken on the cultural-social axis in the study of public health by Biswamoy Pati and Mark Harrison. Cf. Biswamoy Pati and Mark Harrison, *Health, Medicine and Empire: Perspectives on Colonial India*, New Delhi: Orient Longman, 2001.

[20] C. A. Bayly, *Empire and Information: Intelligence Gathering and Social Communication in India, 1780–1870*, Cambridge: Cambridge University Press, 1996, pp. 247–83.

[21] The model of diffusion of Western science to the rest of the world finds best representation in: George Basalla, "The Spread of Western Science," *Science* 156 (May 5, 1967): 611–22. For counterperspectives to Basalla's assertions in the works of historians of science in South Asia, as two representative studies, see, Deepak Kumar, *Science and the Raj, 1857–1905*, Delhi: Oxford University Press, 1995, and Kapil Raj, *Relocating Modern Science: Circulation and the Construction of Knowledge in South Asia and Europe, 1650–1900*, Basingstoke: Palgrave, 2007.

school has exposed the untenable claims of universality of Enlightenment knowledge that was launched in South Asia during colonial rule. They attacked knowledge claims that were seemingly held up to the natives to justify colonial rule, giving a call instead to "provincialize" European thought in developing an understanding of Indian history and society on terms true to Indians. They argued that a particular knowledge system on which Europe held a monopoly did not exhaust all potential sources of positive social change in Indian history. Thus scholars have critiqued nationalism as a derivative discourse and directed attention toward the task of uncovering native visions.[22] Others have sought out imposition, mutual adaptation, and dialectic between Western and Indian institutions and ideologies in the "encounter." They argue that the colonialists could not and did not completely displace Indian systems of thought, and therefore historians must locate the emerging Indian modernity in the outcome of collisions and contestations between European and Indian knowledge in a local context and ultimately in the resilience of the Indian society.

These perspectives are also inhering in elements that limit the possibility of writing a global history of indigo knowledge. All assume that "modernity" was born in the West and subsequently engaged with other knowledges in the non-West. In parallel, others in the subaltern school search for "another reason" in local, native spaces in the colony that if anything is remarkable for its lack of correlation with the Western thought.[23] In different degrees, then, historians in South Asia focus on the analysis of social in the "locality" – ranging from the social space of engagement between colonialists and Indians to the social space of autonomous Indian institutions and imagination. The latter argument is sometimes extended to claim legitimacy for indigenous knowledge in the colony by conceptualizing the existence of alternative modernities and emphasizing the plurality of modern existence. These positions are unhelpful to the task of writing the history of indigo. Indigo knowledge developed simultaneously at several sites. Indigo appeared as a different plant type on numerous continents, and its physiology was shaped distinctly in keeping with local climatic variations. Its universal utility as a

---

[22] Dipesh Chakrabarty, *Provincializing Europe: Postcolonial Thought and Historical Difference*, Princeton, N.J.: Princeton University Press, 2007. See, also, Partha Chatterjee, *Nationalist Thought and the Colonial World: A Derivative Discourse*, London: Zed Books, 1986.

[23] The phrase appears in the title of Gyan Prakash's book. Gyan Prakash, *Another Reason: Science and the Imagination of Modern India*, Princeton, N.J.: Princeton University Press, 1999.

dye plant was realized in different ways by multiple societies. These people put the dye to use for separate purposes on small and large scale. Thus the knowledge of indigo culture was a variable one reflecting the numerous circumstances of its production and use in the world. The knowledge zones and the people possessing them were at the same time also connected with each other at the level of sharing of information and trade relations. This aspect of multisited development of indigo knowledge remained true in the era of passage from the premodern to the modern. It is this aspect of simultaneity of emergence of modern indigo knowledge that is overlooked in the current studies. Rather, the model of a global history that is open to the suggestion that modernity developed at multiple sites simultaneously is more suitable to writing the history of indigo culture.

Indigo Plantations: South Asia and the World

This book seeks out the fundamental connections between the historical space of South Asia and the wider world. It considers whether the emergence of South Asia as a region, a colony, and a nationalist territorial and imaginary space – the dominant focus of historiography so far – was also invaded by impulses from various points in the world that were imbued with additional meanings and possibilities. The history of developing indigo knowledge in Bengal suggests that the Indian subcontinent was influenced by external forces of multiple origin and nature, not only those enabled by British imperialism. Cosmopolitan knowledge of multiple lineage, global trade, and the movement and flows of peoples, ideas, and institutions contributed to the making of knowledge of indigo culture during the colonial times. The imperial framework and colonial relations were never irrelevant to this history. In fact, if anything, they were foundational. The needs of the British Empire led to the launching of Bengal indigo plantations. The profit-seeking motives of the European entrepreneurs who immigrated to the colony, the facilitation by the institutional space of the colonial state, and the remittance requirements of the East India Company sustained the colonial indigo industry. But none of these delimited the frontiers of knowledge determinants for indigo culture. As the indigo industry ebbed and flowed from the last quarter of the eighteenth century, it drew on information from both varied external sources and local traditions, developed its own methods in the local context of the colony, embraced new institutions of knowledge making, and adjusted to the demands of knowledge sanctified by consumers in diverse markets of the world. The categories of Western, foreign, imperial, colonial, and local only go so far to explain the situatedness of this body of knowledge without accounting for all of the contributing factors. A focus on the composite nature of this process of knowledge making clarifies that the colony was a space interconnected with multiple other spaces in the world and that colonial Bengal was a participant in a historical process unfolding on a scale that was not confined within the territorial limits of the Raj or even the formal margins of the larger British Empire.

The use of the category of "world" in this study is precisely meant to emphasize the aspect of wide dispersal of points of influence on Bengal indigo's history. It does not imply that the history of indigo was marked by any sort of universal homogeneity or commonality in patterns.[24] As with "global," "transnational," or "international," the use of "world" as a framework fundamentally represents a common historiographical stand that urges shifting of attention away from the "national."[25] The "nation" has defined the project of history writing in spatial terms for a long time. This could not be truer for South Asian historiography.[26] Regardless of the fact that some historical studies have focused on international trade, religion, and ocean, for instance, whose ambits exceed national frontiers, historians have by and large continued to pivot their attention on the nation as the predominant space of origination or impact. This book seeks to move analysis beyond the current understandings in the historiography that imply that the history of indigo was delimited by influences that could be best categorized as imperial and colonial or native and national. Instead, this study highlights the multiple loci of the history of Bengal indigo.

A gesture forward toward two specific moments in the history of indigo puts the basis of this book's contention in broad relief. One was the global dispersal of an early indigo text authored by Jean Baptise

---

[24] A. G. Hopkins, "Introduction," In A. G. Hopkins (ed.), *Global History: Interactions between the Universal and the Local,* London: Palgrave Macmillan, 2006, pp. 1–38.

[25] "AHR Conversation: On Transnational History," participated in by C. A. Bayly, Sven Beckert, Matthew Connelly, Isabel Hofmeyr, Wendy Kozol, and Patricia Seed, *American Historical Review* III No. 5 (December 2006): 1140–64.

[26] The rise of the subaltern school in the 1980s was based on a critique of the nationalist historiography in South Asia. The subalterns critiqued "nation" as a derivative form and sought to recover other subjectivities by working outside the framework of Western teleologies. As a methodological strategy the subalterns often directed research away from the larger structures ordained by colonialism and toward the autonomous, the individual, and the episodic – "the fragment." In effect such trends minimized the treatment of histories on a large spatial scale.

Labat, the famous seventeenth-century French naturalist. This text was composed at the beginning of the era of indigo plantations. It became an important medium in the global movement of indigo manufacturing knowledge from the Caribbean, in particular from the indigo-growing regions of the French Empire to those of the British Empire. The seamless passage of Labat's work through historical times reveals the openness of political formations, territories, and knowledge systems to information contained in the indigo treatise. A second case in point was the birth of synthetic indigo in 1897, which ultimately caused the demise of plant indigo and brought the end of indigo plantations. A product of the German dye industry, synthetic indigo was an "externality" to the nineteenth century's varied imperial systems. It was rather a product of German industrial history, and its expansion was riveted on capturing major textile markets of Germany's imperial rivals like Britain and other Western textile manufacturing nations.[27] Its ultimate victim was agricultural indigo, which was produced by the major colonial systems of the world. It was thus able to interrupt the arrangement in which European colonies in Asia, Africa, and America produced the indigo dye, and dyers and printers in the West consumed it. It inaugurated a new division in production and consumption. Synthetic indigo captured not only the erstwhile markets of agricultural indigo, but also the new emerging markets for the blue dye, in particular those developing in the Far East. Thus the putative imperial motifs do not explain the birth and expansion of German indigo. The history of its origin in Germany and global expansion – of which the death of agricultural indigo was but a counterpart – requires an analytical ambit beyond those involving imperial systems and local colonialisms alone. It equally requires consideration of dynamics of global scaling.

First, let us turn attention to Jean Baptiste Labat.[28] Labat was born in Paris in 1663 and joined the Dominican order. Although he was a Catholic theologian, preaching was not his only vocation. He was a man respected for his learning and would go on to travel overseas and earn a

---

[27] Even though Anthony Travis has suggested that the monopoly by English traders and merchants on the world's supply of plant indigo might have motivated the search by the Germans for a synthetic substitute, the fact remains that the birth of synthetic indigo followed its own trajectory of industrial history and innovation in the national state context of Germany. Anthony S. Travis. *The Rainbow Makers: The Origins of the Synthetic Dyestuffs Industry in Western Europe*, Bethlehem, Pa.: Lehigh University Press, 1993.

[28] Lafcadio Hearn, *Two Years in the French West Indies*, New York: Harper and Brothers, 1890, pp. 157–83; Robert Bracey, "Jean Baptiste Labat," *New Blackfriars* 5 Issue 151 (June 1924): pp. 136–43.

reputation as a preacher, as a successful colonialist, and later as a writer of travelogues and compendia. He had his calling after he heard that the French colonies in the Antilles were short of preachers on account of the ravages of yellow fever. He applied for permission to go to the Antilles and, permission in hand, arranged for passage to the Caribbean. He reached Martinique in 1693. This marked the beginning of a long sojourn of twelve years in the Caribbean.

It is Labat's first of its kind account of detailed cultivation and manufacturing of indigo that we are concerned with here.[29] Labat was quite versatile in his pursuits in the Caribbean. He helped turn the Dominicans' Martinique sugar plantation into a profit-making unit, assisted a French governor of another island in building defensive fortifications against attacks from the English, and even participated in combat against the English and the Spaniards. He also spent a lot of time traveling, observing local plantations and manufactories. His handbook on sugar manufacturing was considered an indispensable aid by generations of manufacturers. But it was his key composition on indigo manufacturing in the French colony that won him a reputation far beyond the lands of the French Empire. By the late seventeenth century the French colonies had raced ahead of the English colonies in the Antilles in indigo manufacturing. Martinique was indeed a leading center for indigo plantations. Labat based his treatise on the art of indigo manufacturing on what he observed in the Macouba parish of Martinique. This was published in 1722, several years after his return to France, as part of an account of his travels through the Caribbean.

Labat's account of indigo manufacturing spread afar in a way he would never have imagined. Although written primarily with an audience composed of French countrymen and colonialists in mind, the text readily found its way into the imagination and thought of literati and practitioners in the wider world immediately after its publication in Paris. Perhaps the timing was right. The work found relevance at a time when indigo production was expanding worldwide. It was quickly translated into major European languages. A previous combatant against the English, Labat likely may not have fancied its reaching the English. Regardless, an English translation of the section on indigo appeared in London in

[29] Labat was preceded by another French Catholic humanist, Jean Baptiste Du Tertre, who traveled to the Antilles in 1640. Du Tertre also provided an account of indigo that was published in France in 1671. The distribution of Du Tertre's text was somewhat limited at that time, although it became popular among the French colonialists and in the French metropolis.

1731 as part of Phillip Miller's *Gardener's Dictionary*. The latter text and its rendition by other planters in local journals in the American South formed the basis of information as South Carolinians embraced indigo in the 1740s.[30]

Two points are worth emphasizing. First, the description of indigo culture by Labat floated in competing European imperial networks and their colonies. Second, and this is more specifically relevant for the South Asian story, as the Indian subcontinent started opening to immigrant planters from the last quarter of the eighteenth century, those from the British Empire extending into South Carolina could have gained access to the English translations of Labat. Other planters entering India with prior experience in other European colonies – and there were many in the early phase – had also been likely in a position to draw on textual knowledge of indigo culture of the type represented in Labat's account. Knowledge, it would seem, did not follow the bounds of political frontiers imposed by several imperial and colonial systems. The migration of textual knowledge ascribable to Labat was not the exception but the rule, as this study will show. Other French expositions on indigo by writers and planters like Elias Monnereau of Saint Domingue and De Cossigny de Palma of Mauritius, to mention two other prominent cases, similarly made it to the wider world including colonial South Asia. It is the contention of the present study that the history of indigo manufacturing in colonial South Asia cannot be complete without a consideration of these larger, if amorphous, knowledge networks.

The second important global moment chosen for this early exposition was in 1897, the year the German synthetic dye company, BASF, launched its version of synthetic indigo on the market. The arrival of synthetic indigo created a new cast of adversaries on the side of synthetic and natural indigo, respectively, that cut across the familiar fault lines between imperial and colonial or between colonial and native interests, further illustrating the limitations of such categories for understanding the history of indigo. Until then the global market for the blue dye had been supplied by agricultural indigo. Three major imperial systems dominated that market with British India holding the top position as the biggest supplier of Bengal indigo, followed by the Dutch Java, and Central

---

[30] Miller reproduced Labat's account. See the entry "anil" under *AN* in the dictionary. Philip Miller, *The Gardener's Dictionary in Two Volumes*, London: printed by the author, 1743, no page number; David L Coon, "Eliza Lucas Pinckney and the Reintroduction of Indigo Culture in South Carolina," *Journal of Southern History* 42 No. 1 (February 1976): 61–76.

America. The German dye company BASF first interrupted this imperial social formation of the world. Others followed suit and soon several German companies and their subsidiaries in other European nations
started to manufacture the synthetic substitute, which ate into natural
indigo's share of the market. These forces caused the beginning of a slide
in the fortunes of natural indigo that proved to be irreversible except for
short phases, until the natural indigo industry was completely obliterated
at the end of World War I.

The challenge posed by synthetic indigo to the natural dye was of a
global nature that can scarcely be explained on the basis of the dialectics
of colonial relations. There is little doubt that colonialism had led to the
birth of the indigo plantations in Bengal. Colonial power secured the entry
of indigo into South Asia, and the colonial state's land settlement policy,
revenue policy, commercial policy, and laws relating to the use of native
labor set the basic parameters within which indigo culture developed.
But colonialism also tied the destiny of indigo with a worldwide spread
of consumers. The fact that it was consumed by a dispersal of dyers and
printers in the entire world, not only by those in the British home market, meant that global trends in dye consumption could make an impact
on the colonial industry. Indeed in the late nineteenth century colonial
indigo's future was caught up in the global process of transition from natural to synthetic dyes. Emerging outside the orbit of British imperialism
synthetic indigo threatened the investments of British expatriate planters
in India. But as dyers and printers made a switch to synthetic indigo –
including those at home in Britain – the national government could do
little to forbid them. Imperial Britain could do far less to stop the tide of
a global trend. Thus neither an exclusive focus on imperial Britain nor
one on local colonialism can adequately explain the history of struggle
between the German synthetic indigo and the plant indigo of Bengal.
This story, rather, needs explanation on the basis of a broad template that
weaves the study of the imperial-colonial relationship along with that of
the colony's connections with the world, and the transnational processes
that touched the Indian subcontinent.

"A Product of Nature": Science-Nature Relationship

A good part of the book addresses the "science" of indigo culture in a
correlation with cross-cutting movements of knowledge, the local context of peasant production, land structure in colonial Bengal, institutions
that facilitated exchange such as market and trade networks, and the

contingencies introduced by the institutional and disciplinary framework of modern science. The focus of the study accordingly shifts among
agriculture, colonial relations, market, and scientific institutions. But
a running thread through the analysis is the focus on science as a tool
in pushing at the limits of "improvement." What was the best indigo?
What was humanly possible in terms of making the very best indigo? The
answers to these questions were never a "given." As a matter of fact, a
resolution of these complex issues was very much historically contingent.
Philosophers countered each other in texts, market players asserted different viewpoints in an exchange context, and scientists disagreed with
each other. They were all in one way or another engaged in the creation
and validation of knowledge about indigo.

The task of improving indigo through science on the plantations drew
upon the traditions of looking at nature in specific ways or even objec-
tifying it. These traditions were implicit within the informal practical
knowledge and the later formal goal-directed laboratory science as the
plantations evolved from the early modern to the modern period. Later,
scientific practice predetermined the conjuring of indigo as an "object"
with its own characteristics as well as the understandings of what
indigo was, how it could be enhanced, and to what extent. Other ideas
of improvement emerged with the flow of craft practices, experiments,
and trials. The growing understanding of the material artifact influenced
the agenda of enhancement subsequently. And changing externalities of
trade, exchange, and market impinged on the improvement process to
affirm or challenge what constituted real improvement in the making of a
commodity. Thus the developing indigo science depended on the multiple
meanings of what constituted a valid scientific practice, object, nature,
and commodity, meanings that were all historically contingent and open
to definition by dominant forces.[31]

The last third of the book addresses the science-nature relationship
by studying the imputing of "natural" attributes to agricultural indigo
by forces around the plantations. For European planters and scientists
in colonial Bengal, who were trying to improve the natural dye to defeat
synthetic's competition, the agricultural indigo's claim to be a product of
nature was elemental. It was often made part of a central rhetoric in its

---

[31] For a critical perspective on how objects become valid objects of scientific analysis and
fade away, see, Lorraine Daston, "Introduction: The Coming into Being of Scientific
Objects," in Lorraine Daston (ed.), *Biographies of Scientific Objects*, Chicago: University
of Chicago Press, 2000, pp. 1–14.

defense in the market. Thus in 1907, one of the advocates of the industry, Keith MacDonald, said:

Nothing made in a chemical laboratory will even come near what is created in the laboratory of nature. The incessant work going on in the cells of the plant – God's laboratory – can't be equalled by any human agency, which, at best, can only be an approximate imitation, without substance; the ghost and shadow of what has stood the test of … thousands of years. The dye which coloured the ribbons that bound the mummies deposited in the tombs of Egypt more than 2,000 years ago, that coloured the beautiful carpets of Achilles, 1,100 BC and has given luster to the magnificent colours of Oriental textile fabrics for countless generations, is not going to succumb ….[32]

Indigo, the blue dye, did not automatically ooze out of the plant. Rather planters used a streamlined manufacturing process that involved systematic processing of leaves in modern, large-scale establishments on the plantations – also called "factories" – to turn indigo into a dye. European planters in Bengal embraced mechanization, steam power, and the use of chemical substances through the course of the nineteenth century at a faster pace than ever before. All of the above meant a progressively more systematic, efficient, and intrusive – and hardly natural – method of extracting the dye. The apparent disdain for the "chemical laboratory" in MacDonald's verdict apparently did not reflect any disavowal of science. It was more a critique of the factory system of chemical dye manufacturing, which in this view seemed further removed from nature than the planters' layout of land, farming, and manufacture of the agricultural commodity.

What was at stake in these debates was the definition of "nature" as embodied in indigo. The indigo dye hardly seemed to be in any way "prior" to human intervention. The Bengal planters' production of the blue dye involved a clear human intervention with nature's scheme of things.[33] In the final leg of scientific experiments the dominant view in the

---

[32] Letter of Keith MacDonald to the editor, *Indian Planters' Gazette*, December 7, 1907, p. 704. The holdings of this journal are available at the National Agricultural Library, Beltsville, Maryland, USA.

[33] The subject of human-nature interaction forms the cornerstone of environmental history. Nature in these studies frequently appears as "constructed" nature or socially "produced" nature. The concept of pristine, untouched nature at some point much deeper in history is not completely ignored, though. If anything, it appears as a constant referent in illustrating the impact of social action on nature. The pristine nature also has a conceptual presence for analytical purposes of highlighting that ecosystems, environments, or the natural world has a history of its own outside human influence. See Donald Worster's treatment of the "arcadian ideal" in ecological worldviews, whether in Gilbert White's

imperial-colonial world held that indigo dye had reached a "natural" limit to its improvement, thus harking back to the notion of "natural order" that was seen as standing in the way of introducing further changes to the natural dye. These views attributed agency to nature, which purportedly had set final limits on the human ability to act on nature. The multiply implied meanings of nature in the claims of indigo planters and dye scientists were relevant in validating a science for the natural dye and later caused its abandonment. The examination of these meanings of nature throws light on the historical question as to when and how the defenders of natural indigo quit the fight with synthetic indigo and let natural dye meet its final demise.

The study of conceptualizations of nature affords an opportunity to examine "society-nature" or "culture-nature" relationships, an important theme in environmental history and science and technology studies (STS). The study of this important relationship has received due attention by environmental historians and STS scholars, who have frequently discussed the extent to which nature could be taken to be a real, material actor or the end product of human actions and envisioning exclusively. "Nature" is treated in such discussions either in its physical form such as landscape, body, and else, or as natural knowledge. Environmental historians have importantly argued that nature was never a given and thus drawn our attention to the relationship between nature and the historical conceptions of nature.[34] STS scholars have clarified the ways scientists generated truth about the natural world, thus conjuring science as a representation of nature, not any

eighteenth-century history of Selborne, England, or Henry David Thoreau's depiction of the history of Concord, Massachusetts. Worster was motivated to study ecological sciences to understand nature at the level of human-nature interactions. In the preface to the new edition Worster acknowledges in an implied way his intellectual agenda of uncovering the autonomy of nature, referring to one of the fundamental questions of the times as "whether nature has an order, a pattern, that we humans are bound to understand and respect and preserve." Donald Worster, *Nature's Economy: A History of Ecological Ideas*, Cambridge: Cambridge University Press, 1988, first published in 1977, p. ix. See also William Cronon's distinction between "first" and "second" nature, in which he identified the former with some sort of a prehuman nature that was later changed through human interaction toward meeting the ends of society. William Cronon, *Nature's Metropolis: Chicago and the Great West*, New York: W. W. Norton, 1991, p. 56, and *passim*.

[34] The subject has a rich historiography and constitutes the very core of the program of environmental history. For a review of the theoretical stakes in this debate in light of very recent scholarship, see, Kristin Adal, "The Problematic Nature of Nature: The Post-Constructivist Challenge to Environmental History," *History and Theory, Theme Issue* 42 (December 2003): 60–74.

reflection of an ultimate reality.[35] Within the latter tradition a few have conceptualized the existence of "networks" within which, they insist, nature, culture, and science are so intimately lodged that it makes no sense to treat any one as the causative prior.[36]

In this *longue duree* study of developing indigo culture the following chapters analyze indigo's improvement in multiple contexts. Chapter 1 connects the origins of indigo plantations in colonial South Asia with the prior and parallel history of indigo in the Western Hemisphere. It demonstrates that the knowledge undergirding the indigo plantations in South Asia was of a composite nature and was also evolving, a knowledge whose genealogy extended to multiple pasts. The second chapter delves into the agricultural history of colonial Bengal to illustrate how the unique system of indigo manufacturing that took root on the Indian subcontinent involved indigo cultivation on peasant plots. The larger focus of the chapter lies in exploring the interaction between a transnational, modernizing agricultural science in the nineteenth century and local forces of landscape and land structure in colonial India. The third chapter connects the global history of transition toward synthetic dyes with the colonial history of agricultural indigo production in India. In doing so it makes the case that colonial developments had an external arena or context that cannot be adequately captured by reference to the imperial framework and its broader networks alone. Colonial trade in indigo had economically fused the Indian subcontinent with consumers in many national markets. Through focus on the British dye market as a national market it lays out the parameters on which natural indigo had to

---

[35] For historically grounded studies of science embedded in society and social relations, see, Steven Shapin and Simon Schaffer, *Leviathan and the Air Pump: Hobbes, Boyle, and the Experimental Life*, Princeton, N.J.: Princeton University Press, 1985; Steven Shapin, *A Social History of Truth: Civility and Science in Seventeenth Century England*, Chicago: University of Chicago Press, 1994. These and similar studies drew on the methodological contributions of sociologists of the Edinburgh school, who emphasized the symmetry that existed between truth and falsity in the way of creation of scientific "facts."

[36] Bruno Latour, and following him, Donna Haraway completely fuse society and nature in their analysis. Both maintain that society/nature dualism is a product of Enlightenment modernity. Latour lends agency to microbes alongside followers of Louis Pasteur, the bacteriologist, in his study of the pasteurization of French society in the last quarter of the nineteenth century. "There are not only 'social' relations, relations between man and man," Latour concluded, adding, "society is not just made up of men, for everywhere microbes intervene and act." Bruno Latour, *The Pasteurization of France*, translated by Alan Sheridan and John Law, Cambridge, Mass.: Harvard University Press, 1988, p. 35. Haraway blends society/nature in what she calls "hybrids." Donna Haraway, *Simians, Cyborgs and Women*, New York: Routledge, 1991; *Modest Witness @ Second Millennium*, New York: Routledge, 1997.

seek validation in the marketplace. Chapter 4 spotlights the entanglement of the developing indigo science with colonial agricultural institutions, which, in turn, were influenced by a program of colonial agricultural development and rising nationalism. Chapter 5 unravels the planters' questioning of specific aspects of reductionist science and a corresponding rationalization in the market in which the dominant criteria of purity and cheapness had sidelined natural indigo. At the same time it also shows the planters as eager apostles of another arm of the same reductionist science in embracing Mendelian selection. But in the end, nothing seemed to work for the planters and their agricultural indigo as consumers worldwide kept switching to synthetic indigo. Even as natural indigo was gasping for breath, the outbreak of the First World War provided a new lease on life to the plant indigo program. As supplies of German synthetic indigo dwindled, a new effort was made to revive the natural indigo industry. A corresponding science of natural indigo also arose that fundamentally tried to turn out natural indigo in a form similar to synthetic indigo. While these efforts were still under way, the end of the war saw all nations reembracing synthetic indigo and many starting the production of synthetic indigo for their national markets. The producers of natural indigo saw the writing on the wall and stopped all efforts to improve indigo scientifically. In that context the last chapter analyzes the standard of rationalization that was held up to the agricultural dye in settling a lasting meaning of "improvement."

The World of Indigo Plantations:
Diasporas and Knowledge

The indigo planters of European descent extended their sway over the Lower Bengal districts and the Tirhut region of Bengal in the last quarter of the eighteenth century. The earliest origins of the plantation system are hazy, and different claims have persevered among sources claiming priority in patronage of indigo manufacturing in the colony. By one account Louis Bonnaud, a French national, took modern indigo culture to Bengal in 1777. These were the times when the English East India Company had not yet embraced the policy of privileging English entrepreneurs over other Europeans in the settlement of the colony. Another account claims that distinctly European methods of indigo culture were introduced by a colonial state official, the first collector of Tirhut, Francois Grand, in 1785. Grand had claimed, "I introduced the manufacturing of indigo after the European manner, encouraged the establishment of indigo works and plantations, and erected three at my own expense."[1] While it is difficult to establish positively the veracity of such declarations, they nonetheless accurately reflect a general trend. Initiatives at introducing novelty in crops and cultivation by colonial officials like Grand were routine. So was the "private trade" of company officials and other Europeans

---

[1] For a reference to Louis Bonnaud as a pioneer of indigo see, Minden Wilson, *History of Behar Indigo Factories, Reminiscences of Behar, Tirhoot and Its Inhabitants of the Past*, Calcutta: Calcutta Oriental Printing Company, 1908, pp. 65–72; John Phipps, *A Series of Treatises on the Principal Products of Bengal No. 1, Indigo*, Calcutta: Baptist Mission Press, 1832, p. 8. The mention of Francois Grand appears in a later account: T. R. Filgate, "The Bihar Planters' Association, Ltd.," Arnold Wright (ed.), compiled by Somerset Playne, *Bengal and Assam Behar and Orissa: Their History, People, Commerce, and Industrial Resources*, London: Foreign and Colonial Compiling and Publishing Company, 1917, pp. 268–351, quote on p. 268.

licensed to be residents in the colony, who deployed their income privately into lucrative manufacture and trade in colonial commodities like indigo. Francois Grand may just have been one of the earliest entrepreneurs trying to make a fortune out of colonial commerce in indigo.

The modern culture of indigo seems to have developed at a number of sites in the Caribbean, Spanish Central America, South Carolina, and Bengal between the seventeenth and nineteenth centuries. Colonialists and locals in these different spaces contributed to the development of the art and science of indigo cultivation and the manufacture of the blue dye whose knowledge and practice ebbed and flowed with diasporas of planters and dispersal of knowledge that was already made concrete in manuals and encyclopedias. A broad linearity in its history can be traced by following the odyssey of indigo through the Caribbean, the Americas, and colonial South Asia. But current historiographical frameworks that simply assume that a fully mature knowledge of indigo culture was "introduced" into South Asia from the West are not tenable in light of the fluidity and multidirectionality in the knowledge system.[2] South Asia's contacts with the rest of the world were reinforced in the early modern period as new global structures ensured, according to J. F. Richards, that "human societies shared in and were affected by several worldwide processes of change." Indeed, he and other scholars have argued that South Asia was part of this new globality and shared some of the worldwide "traits of early modernity."[3] The Bengal plantations were very much a

---

[2] The Indian subcontinent was the major producer and exporter of indigo to Europe until the seventeenth century. The seventeenth-century export of indigo from the subcontinent by Europeans in the early modern period was predicated on widespread production of indigo in several specific regions of the Indian subcontinent such as Gujarat, Rajasthan, and the Coromandel Coast. Archaeological excavations have documented "premodern" production of indigo in eastern Rajasthan near Bayana. Indigo from this town was a major export item and may have made its way to Lahore, from where Central Asian caravans carried it to distant lands. European and Armenian merchants purchased indigo from Agra, another important Indian mart, and transported it to the markets of Aleppo, an important seventeenth-century entrepôt trade center. Iqtidar Alam Khan, "Pre-modern Indigo Vats of Bayana," *Journal of Islamic Environmental Design Research Center,* (1986): 92–8; K. K. Tivedi, "Innovation and Change in Indigo Production in Bayana, Eastern Rajasthan," *Studies in History* 10 No. 1 n.s. (1994): 53–79, see p. 68. The accounts of the Dutch traveler Francis Pelsaert for Gujarat and Rajasthan in the early seventeenth century, of Phillipus Baldeus for Gujarat in the late seventeenth century, and of Herbert de Jager for the territories between Malabar and Coromandel further confirm the manufacture of dye in these regions before the proliferation of European manufacturing methods that are the center of analysis here.

[3] J. F. Richards has argued for a global history of the early modern period and the place of South Asia within the newly established large-scale global processes. John F. Richards, "Early Modern India and World History," *Journal of World History* 8 No. 2 (1997):

part of this broader planetary process tied to movement and flows of knowledge.

The Birth of Caribbean Indigo

The Caribbean era of indigo dawned amid the unfolding Atlantic World economic rivalries among Spanish, British, French, and the Dutch. In the early decades of the seventeenth century, the Antilles endured relentless wars, conquest, and domination by incoming European powers leading to the foundation of a slave-based plantation society and the creation of the "Atlantic system" of manufacture and trade.[4] In the tropical environments, European settlers set themselves to the cultivation of various staples that could not be grown in temperate Europe. Impetus for the endeavor arose from the growth of a new culture of mass consumption in Europe spurring a huge demand for products like cacao, coffee, cotton, indigo, and sugar. Since 1492 the Spaniards had considered the Caribbean to be their strategic backyard, which had to be jealously policed because it lay on routes leading to their imperial territories in Central America. However, in the early decades of the seventeenth century, as Spanish control over the region dwindled, the previous culture of piracy and looting by privateers and buccaneers started to give way to sedentary plantations by a host of new European powers. From an agricultural perspective, the first significant English settlements arose in St. Christopher (1624), followed by Barbados (1627), and later Jamaica (1655). The French also occupied a part of the island of St. Christopher, and then went on to acquire Guadeloupe (1634) and Martinique (1635), and still later the western part of the island of Hispaniola in the second half of the seventeenth century, subsequently calling it Saint Domingue. Dutch traders were also active in the region, and from their outpost at Curacao they visited separate settlements and bought various kinds of merchandise, including indigo, for sale in Europe. As profits from Caribbean plantations rose, mercantilist nation-states in Europe took measures to extend their formal control over these territories. Administrators were

197–209, quote on p. 203; see also Andre Gunder Frank, "India in the World Economy, 1400–1750," *Economic and Political Weekly* (27 July 1996): 50–64.
[4] Thomas Coke, *A History of the West Indies*, London: Frank Cass, 1971; John G. Clarke, *La Rochelle and the Atlantic Economy during the Eighteenth Century*, Baltimore: Johns Hopkins University Press, 1981; Herbert S Klein, *African Slavery in Latin America and the Caribbean*, New York: Oxford University Press, 1990; P. P. Courtenay, *Plantation Agriculture*, New York: Praeger, 1969; Richard S. Dunn, *Sugar and Slaves: The Rise of the Planter Class in the English West Indies, 1624–1713*, New York: W. W. Norton, 1972.

appointed, local assemblies of expatriate populations recognized, and national armies were sent recurrently to protect these territories with the intent to corner trade in commodities to the benefit of the mother country. Beyond the gaze of mercantilist state officials "illegal" trade by merchants in their private capacity also added to the flurry of commercial activity in the Caribbean.[5]

In Barbados the planters first turned to indigo as the price of other staples like tobacco and cotton plummeted in the third decade of the seventeenth century while that of indigo rose sharply. As Robert C. Batie has noted, a steep rise in the price of indigo in one key market, the Netherlands, which rose from 4.43 guilders a pound in 1638 to 8.10 guilders in 1640, explains the impact of international price levels on changes wrought on Barbadian plantations. The best years of Barbadian indigo production may have been between 1638 and 1643. The meteoric rise of indigo in the Barbadian economy was not destined to last long as sustained depression in the price of indigo in turn made the planters turn to sugar rather abruptly in 1643.[6]

In the second half of the seventeenth century, the indigo plantations spread to other West Indian islands. In Richard Dunn's description of plantations in Jamaica, the 1660s seem to be very much the decade of cacao and indigo. The later history of the Jamaican planter Bryan Edwards also indicates that in 1673 the primary produce of the island included "cacao, indigo, and hides."[7] Around the same time, indigo estates were also spreading in the French colonies of Guadeloupe and Martinique. Later, at the turn of the eighteenth century, Saint Domingue became the predominant producer of indigo and remained so through the course of the century.[8] As early as 1713, as many as 1,000 indigo works existed

---

[5] The contraband trade is discussed in Richard Pares, *War and Trade in the West Indies, 1739–63*, London: Routledge, 1963; more recently, the discussion of contraband trade with reference to free coloreds in Saint Domingue appears in John Garrigus, *Before Haiti: Race and Citizenship in French Saint Domingue*, New York: Palgrave, 2006.

[6] Hilary McD. Beckles, *White Servitude and Black Slavery in Barbados, 1627–1715*, Knoxville: University of Tennessee Press, 1989, pp. 23–7; Robert C. Batie, "Why Sugar?: Economic Cycles and the Changing of Staples on the English and French Antilles, 1624–54," *Journal of Caribbean History* 8–9 (1976): 1–42.

[7] Richard S. Dunn, *Sugar and Slaves*, pp. 168–70; Bryan Edwards, *The History Civil and Commercial of the British Colonies in the West Indies, to which is added An Historical Survey of the French Colony in the Island of St. Domingue*, London: B. Crosby, 1798, pp. 77, 81.

[8] John D. Garrigus, *Before Haiti: Race and Citizenship in French Saint Domingue*, New York: Palgrave Macmillan, 2006, see especially pp. 21–50.

in Saint Domingue, many under the control of free colored people in its southern parts.[9] By another contemporary account there were 3,160 indigo plantations on the island in 1790, far exceeding the number of 793 plantations for sugar, a commodity whose importance in the Caribbean has been well recognized.

While it is true that the changing nature of staples reflected a response to price shift, the onset of indigo production across Antillean islands was also predicated on the acquisition of new skills. As noted by Robert C. Batie, the switch to indigo and later to sugar in Jamaica was not an automatic process, but was marked by initial hesitation on the part of planters. Indigo production required new machinery, more slave labor, and new "knowledge," and planters ruminated for a while before leaving the safety of a trade with which they were familiar. The manufacture of indigo required information and training to oversee separate processes like soaking of leaves, beating, and drying of the dye, each of which involved distinct physical and chemical processes. The well-documented switch to indigo trade only confirms that the planters were willing to reequip themselves and to cross the threshold in acquiring skills and machinery as long as the price was right. But it stops well short of explaining the mediating world of knowledge, skills, and preparatory work in methods of cultivation and production that made the transition to indigo actually possible.

The first indigo plantations in the Caribbean set up by the English drew on the knowledge lodged in local native practices and within the trade networks dominated by the Spaniards in the "greater Caribbean." Since the time of Columbus, the island of Hispaniola had served as a Spanish base in the Greater Antilles from which the former led military attacks to conquer territories in Central America. The military, trade, and political linkages across the Caribbean remained in place until the early sixteenth century. Indigo was a key commodity in this part of the Spanish world of politics and commerce. Spanish traders exported indigo from the Greater Antilles and Tierra Firme coast of central America to Seville on the Spanish mainland. In the 1560s, the Spaniards began to organize the production of indigo for export from the Central American territories. The interior valleys of Honduras and El Salvador were the major centers

---

[9] John D. Garrigus, *Before Haiti*, p. 30; Bryan Edwards, *History Civil and Commercial of the British Colonies in the West Indies, to which is added An Historical Survey of the French Colony in the Island of St. Domingue*, p. 361.

of production for Spanish Guatemala indigo in the sixteenth century.[10] In the Caribbean half of this Spanish indigo empire native populations were similarly already knowledgeable about methods of indigo cultivation and production. The new incoming waves of Europeans entered into an enduring period of "interaction" with this knowledge regime living through the age of privateers and pirates, just as the Spaniards had in Central America. Some tentative answers about this process of "engagement" come from the domain of privateering in the Elizabethan era Caribbean. Among others, Dunn has provided clues by pointing toward the period of piratical economy in the Caribbean islands prior to their formal settlement. He points to the traffic of English sailors in the region who went there to trade with the Spanish colonists in their private capacity or to plunder their goods and slaves. And indigo was one principal commodity of that plunder and trade. Thus he argues that the Englishmen were "no babes in the woods" when they set up their plantations in the 1620s. Rather, they were "heirs" to the prior regime of plunder and conquest that had lasted for more than half a century and drawn English trade and capital to the region.[11] The knowledge of indigo culture seeped from the prior regime into the formal plantations as the latter began to be set up in the Caribbean.

The potent resources of the global British Empire actively assisted the accumulation of knowledge and wares for indigo development in the English West Indies. In fact, one gets a sense of deliberate plans by English capitalists to "move" knowledge from the East to the tropical climes of the New World. The English East India Company, for example, had established its "factory" at Surat in Gujarat, from where it exported Indian goods. Indigo was the primary commodity of export in 1620. While indigo export from Gujarat was anything but trivial for the entire century,[12] after midcentury indigo from other places in the West was becoming more important. English factory records disclose a shifting imperial emphasis from South Asia to the New World as far as production

---

[10] Dauril Alden, "The Growth and Decline of Indigo Production in Colonial Brazil: A Study In Comparative Economic History," *Journal of Economic History*, 25 No. 1 (March 1965): 35–60, especially, pp. 39–40.

[11] Dunn, *Sugar and Slaves*, pp. 10–11; Kris Lane, *Pillaging the Empire: Piracy in the Americas, 1500–1750*, Armonk, N.Y.: M. E. Sharpe, 1998; Miles Ogborn, *Global Lives: Britain and the World, 1550–1800*, New York: Cambridge University Press, 1998, pp. 16–46.

[12] Om Prakash, *New Cambridge History of India: European Commercial Enterprise in Pre-Colonial India*, New York: Cambridge University Press, 1998, pp. 226–7.

of indigo was concerned.[13] One series of correspondence among factory records from 1662 to 1663 reveals an effort by English company officials to introduce Indian methods of indigo production in the island protectorate of St. Helena in the South Atlantic. This reflected a general desire to shift the pivot of indigo production westward. The New World production offered obvious advantages in proximity to Western markets. Besides, South Asian export was based on English purchase of indigo from Indian peasants who were the direct producers of the dye. In the West Indies, in contrast, the production of indigo under the direct supervision of European planters afforded more control to the manufacturers. Reflecting the general shifting emphasis away from the Indian subcontinent, a letter by a colonialist proposed to send indigo seeds from the East India Company's stocks in Surat, a town on the west coast of India, to St. Helena. It also promised to look into the possibility of sending across a "black," presumably an Indian native, "that hath knowledge how to sow it and afterward worke it to perfection." The effort did not materialize in the end as colonists in St. Helena conceded that they had been unable to find a local person to sow the seeds procured from India in the appropriate way: "Wee cannot by any means procure a black to sew it."[14]

Modular Texts and Their Historical Mobility

The process of formal knowledge generation on indigo received a fillip locally in the West through a trailblazing production of texts. The major initiative here first occurred among French naturalists who traveled in the French West Indies in the seventeenth and eighteenth centuries. Their contributions were immense in terms of providing input for subsequent writings, whether encyclopedias of commercial products, gardener's texts, or pure scientific works. By writing about what they saw on indigo plantations, the naturalists transformed the informal, tacitly known, and widely practiced craft into concrete knowledge that was now textual in form. In other words, they committed experiences circulating in the indigo world, those resting with "old hands" and "native" customs, into tangible, described knowledge. They made this knowledge clearer and more legible to current practitioners and transportable and helpful in enlisting

---

[13] William Foster, *The English Factories in India,* Oxford: Clarendon Press, 13 vols., 1906–23.

[14] William Foster, *The English Factories in India,* Oxford: Clarendon Press, vol. 11, pp. 186–214. British Library, Oriental and Indian Collections, OIR 354.54P.

potential entrants. In addition, and very importantly for the future development of the "science" of indigo manufacture, they collected plant materials and cataloged them for later reference. Their reproduction of the tropical plant and culture of indigo production served philosophers in the distant, temperate metropolis where the "wise men" did not have access to the plant locally. With time, the travelers' inputs were incorporated into important scientific texts and practical agriculturist guides.

The first two significant travelers from the French metropolis who touched upon the subject of Antillean indigo culture in a substantive way were both missionary-naturalists. A Catholic humanist of the Dominican order, Jean Baptiste Du Tertre, immigrated to the Antilles in 1640. He described life in Caribbean society and defended mercantilist policies, and he included a description of indigo as part of his narration of flora and fauna in the Caribbean. His study covered all parts of the French territories in the Lesser Antilles including St. Christopher, Guadeloupe, and Martinique. The writings were published when Du Tertre returned to France after a sojourn of eighteen years in the Tropics. He published his most important work, *L'Histoire generale des Antilles habitués par les Francais*, in 1667, a popular text that went into a reprint in 1671.

The next missionary, Jean Baptiste Labat, also of the Dominican order, visited the Antilles between 1693 and 1706 and produced a work that focused more directly on the natural history of the French West Indies than Du Tertre's. Labat resided in Guadeloupe and Martinique and visited Saint Domingue in the early eighteenth century. He published his account of his travels, *Nouveau Voyage aux isles Francoises de l'Amerique*, which contained a detailed account of indigo cultivation and manufacture as practiced in the Macouba parish of Martinique, in 1722.

Labat's indigo account turned out to be the first widely dispersed comprehensive explanation of indigo processes of the era, and it also traveled seamlessly across historical times. It was translated into a Dutch rendition just three years later. In 1731 an English translation of the work appeared in London as an entry in Phillip Miller's *Gardener's Dictionary*. As noted earlier, the latter text was a source of information for the South Carolinians when they took to indigo in the 1740s.[15] The text by Labat was also translated into German and appeared in multiple editions in the

---

[15] Miller reproduced Labat's account. See the entry "anil" under *AN* in the dictionary. Philip Miller, *The Gardener's Dictionary in Two Volumes*, London: printed by the author, 1743, no page number; David L Coon, "Eliza Lucas Pinckney and the Reintroduction of Indigo Culture in South Carolina," *Journal of Southern History* 42 No. 1 (February 1976): 61–76.

German language from 1768. It likely formed one source of information on indigo culture that appeared in editions of the key eighteenth-
century encyclopedia of commerce in Germany composed by Carl
Günther Ludovici. And, in a further illustration of fluidity of knowledge,
the detailed description of indigo culture provided by Ludovici was translated and published in Calcutta as part of a comprehensive collection
of indigo works at the end of the eighteenth century. Thus an advertisement in an important Calcutta newspaper announced that the publisher
R. Nowland, Calcutta literatus, would be selling a practical manual on
the culture of indigo beginning next year that would include translations
of Ludovici's work, among others. Nowland was trying to attract the
attention of proprietors of indigo factories and their controlling banks in
the colonial Indian metropolis to whom the name of Ludovici was apparently not foreign and for whom his work on indigo would have seemed
a bargain.[16]

Indigo Culture in Saint Domingue and the Tract of Elias Monnereau (1736)

Attesting to the fact that knowledge of indigo culture had achieved a
high degree of specification in the French colonies in the West Indies,
a detailed account of its cultivation and manufacture was published
in Saint Domingue in 1736. The author, Elias Monnereau, was a local
planter of several years' experience. He emphasized that his work was
a product of direct, unmediated experience and deep understanding of
the processes of production, in contrast to accounts of "different travel-
lers" that were based either on those authors' recounting of "testimony
of others" or comprised the observations of "mere spectators." He bran-
dished his production as the most credible account of indigo culture
to date. Smugness aside, Monnereau's claims had some basis. By the
end of the first quarter of the eighteenth century Saint Domingue was
the Caribbean island with the longest history in indigo cultivation. The
French island had also won acclaim as the dominant producer of the finest quality of indigo in the world. Monnereau's long and accomplished
experience in indigo culture was a part of the deep and exceptional history of indigo in Saint Domingue. As he said, he wished for that experience to serve the interest of his French countrymen and fellow colonists

---

[16] *Calcutta Gazette*, April 18, 1793. The holdings of the newspaper are available among the
Oriental Collections of the British Library, SM 128.

specifically.[17] But little did he know that Saint Domingue indigo production that his work described would have a lasting influence on indigo
plantations on the West Indies islands of France's key rival, Britain.
Even more importantly, as later sections will discuss, Monnereau's
work would be translated to reach the wider English-speaking world
and later have an abiding influence on indigo culture in Bengal, the
colony that was acknowledged to be the jewel in the British crown.

In the book entitled *The Complete Indigo Maker* (Fig. 1) Monnereau
described three primary varieties of the indigo plant in Saint Domingue:
real indigo, bastard indigo, and Guatemala indigo. The first was particularly vulnerable to weather conditions and to attack by pests even
though it was easier to manufacture and gave color of a good quality.
Thus, even though it yielded the best and maximum output of indigo, it
presented planters with many challenges during cultivation. Describing
its vulnerabilities, Monnereau said, "The wind, rain, the sun, all conspire
to destroy it; and even the earth where it grows seems to deny it assistance." The burning fly and caterpillars attacked the plant at a later stage
in its growth. The bastard indigo was hardier as it grew in all types of soil
and was less vulnerable to the countless insects endemic to the Tropics.
The plant stayed in the field longer than the real indigo variety, but at
least the chance of losing the crop to weather and insects were lower.[18]
The Guatemala variety was particularly difficult to run through the
manufacturing process and, thus, not favored. There also existed other
minor varieties like the wild indigo and Indigo Marry.

For all indigo varieties cultivation was labor intensive. It involved multiple rounds of clearing and loosening of the land to prepare it for sowing
in the month of December. Slave labor was used to dig holes five or six
inches apart in soil with a hoe, to drop seeds with a "gourd rake," and to
cover the punched seed with soil. Monnereau thought that seven or eight
seeds were sufficient for each hole, though he cites his predecessor Labat
as recommending up to eleven or thirteen. Wet soil was of paramount
importance for the seed to germinate. But for large plantations it was

---

[17] Elias Monnereau, *The Complete Indigo Maker: Containing an Account of the Indigo
Plant; Its Description, Culture, Preparation, and Manufacture, to Which Is Added a
Treatise on the Culture of Coffee*, translated from the French of Elias Monnereau, a
planter in Saint Domingue, London: P. Elmsly, 1769. British Library, T36433. See the
preface for the author's reflections on his purpose for writing the account, pp. v–x.

[18] It is likely that the "bastard" variety was the product of haphazard crossing with local,
acclimatized varieties growing in the wild. The hardiness and resistance of bastard indigo
against local pests were the mark of inheritance of locally developed resilience in the
local environs.

THE COMPLETE

INDIGO-MAKER.

CONTAINING,

An Accurate Account of the INDIGO PLANT;
Its Defcription, Culture, Preparation, and
Manufacture.

WITH

Œconomical RULES and neceffary DIRECTIONS for a
Planter how to manage a Plantation, and employ his
Negroes to the beft Advantage.

To which is added, A

TREATISE on the CULTURE of COFFEE.

Tranflated from the FRENCH of

ELIAS MONNEREAU,
PLANTER in ST. DOMINGO.

LONDON:

Printed for P. ELMSLY, in the Strand.

MDCCLXIX. 1741

FIGURE 1. Front cover of Elias Monnereau's *Indigo Maker* 1769. (c) The British Library Board.

difficult to accomplish sowing over the entire estate in the small window of time while the ground was still wet. To maximize acreage, planters sometimes took a calculated risk by indulging in "dry planting" with the hope that following rains would help the seedlings to sprout. Sometimes the gamble did not pay off, leading to a complete loss of that portion of planted land. But planting was generally a long stretched-out process lasting well into May. The disadvantage of late planting was that the crop in these cases did not permit as many cuttings of the plant as was possible with early planting. Repeated weeding was absolutely necessary because the tropical climate and rich soil encouraged plentiful weeds that could choke off the young seedling. Weeding every fortnight was strongly recommended until the plant had reached a respectable height to overcome the competition of weeds for nutrients and sunlight.

The manufacturing process was well-defined and divided into three distinct stages, which required skillful supervision. The operations were assigned to take place in three different vats made of stone or wood. The harvested leaves were packed in bundles and transported to the factory site, where slaves loosely spread them into the first vat, called the steeper. Water was run into the vat; covered leaves were then allowed to undergo fermentation. A colder day prolonged the period of fermentation beyond the normal completion time of twelve to fifteen hours. But sometimes the period of fermentation could extend up to thirty hours, depending upon multiple variables, and it is here that the keen eye and experience of the planters were most useful in navigating the operation. Some planters preferred to store water in a reservoir to warm it with the Sun's rays before running it into the steeper. This additional preparatory stage hastened the process of fermentation. The "critical time of the dissolution" of the herb in the steeper also depended on the quality of the crop. A skillful planter bore this fact in mind while watching the changing color of the water, its drossy nature, and the content of suspended matter in the liquor while determining the terminal "just point of fermentation." A crop grown under excessive rains produced imperfect grain, and so did a crop grown under drought conditions. Thus, in these cases, the imperfect nature of suspended matter did not indicate imperfect fermentation. Similarly, the first cutting from the crop often produced "false grain," and a crop harvested hurriedly after ravages of caterpillar produced drossy water, neither of which denoted that the process had been compromised. The planter watched the pace of dissolution in the steeping vat by periodically extracting water from the middle parts of the vat through an aperture and placing it in a "proof cup" to gauge the quantity and state of its suspended matters. The changing color of liquor in the vat also reflected the pace of fermentation. Bluish water with a green background or brownish tincture indicated that the vat had "rotted too much" while reddish or yellowish green color meant that the vat required additional fermentation of a couple of hours. The "color of coniac brandy" [*sic*] announced to the planter that the optimal point had been obtained to leave him secure in the belief that he had "extracted the quintessence."

In the next stage, the fermented liquid was transferred to the beater, a vat that was deeper in dimension than the steeping vat and that lay beside the steeper and below it on a gradient. Underlining the importance of this stage, Monnereau called beating "the principal operation," one "which alone can bring Indigo to perfection, or destroy it."[19] The workers subjected liquid to violent shaking, scooping, and stirring with the help of bottomless buckets attached to a long piece of timber. The process was continued "till such time as the salts and other particles of the plant are united and blended together."[20] The completion of this supposed collusion had to be monitored and regulated. The key again was to reach the optimal point when all the color available in the plant had been extracted. Any less beating than the optimal would leave the color suspended in liquor that did not settle to the bottom of the tank and thus escaped with runoff effluents. And overbeating caused the color to redissolve and become unrecoverable, an event that was, according to Monnereau, "an accident that is without a remedy." Thus, planters regularly visited the vat and used the proof cup to check completion of the process of formation of color. When they witnessed that the sedimentation started forming round grains like sand that rolled over each other, leaving clear water to rise to the top, they stopped beating. The beating stage was also crucial because it afforded an opportunity to discover and correct any mistakes made during the stage of fermentation. "The fixed point of fermentation is very seldom hit upon," warned Monnereau. Thus he advised the planters to watch for signs of insufficient fermentation, often indicated by excessive froth at the top of the vat that was quickly dissipated with a sprinkling of oil. If that was the case, he asked planters to prolong beating. In the case of excessive fermentation, the grains appeared very soon after the start of the beating process. The water, however, did not become clear in this case, but rather stayed muddy. In such

[19] *Indigo Maker*, p. 39.
[20] *Indigo Maker*, p. 24.

an instance of proven overfermentation, the planters were advised to stop beating as soon as early grains had formed.

After beating, the vat was left to rest for a couple of hours for the grains to settle to the bottom of the tank. The beating vat contained three holes at different heights on one of the sides to let off water. These were unplugged one after another beginning at the top until only the dye remained at the bottom. The latter flowed out of the beating tank into the receiver. A broom was often used to sweep the floor of the beater to collect all of the dye. The indigo was then transferred into linen sacks and hung for a day to allow the remaining water to drip away. It was then spread in cases and exposed to the sun to dry. It was cut into one-inch squares, and these cakes were then further dried to give them the required hardness. Some planters preferred to dry indigo in the shade, a much longer process that extended up to six weeks.

The reference to previous works on indigo in Monnereau's tract attests to the fluidity of information and to the historical reality of the multiple lineages of indigo culture that had consolidated itself in eighteenth-century Caribbean. Monnereau made effective use of information provided by the naturalist Jean Baptiste Labat, his fellow countryman, whose work would have been easily accessible to him. In fact, he entertained high regard for Labat's work as a naturalist and endorsed his morphological description of the indigo plant while criticizing the flippant allusions to the plant by other writers who had described it in general terms as a shrub or even, inaccurately, called it similar to hemp. Displaying the cosmopolitan basis of knowledge on indigo in Saint Domingue, Monnereau also nodded in the direction of the seventeenth-century account of indigo manufacturing in India by the French traveler Jean Baptiste Tavernier. Tavernier had made multiple voyages to the East and Southeast Asia between 1640 and 1667. On these extended voyages he had undertaken extended travels on the Indian subcontinent and even visited the court of the esteemed Mughals. Although the primary commercial motives on these voyages were collection and sale of precious stones from the Golconda mines in southern India, his travelogue also described methods of indigo manufacture in India, among other topics. Monnereau used Tavernier's account in support of his theory that leaves of the indigo plant, and not stalks, contained the color-yielding principle. Although he was generally disparaging of the manufacturing technique used by Indian manufacturers, while highlighting the superiority of Saint Domingue practices, he cited the Indian practice to reemphasize the point that if color was not exclusively lodged in the leaves the Indians would not spend so much time separating leaves from the stalk.[21]

There is no doubt that the knowledge about indigo manufacturing in Saint Domingue that Monnereau so well described would have influenced indigo culture in the English colony of Jamaica over the course of the eighteenth century, where major planters had moved to sugar and cotton production,[22] and vice versa. This happened particularly as a result of the thriving illicit trade that enabled separate trading communities to transmit key information. Some evidence of such an overarching informational order can be deduced from the similarities in the specifics of the culture of indigo production in the two adjacent imperial systems. But there is also some direct evidence of exchange of information between French and English colonies. Bryan Edwards's history of Jamaica refers to the mercantile body, the Chamber of Agriculture in Hispaniola, that had apparently "made repeated experiments" on indigo and published its results for "the benefit of the public." Apparently this publication was available in Jamaica or to Edwards himself because he cites this report on the determination of the "true point of fermentation," the terminal point

[21] Elias Monnereau, *Complete Indigo Maker.* pp. 4–5. It is not certain whether Monnereau actually saw the original account of Tavernier. But he had certainly seen excerpts from Tavernier's *Travels in India* describing indigo manufacturing, which had been incorporated into an important text on drugs in France. Jean Baptiste Tavernier, *Travels in India by Jean-Baptiste Tavernier, Baron of Aubonne,* translated from the original French edition of 1676 by V. Ball, second edition, edited by William Crooke, 2 vols., London: Oxford University Press, 1925; see vol. 2, *Concerning Indigo,* on pp. 8–12. The reference to leaves being the only color-bearing part appears on p. 10, and the reference to the manufacturing process and drying on pp. 10–12. Tavernier refers to the practice of soaking of leaves in water in a single large tank "80 to 100 paces in circuit," stirring in the tank for days, allowing the "slime" to settle over additional extra days, and then collecting the dye after draining off water and drying in the open sunshine.

[22] The historical accounts of the English West Indies by Bryan Edwards and Edward Long, compiled late in the eighteenth century, note the decline of indigo. Bryan Edwards expresses astonishment at the decline of the indigo industry, considering "the cheap apparatus attending the manufacture" and "the small number of negroes requisite for its culture." Obviously Edwards was speaking in relative terms comparing the cheap apparatus with the more elaborate machinery for sugar and the huge labor demands of cotton plantations. Similarly, Long seems almost to rue the decline of indigo cultivation in Jamaica, where the number of indigo works had declined from sixty in 1672 to a mere twenty at the time of his writing. Bryan Edwards, *The History Civil and Commercial of the British Colonies in the West Indies, to which is added An Historical Survey of the French Colony in the Island of St. Domingue,* London: B. Crosby, 1798, pp. 239–40; Edward Long, *The History of Jamaica, or a General Survey of the Antient and Modern State of the Island with Reflections on its Situation, Settlements, Inhabitants, Climate, Products, Commerce, Laws, and Government,* vol. III, London: T Lowndes, 1774, pp. 679, 681.

of maturity after which the liquor had to be immediately turned over to beating. The method involved dipping a pen in the fermentation vat at intervals of every half-hour and making a few strokes. The planters were advised to stop fermenting the moment the liquor started turning color-less on disturbance. Even though he had never tried it, Edwards found the method suggested in the report unconvincing: "It is astonishing that an experiment so simple in itself, if it answers, should have been for so many years unknown to the indigo planters in general." As a mere narrator, though, he found the alternative method from one "Mr. Lediard" more convincing. This method required making apertures in the lower part of the vat to let a small stream of liquor continually ooze out through it. According to this prescription, fermentation should be stopped a few hours into fermentation when the liquor started turning from green to a copper-like color at the bottom of the vat. Although Edwards does not elaborate, it can be said with certainty that the "Mr. Lediard" alluded to is John Ledyard, an experienced dyer of Milksham, Wiltshire, in England. John Ledyard composed an account of appropriate methods of produc-ing indigo that was published first in Bath in 1773, and then at Devizes in 1776. It was a text directed to English planters based in Carolina. This publication also included letters between Ledyard and a Carolina based planter that were republished from the *Carolina Gazette*. Edwards's information originated from the world of indigo manufacturers and traders existing between the American South and England, further evi-dence of the expanding circumference of knowledge pertaining to indigo manufacture.[23]

There is little doubt that the most perfect West Indian system of indigo manufacture was established in the French Saint Domingue in the eighteenth century. This fact is evident in the reading of the accounts of indigo production in the British West Indies by Edwards and Long that "metropolize" the French methods of cultivation and culture in Hispaniola. Those accounts are often written with a sense of defer-ence to the ideal type in Saint Domingue and are replete with numerous

---

[23] Bryan Edwards, *The History Civil and Commercial of the British Colonies in the West Indies*, pp. 237–8; the actual designation of the institution was "Chamber of Agriculture." *Cf.* James McClellan, *Colonialism and Science: Saint Domingue in the Old Regime*, Baltimore: Johns Hopkins University Press, 1992, pp. 44–5; *Methods for Improving the Manufacture of Indigo: Originally Submitted to the Consideration of the Carolina Planters; and Now Published for the Benefit of all the British Colonies, Whose Situation is Favorable to the Culture of Indigo. To Which are Added Several Public and Private Letters, Relating to the Same Subject, by an Experienced Dyer*, Devizes: T. Burrough, 1776.

references to French colonialists. Some of the differences in the culture in English colonies, however, are attributable to the inertia of old, conventional practices and factors of climate, terrain, and particular varieties of indigo grown in different territories. In other words, in the world of interregional knowledge flows "differences" in local practices were a norm. For example, it was routine in Jamaica to dry the indigo after manufacturing in the shade and not under direct sunlight. Edward Long expressly warned against "exposure to the sun, which would be very harmful to the colour of the dye." In contrast, a 1770 account of indigo culture in Saint Domingue put forth sun drying as a standard practice although it mentioned that drying in shade, though time-consuming and requiring ample space to store a large inventory over the long period that the slow shade drying would require, might indeed yield a better dye. It is likely that the large-scale production of indigo in the French colony made shadow drying impractical. It is also likely that the practice of sun versus shade drying reflected the specific properties of indigo varieties grown in respective colonies, their soil and climate conditions. Similarly, the apparatus used for manufacturing in the English colonies was not as elaborate as that used in French colonies. Thus, the Jamaicans used only two vats, the "steeper" and the "battery." After the beating process was over, the dye was simply scooped out of the battery. In Saint Domingue, the indigo apparatus comprised three different vats: a reposer or a receptacle, a third vat, was added to the steeping vat and a beating vat or battery, as also described by De Beauvais-Raseau, a militiaman in Saint Domingue, who composed a treatise on indigo and published it in France.[24] It is likely that the lack of attention to the apparatus in the English colonies was the result of the relative decline of the industry itself. The remaining indigo works in the English colonies were not drawing the major part of English colonial capital.

De Beauvais-Raseau's Account (1770)
of Improvements in the Caribbean

There is evidence that considerable improvement in methods of indigo culture was secured all across the Caribbean in the second half of the

---

[24] Edward Long, *The History of Jamaica*, p. 678; De Beauvais-Raseau, *L'Art de L'Indigotier*, Paris: L. F. Delatour, 1770, translated by Richard Nowland, *A Treatise on Indigo*, Calcutta: James White, 1794, pp. 5–64; for specific references to sun drying, see, pp. 22–4, 64; for reference to the third vat, see, pp. 53–4; henceforth, cited as *A Treatise on Indigo*.

eighteenth century.[25] Indigo tracts from this period explicitly stress the importance of maintaining a high quality of water used in fermentation tanks. "The softest water answers best for the purpose," counseled Long, expressing a new carefulness in this regard that contrasted with Monnereau's neglect of this aspect in his account for Saint Domingue four decades before. De Beauvais-Raseau, also in the later part of the eighteenth century, was absolutely insistent that planters use good quality water, saying that "the quality of water made use of, have a great influence on the manufacture of Indigo." The water from rivers and streams that were not too cold, nor too hard, and clear was best for "dissolving the plant." Well water in the region often had salinity and he advised planters to avoid it if they could. He had apparently studied the impact of saline water on the final product, which could make the final dye "a bad acquisition." Such indigo would have a fine appearance, but on exposure to air the salinity in the indigo would attract dampness and the product would then be less attractive to buyers who preferred solid cakes. He also warned against the use of stream water if it had sediments that would "suspend the activity" and spoil the indigo "very considerably." Those planters who stored water in a reservoir he urged to guard against insects' breeding in the tank, which could spoil the indigo. The importance given to the quality of water is evident in De Beauvais-Raseau's reminder to planters wishing to set up an indigo work: "When you design to build Indigo works, you should above all things examine the place; observe if it is possible to convey the water of any river or stream to the vats; for if you have not this convenience, you must independently dig wells or make reservoirs ... without which the finest works would become useless.

[25] New emerging historiography has asserted that the Caribbean basin retained primacy in the production of indigo until the end of the eighteenth century against the grain of the general impression that the region had moved to other commodities at the cost of indigo. For Saint Domingue, John Garrigus's innovative research has argued that if one moves away from official trade statistics the impression that the primacy of indigo was challenged by sugar and coffee does not hold. He argues instead that indigo continued to be dominant using a new set of sources on contraband trade in indigo by English and Dutch interlopers as well as French merchants from the geographically distant, mountainous southern coast of the peninsula, which escaped inclusion in official statistics. Garrigus has also argued for indigo production by the "free men of color" in the colony. In one parish alone, for which records are explicit, the expansion of indigo between 1750 and 1787 far exceeded that of cotton, coffee, and sugar. John D. Garrigus, *Before Haiti: Race and Citizenship in French Saint Domingue*, New York: Palgrave Macmillan, 2006; "Blue and Brown: Contraband Indigo and the Rise of a Free Colored Planter Class in French Saint-Domingue," *Americas* L2 (October 1993): 233–63; Richard Pares, *War and Trade in the West Indies, 1739–63*, London: Routledge, 1963.

When you have made sure of this essential point, you may begin your buildings [for manufacturing]."[26]

The most significant aspect of indigo manufacturing in the latter part of the eighteenth century may have been the rising use of limewater and other alkali substitutes as a precipitant. Both Long and Edwards pointed to the use of lime in the beating vat in the English side of the Caribbean. Bryan Edwards actually described addition of a separate lime vat in the vat series and suggested positioning of an outlet at least eight inches from the bottom of this cistern containing the alkali so that only clear limewater and not any of the solid lime residues made their way into the fermentation vat. De Beauvais-Raseau also discussed the use of precipitates in the French Caribbean and wondered about their efficacy, a pointer perhaps to the recent emergence of its use. He cited other authorities to the effect that in the preparation of indigo in India quick lime (*chunam*) was used, whereas planters in Carolina used limewater. He conceded that alkalis had a positive effect on the formation of the grain during manufacturing. Apparently, as per Raseau's account, the Saint Domingue planters had tried lime, but their initial trials had not been very successful: "The first trial with lime, not having been executed with exactitude and science, produced an Indigo which discouraged the Artists from renewing the trial." Its substitute, urine, was known to catalyze the formation of dye grains in Jamaica as long as the addition was "in proportion to the perfection of the fermentation and beating." In the latter case he regretted that "it does not appear that any effort has been made to render that knowledge advantageous." All these meant that "the question upon the discovery of the true precipitant, thus remains undecided; but there is every reason to believe that an able Chymist would succeed and resolve it, if his operations were to be seconded by all Proprietors of Manufactories to whom the knowledge of it would be so very interesting." By all indications, the planters in the French colony were hesitant to use alkalis, a practice that in the opinion of De Beauvais-Raseau awaited "greater experiments and better trials" for its validation.[27]

The production of texts was meanwhile catching up at all levels from the colony to the metropolis in an attempt to streamline the vat processes. There is evidence of efforts to write manuals for the practitioners at even the most local levels. While such accounts that would have been written or published for limited circulation did not enjoy any remarkable

---

[26] Edward Long, *The History of Jamaica*, p. 676; *A Treatise on Indigo*, pp. 14, 55–6.
[27] *A Treatise on Indigo*, pp. 7–8.

longevity, fragmentary references at least allude to the fact that they were created. Thus the historian John Garrigus's study of the Raimond family in southern Saint Domingue, a family of free coloreds engaged in indigo plantations, discloses such an effort by one Julien Raimond. Raimond is reported to have "subscribed 200 livres for the publication of new information about the distillation of indigo dye." This amount was minuscule, especially compared to his fortune totaling 202,000 livres in 1782, but still reveals efforts to produce texts for local use.[28]

De Beuavais-Raseau's account, fundamentally a practical manual, made an additional effort to explain theoretically the nature of vat processes within a broader "theory of fermentation." Texts like De Beauvais-Raseau's scored over local texts by their superior ability to explain the processes, based as they were on their authors' ability to move between the colony and the metropolis and to meld the knowledge between praxis and philosophical speculation in the two zones. These compilers of "treatises" successfully combined their understanding of vat processes born out of witnessing manufacturing processes firsthand in the colony with the insights of savants and philosophers in the metropolis. Thus De Beauvais-Raseau's *Treatise* explained the transformations in the indigo vats as being composed of "three revolutions." The author offered both the theory that underlay these transformations and their correspondence with observable changes. "The theory of this Manufacture is founded upon the fermentation of vegetables that are subject to pass from a hot or spirituous state, to an acid one, and from this to a putrid state." But since the acidic stage was so fleeting the planters were counseled only to concern themselves with changes in spirituous states. The herb passed from the fully fermented state, or "most spirituous state," to the state of putrefaction or excessive fermentation very suddenly. The putrid state of the liquid meant a partial to complete loss of quality and quantity of the dye, and thus all precautions had to be taken to preempt the onset of putrefaction in the fermentation vat. The task of fermentation was to develop all "active and passive principles" that would go on to form the grain. But, when not handled properly, fermentation ended up destroying the active principles, while separating other unnecessary ingredients from the leaf that remained suspended in water and could not be processed into color during beating. Beating accomplished "the union of the particles and the formation of the grain." An optimal beating not only caused a speedy

[28] John Garrigus, "Blue and Brown: Contraband Indigo and the Rise of a Free Colored Planter Class in French Saint-Domingue," p. 249.

union of principles but also ensured that "all parts proper for the fecula meet, adhere and concenter [*sic*], forming small bodies of different sizes," in other words, form plentiful, healthy grain.

But the knowledge was not yet fully mature and did not give practitioners any significant control over the flow of manufacturing process. De Beauvais-Raseau sounded a warning along these lines, saying, "In this art no specific rule is laid down for the time of fermentation, nor for that of beating." The inherent contingencies of the process "render[ed] this art variable, obscure and subject to many errors, and failures during the process."[29] He asked planters to be observant to notice "the just point" at which fermentation and beating were completed. There were signs and indications to react to; an important visual marker was that a fermented vat at perfect maturity had a deep green color. He also advised that water should be drawn from various parts of the vat and placed in the proof cup and observed after a little shaking. If the fermentation was complete and adequate, the silver cup would receive grains of the dye on its bottom as stirring caused the same physical effect in the silver cup that beating obtained in the second vat. The planter was called upon to "seize the moment" as any prolonged fermentation after this point would cause loss of dye. Similarly, optimal beating would be indicated by water at the top that was amber in color and clear overall. On the other hand, if beating was exceeded, the water appeared to be "troubled" and brown or blue in color. The indications of spoilage after the fact could also be useful as they would enable the "artist" to recalibrate the length of time for processes in subsequent vats. Thus it was useful to bear in mind that a final, dried indigo, if friable, indicated that either fermentation or beating had been inadequate. And a black or slaty color paste of indigo or an indication of grains that were not coalescing meant that either fermentation or beating had been exceeded.

Guatemala Indigo: Peasants and Capitalists

On the far western rim of the Caribbean basin Spanish Central America emerged as the producer of "Guatemala indigo." The local variant of the indigo plant, called *xiquilite*, grew wild in the area, and the Spaniards first began to depend on local Indians to collect the plant and process the dye that they then acquired to be sold to overseas markets as they extended their sway in the region. Spanish export of indigo from

[29] *A Treatise on Indigo*, p. 6.

the region began as early as the mid-sixteenth century. Direct Spanish involvement with indigo production started in the early seventeenth century and was reflected in the establishment of indigo estates (*haciendas*) and dye works (*obrajes*) that over the course of the seventeenth century continued to coexist with peasant units that were dominated by Indians, mulattoes, and the Hispanicized ladino peasantry. With direct Spanish involvement in production El Salvador became the primary region for the production of indigo; western Nicaragua and southwestern coastal parts of Guatemala were the two other axes of the production zone.[30] In the latter half of the eighteenth century, "Guatemala indigo," still primarily exported out of the El Salvador region, won a reputation for its high quality and was counted along with French and Carolina indigo among the principal dyes for an industrializing Europe.[31]

The widely acknowledged "best quality dye" from Guatemala in the eighteenth century was produced with processes that shared common elements with those in vogue elsewhere in the West Indies. The early pre-Hispanic history of indigo production and the period of engagement between the native and Spanish systems has not been explored by historians so far. But in the last quarter of the eighteenth century, the process of manufacturing consisted of putting the leaves into large stone cisterns or vats, called *pilas*. They could be soaked for as long as twenty-four hours. Some practitioners were also known to use warm water in the fermentation vat. The beating process in the next vat was supported by water wheels or animals. The importance of determining the "just point" was realized here as well, just as in the Caribbean traditions. In the historian Macleod's description, a specific group of professionals had specialized in the art of determining the finishing point on the basis of observations; they were called *punteros*, or point watchers.[32] Jose Antonio Fernandez Molina indicates the existence of an alternative method of dye production

---

[30] David Browning, *El Salvador: Landscape and Society*, Oxford: Clarendon Press, 1971, pp. 66–77; Murdo MacLeod, *Spanish Central America: A Socioeconomic History, 1520–1720*, Berkeley: University of California Press, pp. 176–203. Since the first Spanish exports of indigo had been sourced from the Guatemala region, the name "Guatemala indigo" stuck with the Central American indigo.

[31] Jose Antonio Fernandez Molina, "Colouring the World in Blue: The Indigo Boom and the Central American Market, 1750–1810," Ph.D. Thesis, University of Texas at Austin, 1992. Molina identifies Central America's period of indigo boom as extending from 1760 to 1792, one that coincided with its "strong linkage to the world market" (p. 4). See also, Molina's description of the coexistence of estates and peasant production units in the region (chapter 2) and his graph for prices in the London market in 1775 (p. 48).

[32] Murdo MacLeod, *Spanish Central America*, p. 180.

in which indigo leaves were boiled and stirred in large pots by peasants. It was claimed that indigo produced through this process was of the best quality. The large quantity of firewood required for heating likely pre-empted the introduction of this technique into large estates, where the scale of operation was exponentially larger than on peasant units.[33]

"It is not clear how the colonists learned vat-processing technology," Jose Molina has said, acknowledging inability to state definitively how the indigo-making techniques took a mature shape in Central America.[34] In the absence of definitive sources it can be speculated that trade networks of the Atlantic system as well as the larger early modern global networks extending from India to the Western Hemisphere had ensured that separate trade zones of indigo manufacturing were not as isolated as we have so far assumed them to be. The Portuguese were connected with the indigo manufacturers on the Indian subcontinent who knew the practice of soaking and beating of leaves. They had imported their first batch of indigo from India in 1498. The Spanish monarchy was intent upon learning the art of indigo manufacture. Within such a context of trade links and local desire in the Iberian Peninsula to master the art of indigo manufacturing, it was likely that information about it moved across the globe through networks of traders and travelers who might have carried information and even texts. Networks similar to the ones that made Monnereau in Saint Domingue aware of premodern indigo culture on the Indian subcontinent were active in Spanish America as well. Further, coastal Central America, which was the extreme western rim of the Caribbean, would also have received pertinent information from the Antillean islands and their systems of indigo manufacture in an age when piracy and private trade interests often trumped the attempt by mercantilist governments to monopolize and check the spread of key trade information.

Carolina Indigo: Empire and Market

The mid-eighteenth century also witnessed South Carolina's emergence as an important producer of indigo in response to market forces and British imperial imperatives. The huge demand for the dye in Britain was

---

[33] Jose Antonio Fernandez Molina, "Colouring the World in Blue," pp. 127–8. A similar "hot water extraction process" was also used by natives in India and even promoted by the European naturalist William Roxburgh at the end of the eighteenth century on the Indian subcontinent. This will be discussed later.

[34] Jose Antonio Fernandez Molina, "Colouring the World in Blue," p. 111.

not completely met by French and Spanish supplies, and this gap between demand and supply opened up a new opportunity for the Carolina planters. The English Parliament proactively encouraged indigo cultivation by its British subjects in the American colonies with an eye toward securing supply lines for the textile industry at home. It passed a law allowing payment of a bounty of six pence per pound for indigo produced in the Carolinas if it were directly sent to Britain. The Carolina indigo went some way toward meeting the metropolitan demand and replacing erstwhile English Jamaica planters as important colonial suppliers to the home market.[35]

The growth of indigo plantations in Carolina has also been explained in terms of the necessity of "stabilizing the Carolina economy,"[36] in other words, in terms of the contingencies of local economic factors. The very first settlers had taken indigo seeds to Carolina from Barbados in 1670. Specific plantations of colonists were raising indigo the following year. In the early colonial economy, where settlers tried an array of crops and other products, the Carolinians moved away from indigo to rice at the end of the seventeenth century. In the first three decades of the eighteenth century, rice emerged as the principal staple and item of export from South Carolina. The instability in the market for rice from 1730 onward made Carolinians look for other commodities. Eventually they settled on indigo, again given the huge demand for it in Britain. In 1737, the local assembly enacted a law to pay a bounty of a shilling for every pound of indigo exported out of Carolina in an effort to promote the staple, although the bounty was withdrawn in the subsequent year in the face of opposition from English planters growing other crops.

The initiatives of Eliza Lucas and her husband, Charles Pinckney, in the expansion of indigo plantations in South Carolina are representative of the process of the flow of information facilitated by networks across continents. Eliza Lucas's father, George Lucas, the governor of Antigua in the West Indies, owned plantations at Wappoo in South Carolina. It is

[35] At mid-eighteenth-century mark the demand for indigo in Britain stood at about 500,000 pounds per annum. Edward Long's later text, cited earlier, gives the export figures of indigo from Carolina to Britain in the same year as 200,000 pounds, which amounts to less than half of the total requirement in Britain. *Cf.* Edward Long, *The History of Jamaica*, p. 680.

[36] Virginia Gail Jelatis, "Tangled up in Blue: Indigo Culture and Economy in South Carolina, 1747–1800," Unpublished Ph.D. dissertation, University of Minnesota, 1999; see, pp. 1–41 for a description of early consolidation of the indigo industry in South Carolina; quote on p. 11.

on these plantations that Eliza Lucas ventured into indigo production in the early 1740s. Her father sent her seeds from the West Indies and later an indigo expert from the English colony of Montserrat. The movement of a trained hand from the Caribbean secured for Lucas the "tacit knowledge" considered important by science and technology studies scholars for viably transplanting technoscientific systems, the system of indigo manufacturing through vat processes, in this case.[37] The knowledge received from elsewhere was also improved upon and perfected locally. Lucas was in correspondence with another South Carolinian planter, Andrew Deveaux, who conducted a series of indigo experiments of his own. Lucas's husband was an astute observer of the cultivation practices and contributed several articles on the subject to the local *Carolina Gazette*.[38]

In addition, the consolidation of indigo plantations in South Carolina would imbricate itself with a second layer of knowledge circulation involving the British metropole. Merchants, planters, gardeners, and dyers with connections to Britain and the American mainland augmented prevailing knowledge systems undergirding the indigo plantations in South Carolina with inputs of their own. One among them was James Crokatt, a successful colonial merchant, who had moved back to England in 1737 but kept his connections with merchants in Carolina well oiled. From the metropolis he published a manual for improving the quality of Carolina indigo that was addressed to the English manufacturers of indigo in America. His efforts stemmed from the realization that Carolina indigo was easily surpassed in quality by French and Spanish supplies in the metropolitan and international markets.[39]

---

[37] The STS understanding of tacit knowledge has arisen from the theoretical foundations of Michael Polanyi: Michael Polanyi, *Personal Knowledge*, London: Routledge & Kegan Paul, 1958, see esp. part 2, "The Tacit Dimension," pp. 69–245; H. M. Collins, "The TEA Set: Tacit Knowledge and Scientific Networks," *Science Studies* 4 No. 2 (April 1974): 165–86; *Changing Order: Replication and Induction in Scientific Practice*, Chicago: The University of Chicago Press, 2nd ed., 1992, pp. 51–78; for the technological context, see, Donald MacKenzie and Graham Spinardi, "Tacit Knowledge, Weapons Design and the Uninvention of Nuclear Weapons," *American Journal of Sociology* 101 No. 1 (July 1995): 44–99.

[38] Virginia Gail Jelatis, "Tangled up in Blue," pp. 26–8; H. Roy Merrens (ed.), *The Colonial South Carolina Scene: Contemporary Views, 1697–1774*, Columbia: University of South Carolina Press, 1977, pp. 144–63.

[39] James Crokatt, *Further Observations Intended for Improving the Culture and Curing of Indigo, &c. in South-Carolina*. London, 1747. This document is available at the British Library in London. IOC 967.c.34.

In a follow-up report Crokatt comprehensively addressed the weaknesses of Carolina indigo in an effort to bolster its prospects. He thought that the indigo from Carolina was better than any from the English colonies of Montserrat and Jamaica, but his effort was directed at making the commodity equal in quality to French indigo, which in his opinion was the very best indigo. For that to happen, he counseled, the planters in Carolina and exporters would have to work together. The weakness of Carolina indigo lay in a brittle or flinty exterior and a less than cured core, which, he opined, was the result of either drying in the Sun or using too little or too much limewater during production in the vats. Some consignments were found to contain "trash" or earth. Some of this seemed to have resulted from the leaves and stalks kept in the fermentation vat longer than required because of the eagerness of planters to maximize the quantity of the dye. Such efforts led to a larger output but severely compromised its quality. Such a product predictably did not fetch a good price in the market. Crokatt wished the planters to focus on quality and the exporters to be selective in buying the best products to ship to England. Otherwise, he warned, they would be deemed "enemies to the country" as they would be leading the Carolina indigo to doom just as the prior indigo from Jamaica grown by English planters had perished in the market.[40]

Crokatt also made himself a conduit in the transmission of ideas on indigo cultivation between wider interimperial networks and Carolina. He was in touch with Philip Miller, a major figure in the intercontinental flow of knowledge of indigo, who had excerpted Labat's French work on indigo for the first time in his *Gardener's Dictionary* of 1731 for the English reading publics. Crokatt supplied seeds of the indigo varieties from Carolina to Miller for planting at his Physic Garden in London for experimentation and trial purposes. But more broadly, through his liaisons with Miller, he made sure that Carolinians were invested in and benefited from the wider information order pertaining to the specifics of the French and British colonial systems of indigo manufacturing. Crokatt included the counsel of Miller and others on specific practices that could help Carolina planters overcome the problem of brittleness. He communicated two key suggestions from informed persons the first of which related to the skillfull use of limewater. The brittleness arising out of the use of an inappropriate amount of limewater was a problem that "can

[40] James Crokatt, *Further Observations Intended for Improving the Culture and Curing of Indigo*, pp. 4–7.

only be rectified by Experience," he wrote. If such skill was not available locally, he advised that the Carolina planters were better off dispensing with its use completely. Not using limewater would mean prolonging the process of precipitation, but in the end, limewater was not indispensible. After all, Crokatt pointed out, the French and the Spanish were known to produce indigo of the very best quality without using any alkalis. The other major recommendation to Carolina planters involved a clear directive to dry the indigo slowly and in the shade; it should be kept out of sunshine and away from wind immediately after leaving the vat. Once it had obtained some solidity, the planters were advised to place the solids in a dry and shaded area. The drying in shade was a typical practice in the English colonies, as opposed to the sun drying prevalent in French colonies.

In South Carolina, the "West Indian system" took root by the mid-eighteenth century.[41] Commentators from the period have discussed the arduous work that went into preparing the field for indigo and in caring for the crop even while asserting that the local soil "was the finest in the world for raising this plant." Writing in *The Gentleman's Magazine,* a Charlestown planter stressed the importance of extra care in preparing the land needed for raising indigo in Carolina. "An *Indigo* field ought to be a perfect garden," he wrote. He argued that in general terms it took three times the amount of labor in Carolina as compared to the time invested by workers in agricultural fields in England. If intense effort was not made in the land, the crop would be destroyed, he emphasized, saying "our negroes are always at the hoe." The article suggested that the planters should use a drill plow and use methods described in Jethro Tull's book on husbandry to cut down on the time and labor for preparing the land and weeding. He praised the French, who had adopted a better tilling and weeding method. But regardless of labor saving mechanisms and purported achievable efficiency, there was no doubt in his mind about the enormous effort that would be needed to work up the lands: "There cannot be too great pains taken to prepare land for *indico* [*sic*], nor to keep it clean, while growing, from weeds and grass; for our lands must be daily and hourly attended, otherwise everything planted in them would be quickly choaked [*sic*]."[42]

[41] According to one estimate, the "golden years" of indigo production in Carolina lay between 1750 and 1775. Virginia Gail Jelatis, "Tangled up in Blue," p. 9.
[42] Letter from a Charlestown planter, dated November 30, 1754, *Gentleman's Magazine,* for May 1755, pp. 201–3.

In the standard Carolina system of manufacturing, planters used a steeper and a beating vat. One account refers to the stacking of the plant in the steeper in a way so as to make the stalks face upward in an effort to submerge the leaves completely.[43] The vats were made out of cypress plank. There is also mention of a separate lime vat that opened into the beating vat and was used for adding lime during beating to facilitate precipitation. As in the other English colonies in the West Indies, the Carolina practice was also to dry indigo after it emerged from the beating vat in the shade, and a separate drying house was provisioned for this purpose, another practice from the English colonies as distinct from the French colonies in the Caribbean.[44]

Even by the third quarter of the eighteenth century, Carolina indigo had failed to close the quality gap with either the French or the Spanish variety in terms of purity or consistency. In a letter dated April 19, 1775, John Ledyard assailed the relatively poor quality of Carolina-manufactured indigo. Some of it was so poor in quality so as to sell for 6 d. per pound. Only some batches of dye from America, he said, were of superior quality and sold at a high price of 6 or 7s. a pound. The superior French indigo, in comparison, regularly fetched between 7s. 6 d. and 8s. a pound, and

---

[43] *A Short Description of the Province of South Carolina, With an Account of the Air, Weather, and Diseases, at Charlestown, 1763*, London: privately published, 1770, p. 94.

[44] *Gentleman's Magazine*, for May 1755, pp. 256–8. Minor differences in manufacturing processes appeared elsewhere on the mainland. Andrew Turnbull, a Scottish physician, and his partner, Sir William Duncan, received a grant to set up a settlement at New Smyrna in East Florida in 1766, which became an important center for export of indigo out of East Florida. The first shipment out of this unit was made in 1770, and between 1771 and 1777, 43,283 pounds of indigo were exported. This East Florida indigo manufacturing at New Smyrna involved the three-vat system, the use of lime, and drying in the shade, the quintessential features of the English West Indian system. Indigo was the chief produce of East Florida, and yet the total production of East Florida was not sufficient to "warrant the call of a large vessel." Sometimes the indigo exporters in East Florida had to move their goods to Charleston to be exported from the port there. Kenneth H Beeson Jr., *Fromajadas and Indigo: The Minorcan Colony in Florida*, Charleston: History Press, 2006. The Dutch traveler Antoine Simon Le Page Du Pratz visited Louisiana a little earlier, between 1718 and 1734, and described the indigo plantations established there by the French settlers. These plantations took their inspiration from practices prevalent in French West Indian colonies; thus, Du Pratz's description of indigo manufacturing near New Orleans mentions the typical three-vat system. But he makes no mention of the practice of adding lime to the beating vat. Similarly he alludes to the practice of drying of indigo under direct sunlight, just as the French colonists did. M. Le Page Du Pratz, *The History of Louisiana or of the Western Parts of Virginia and Carolina*, translated from French, London: T Becket, 1774, pp. 168–71, 191–4. It must also be stated that most other regions were minor producers of indigo compared to South Carolina.

Spanish or Guatemala indigo between 10 s. and 14 s.[45] Ledyard had conducted a few experiments in England in order to get to the bottom of the poor quality of Carolina indigo. He wrote publicly to communicate the results of those trials, claiming that he was acting out of patriotic feelings to help improve the quality of a manufactory produced by fellow countrymen based in the American colonies. But he also justified his purchase of the blue dye produced by French and the Spanish: "A prejudice in favor of my country and its productions I avow: it is inherent in every Briton; and that partial esteem I have had for my countrymen in every part of the British empire, has induced me always to give their manufactures the preference, *when it could be done without injuring trade.*" Obviously a dyer's preference for his trade trumped patriotism for Ledyard, a colorist of forty years' standing in the English metropolis. He pointedly referred to a letter published in the *Carolina Gazette* by "a planter" who had questioned the patriotism of dyers and indigo salters in Britain who were buying foreign indigo. Ledyard confronted this Carolina planter with reasoning based on economic facts: "The reason of my being obliged to buy foreign [indigo] was, that yours was worse in proportion to the price than the foreign, the difference in price not being equal to the difference in goodness." He stated that in his view only political intervention by the government, not market forces, could make the Carolina variety in its current form do better in English markets. "I assure you, Sir, and the rest of your brethren [in Carolina] that nothing less than a stop to the importation of Spanish and French indigo, can induce us to use yours, unless some amendment is made in ... [the quality], or the price proportioned to its just value."[46]

The New Indigo System in Bengal

The global template of the British Empire facilitated the movement of men and skills and eventually the rise of Bengal as the leading producer of indigo in the nineteenth century. As a predominant textile manufacturer

[45] "Extract of a Letter to M. Brewton, Esq;" *Methods for Improving the Manufacture of Indigo: Originally Submitted to the Consideration of the Carolina Planters op. cit.* pp. 47–8.

[46] "To the Author of a Letter in the *Carolina Gazette,*" of August 24, 1772, No. 725, Signed "A Planter," dated May 29, 1773, *Methods for Improving the Manufacture of Indigo: Originally Submitted to the Consideration of the Carolina Planters, op. cit.* pp. 24, 28–9.

Britain was already invested in a global network of production and consumption of the blue dye. As the supply of indigo from South Carolina and the West Indies became uncertain, the former on account of the independence of the American colonies and the diversion of land to rice and cotton, and the latter on account of a major black rebellion in the French Saint Domingue (1791), the different nodes of the empire were vitalized to reinvent another trade and supply regime similar to the previous one. In the changing turn of circumstances in the late eighteenth century it was the newest and the most valuable acquisition of the empire on the Indian subcontinent that seemed to offer an opening. Following the victory at Plassey in 1757, the English East India Company had further acquired political control in Bengal and was poised to become the predominant power on the Indian subcontinent. Planter entrepreneurs from other parts of the empire and elsewhere began to flock to this new tropical zone in the empire. The soil and climate in Bengal was suitable for growing indigo. The company also reckoned indigo to be a suitable item of export on its ships bound for home. Its officials made a deliberate effort to promote the cultivation of indigo in Bengal, and the institutional space created by the mercantile company Raj facilitated the expansion of indigo manufacturing. By the first decade of the nineteenth century Bengal indigo became the world's preeminent blue dye. Between 1786 and 1810, the import of indigo into Europe from India rose from less than half a million pounds to approximately six million pounds.[47]

The importance of indigo to the colonial commerce has been clearly established by economic historians of eighteenth- and early nineteenth-century colonial Bengal, in particular by Benoy Bhushan Chowdhury and Amales Tripathi.[48] They have described the economic necessity of indigo

---

[47] David Macpherson, *The History of the European Commerce with India*, London: Longman et al., 1812, p. 415, cited in Molina, "Colouring the World in Blue," p. 57.

[48] Early history of indigo in India is available in N. K. Sinha, *Economic History of Bengal from Plassey to the Permanent Settlement*, vol. I, Calcutta: K. L. Mukhopadhyay, 1956, pp. 87–8, 216–17; for an early analysis of the political economy context of EIC's trading activities, see, Holden Furber, *John Company at Work: A Study of European Expansion in India in the Late Eighteenth Century*, Cambridge, Mass.: Harvard University Press, 1951; although focused on the indigo rebellions of 1859–62, Blair Kling's brief treatment of EIC's efforts to set up indigo plantations in the late eighteenth century is still valuable. Kling tells us that the company government made deliberate efforts to promote the trade and manufacture of indigo, such as contracting with private traders for a supply of indigo in the initial phase that were carried to England in the company's ships until 1788. It then began a system of giving advances to European entrepreneurs for setting up plantations. This exceptional measure continued until 1802. Imperial figures like Governor General John Shore (1792–8) also

to the "remittance trade" of the East India Company. These historians have also clarified the larger political economy context that facilitated monopoly trade by the company in indigo. Jacques Pouchepadass's study of the indigo peasantry in Bihar in the later part of the nineteenth century has similarly illuminated the entanglement of colonial plantations as a production system with the system of indigenous peasant production.[49] These studies are valuable in clarifying the nature of colonial economy through examination of mercantile commerce and agrarian society. But these economic studies do not address the knowledge dimension of plantations in Bengal. In fact, explicitly or implicitly, these analyses frame "technique" itself as devoid of any dynamic of its own. The science of cultivation and manufacture merely becomes a superstructure to more foundational political, social, and economic forces of exploitation.

Need for a New Analytical Framework

The history of the origin and development of indigo on the Indian subcontinent in the eighteenth and the nineteenth centuries needs an additional perspective focusing on the knowledge dimension of indigo culture in colonial Bengal. The subcontinent's connection to indigo manufacturing was deep.[50] Until the first half of the seventeenth century European companies traded in a large amount of Indian indigo produced in Gujarat, Rajasthan, and the Coromandel Coast. Thus the claim commonly appearing in writings that Europeans "introduced" the system of indigo manufacturing from the Western Hemisphere not only is semantically inaccurate, but also misses an opportunity to analyze the elements of continuity from the earlier period. At the same time, uncritical claims

played a critical role by giving duty protection to Bengal indigo against imports from adjoining provinces. *Cf.* Blair B. Kling, *The Blue Mutiny: The Indigo Disturbances in Bengal, 1859–1862*, pp. 17–18.

[49] Benoy Chowdhury, *Growth of Commercial Agriculture in Bengal, 1757–1900*, Calcutta: India Studies, 1964; Amales Tripathi, *Trade and Finance in Bengal Presidency, 1793–1833*, Calcutta: Oxford University Press, 1979 (first published in 1956); Jacques Pouchepadass's work remains the best account of indigo agriculture and peasantry in colonial Bihar. *Cf.* Jacques Pouchepadass, *Champaran and Gandhi: Planters, Peasants and Gandhian Politics*, Delhi: Oxford University Press, 1999, pp. 1–84.

[50] There is no denying that this connection predated the launch of fresh plantations in Bengal in the 1770s. The medieval Indian roots of indigo have been traced by other historians. Irfan Habib, *The Agrarian System of Mughal India, 1556–1707*, Bombay: Asia Publishing House, 1963, pp. 66–8, and passim; Ishrat Alam, "New Light on Indigo Production Technology during the Sixteenth and Seventeenth Centuries," A. J. Qaisar and S. P. Verma (eds.), *Art and Culture*, Jaipur: Publications Scheme, 1993, pp. 120–6.

of continuity between the earlier systems of indigo production on the Indian subcontinent and later plantations in Bengal tend to be neglectful of important changes to indigo manufacturing that that had accrued in the era of plantations. Blair B. Kling's significant political tract on the Bengal plantations thus noted that "from the seventeenth to the twentieth centuries indigo was a fugitive among industries, wandering from Gujarat in western India to the West Indies and then back to Bengal in eastern India."[51] Kling is apparently correct in spotlighting the global relocation of indigo manufacturing. But he missed to add that the plantation manufacturing that arose in Bengal at the end of the eighteenth century was significantly different from the previous form of indigo manufacturing in the region.

A framework based on flows of knowledge rather than on fixed knowledge systems and their "transfer" across concrete territories provides an accurate means to connect Bengal plantations with earlier systems of indigo manufacturing on the Indian subcontinent and in the Western Hemisphere. The Bengal indigo plantations were a beneficiary of state facilitation, with connections to the colonial economy, but their genealogy in the knowledge dimension was not circumscribed by the territorially bounded nature of the local colonial state or even the formal, territorial limits of the British imperium. Indeed, indigo that began to be produced in Bengal toward the later part of the eighteenth century could claim to have many pasts extending to medieval India as well as to numerous parts of the world more recently in the seventeenth and the eighteenth century. Flows of knowledge between India and the rest of the world during medieval times and knowledge essentially embedded in diasporas of planters moving between European metropolises and the Caribbean, Central America, and South Carolina in the era of plantations constituted the cumulative past of the Bengal plantation.

A closer analysis of a key process of manufacturing, fermentation, for instance, can put into relief the nature of knowledge flows across the physical span of indigo's global information order. In common

---

[51] Blair B. Kling, *The Blue Mutiny: The Indigo Disturbances in Bengal, 1859–1862*, Philadelphia: University of Pennsylvania Press, 1966, p. 15. Jacques Pouchepadass, *Champaran and Gandhi: Planters, Peasants and Gandhian Politics*, Delhi: Oxford University Press, 1999; however, the worldwide production and trade in indigo with an evident India connection went as far back as medieval times and even extended into the classical period; for a description of earlier times, see Jenny Balfour-Paul, *Indigo in the Arab World*, London: Routledge Curzon, 1996; *Indigo*, London: British Museum Press, 1998; Frederick H. Gerber, *Indigo and the Antiquity of Dyeing*, Ormond Beach, Fla.: Gerber, 1977.

understanding the "West Indian system" in the Caribbean had developed the method of fermenting leaves in a steeping vat to perfection. But the art of soaking indigo leaves was an ancient one. In a late eighteenth-century commentary Edward Bancroft traced the practice of drenching leaves to India, where it possibly began "several thousand years ago." Jean Baptiste Tavernier's text describing mid-seventeenth-century practices in India had referred to the practice of soaking of indigo leaves in a large tank, while in Europe itself, woad, the predecessor to *indigofera*, was subjected to "partial and immature fermentation" involving the pounding of leaves followed by drying. Indigo derived from several plants on various continents was variously subjected to drying and pounding, pounding of moist leaves followed by drying, and kneading into round balls of blue dye. These were sold in this form to the dyer and were macerated or fermented at the dyer's house or in private households before use. In mid-eighteenth century natives in Africa were known to "take the well dried leaves of averei, the indigo tree, and reduce them to powder which they put in a very large vessel full of water."[52] Across the rest of the world, from China to Java to Senegal and Egypt, variations on these practices of soaking in boiling water, simply leaving indigo in a heap in open air to sweat, and grinding in a dry state followed by soaking achieved partial fermentation of leaves in multiple ways.[53] In contrast, the planters based in the West Indies loaded indigo into fermenting vats, added water of a known quality, and supervised its fermentation for a stipulated time under the watchful eyes of a trained practitioner before turning the liquid over to the beating vat. Thus it would be erroneous to claim that the art of fermentation originated in the Caribbean. A more accurate description would focus on the prior and coexisting practices in various dispersed geographies of the world and on how information about these process flowed across the globe. When that is done it would

---

[52] Jean Baptiste Tavernier, *Travels in India by Jean-Baptiste Tavernier, Baron of Aubonne*, translated from the original French edition of 1676 by V. Ball, second edition, edited by William Crooke, 2 vols., London: Oxford University Press, 1925, vol. 2, pp. 10–12; Claude Francois Lambert, *A Collection of Curious Observations On the Manners, Customs, Usages, Different Languages, Government, Mythology, Chronology, Ancient and Modern Geography, Ceremonies, Religion, Mechanics, Astronomy, Medicine, Physics, Natural History, Commerce, Arts, and Sciences, of the Several Nations of Asia, Africa, and America*, translated by John Dunn, London, 1750, p. 249. This document is available at the British Library in London. IOR, 571.d.7.

[53] Edward Bancroft, *Experimental Researches containing the Philosophy of Permanent Colours and the Best Means of Producing them, by Dyeing, Calico Printing & c.*, vol. 1, Philadelphia: Thomas Dobson, 1814, pp. 123–6 and notes.

appear that the West Indian planters merely appropriated and developed the practice of fermentation.

The worldly network of indigo knowledge also accommodated internal differences in practices and numerous deviations from the "normative" practice. This aspect of internal variegation comes through in examining the case of addition of lime as a precipitant to the beating vat. This practice varied regionally within the system of manufacture often addressed by the blanket phrase "the West Indian system." The function of lime-water was widely acknowledged as assisting precipitation of the dye. In Jamaica, the addition of lime was routine. The planters were advised to be careful to add the correct amount as any extra alkali would compromise the texture of the dye cake. A Charlestown planter on the American mainland was confident that the addition of lime catalyzed precipitation of color grains and saved labor during beating. But it seems that the use of alkali was not as widespread in the French colonies. De Beauvais-Raseau called the science of use of precipitants indeterminate. But the major theoretician for French indigo, De Cossigny de Palma, a planter based in the Indian Ocean French colony of Mauritius, seemed quite open to the idea of adding precipitants. He acknowledged that the tradition of adding precipitants and the process by which alkalis acted on the liquor were "new and unknown" but still vouched for their efficiency. He recommended the use of precipitant for steering the formation of grain, especially if the crop had been harmed by excessive rain or drought or an error had occurred in the process of fermentation. But the addition of alkali could be helpful even in the case of normal crop output and an optimally completed manufacturing process. [54]

Openness to appreciating the fluidity of accumulating indigo knowledge can help in understanding the broadly dispersed genealogy and cosmopolitan borrowings of indigo knowledge. The transnational knowledge of indigo with all of its internal variegation reached Indian shores in the late eighteenth century at the behest of texts, diasporic planters, and naturalists. It had been developed in the previous centuries not only by practitioners in many tropical plantations, but also by philosophers and naturalists in several Western metropolises. Craft knowledge was perfected in specific territories and allowed to migrate along with practitioners or

[54] Edward Long, *A History of Jamaica*, vol. iii, *op. cit.*, pp. 677–8; *Gentleman's Magazine*, for June 1755, p. 257; Richard Nowland, *A Treatise on Indigo*, p. 7; De Cossigny de Palma, *Memoir Containing an Abridged Treatise of the Cultivation and Manufacture of Indigo*, Calcutta: 1789, first French edition published in 1779, pp. 78–94. Manuscripts Collection, British Library, 459.a.9.

their apprentices. Similarly, productive knowledge was generated on the plantations and in the laboratories of naturalists, planter-savants, and experts. That knowledge would be further changed on the Indian subcontinent during the process of engagement with local knowledge and practice. Attention to the complexities of this constantly changing knowledge system enables a clearer understanding of how the so-called West Indian system developed on the Indian subcontinent.

Indigo Texts and the Indian Subcontinent

Some of the texts on indigo cultivation and manufacturing that made their way into Bengal at the end of the eighteenth century acquired seminal importance as practical manuals for those launching themselves into indigo culture on the subcontinent. Among these authoritative compendiums were those by French authors such as Elias Monnereau and De Cossigny, and the compendium in English made available by Richard Nowland, which included excerpts from several European authors including the French work by De Beauvais-Raseau. These works were translated and made available within the larger context of expansion of imperial control and development of local colonial economy surrounding indigo plantations in Bengal. The need for information about indigo processes had created a market for indigo texts on the subcontinent. Company officials and private publishers, and even individuals in some cases, acted in response to that need and became facilitators in the intercontinental movement of texts on indigo. Authors, translators, and publishers were deliberate agents in creating an informational sprawl. The translator of De Cossigny's magnum opus on indigo manufacturing, Manuel Cantopher, had perceived a need for indigo texts in "the present rising state" of the industry in Lower Bengal. The author also gave a clue that within the European indigo industry in Bengal at that time, native participation and, one may assume, native practices still had a definitive and recognizable presence, alluding to "several persons concerned in it being only under the guidance of natives of India." The persistence of Indian methods had, in fact, prompted Cantopher to undertake the translation work. He thought that while Indian methods of manufacturing indigo were legitimate on account of the "antiquity of their practice," the colonial practitioners "may derive advantage by comparison [of their art] with that laid down by the learned author of the present treatise," that is, De Cossigny. The best practices of indigo manufacturing had not been settled yet. In fact, he pointed out that De Cossigny himself was "far from thinking that he has exhausted the subject," and that there were grounds for further observation and experience to explain what he "has left unascertained." Thus he invited the European planters in India to "add something to the general stock of knowledge." An exchange of information and continual improvement based on further experience were necessary.[55]

Richard Nowland, another Calcutta-based publisher, had also sensed the need for a compendium on indigo in the colony. Interestingly, he cited the prior assertion by Cantopher that Bengal indigo operations were "*only under the guidance of the Indians*" [italics in the original] to justify the production of another treatise in order to expose planters in India to "useful works from Europe." He translated a wide selection of European indigo works, which included not only the French indigo expert De Beauvais-Raseau's and the German Carl Günther Ludovici's, but also many seventeenth-century accounts by Dutch, French, and German Orientalists and travelers to Asia and Africa. Europe's dominant position in eighteenth-century indigo production had made it a veritable clearinghouse of indigo information collected from around the world. In his entrepreneurial drive to cash in on the need for published literature on indigo, Nowland provided a conduit for the passage of that diverse knowledge to India.[56]

All these texts were normative in the sense of privileging the foundational elements of the three- vat system. At the same time some of the authors and publishers implied that this knowledge was "in the making" that had not closed itself off to new information. Nowland clearly stressed this fact: "This art above all others is the most unfixed." Nowland described indigo manufacturing in seventeenth-century India, including the prevalence of the "dry method" that involved drying the indigo leaves, pounding them, and then either soaking or combining soaking and beating, a practice that contrasted with the West Indian method of fermenting fresh indigo leaves in a vat. He also presented information on the Dutch methods of indigo culture in Java and in South Carolina and made some cursory remarks on prevalent methods in China and Madagascar. He noted that the Indian method was "the simplest" and its indigo quality "in general ... preferable to any other," even as he acknowledged that the

---

[55] See the translator's "Introduction" in De Cossigny de Palma, *Memoir Containing an Abridged Treatise of the Cultivation and Manufacture of Indigo*, Calcutta: 1789, first French edition published in 1779, pp. ix–xii.

[56] Richard Nowland, *A Treatise on Indigo*, translated from the French, Calcutta: James White, 1794, pp. v–vii.

European method was the best "in point of accuracy." The final word, he observed, should be left with the European manufacturers. It is also note-worthy that in drawing a contrast between "simple" Indian methods and "accurate" European methods, Nowland also seemed to be representing the broader trend in transition from craft to science in indigo manufactur-ing itself. Some texts by illustrating the flexibility of the vat system might have exhorted planters to a new round of experimentation. This aspect of flexibility comes across most clearly in De Cossigny's text, one based on his experiences as an indigo planter in the French colony of Mauritius. Away from the Antillean Islands, in the southwestern Indian Ocean, De Cossigny had experienced the perils of implementing the Antillean sys-tem. In addressing the subject of drying of indigo under direct sunlight or in shade, respectively, French and English practices in the West Indies, he wrote that "neither of these processes succeed well at the Island of Mauritius, the one being too quick, and the other too slow, that in the shade [also] generating insects in the lower parts of the fecula."[57]

De Cossigny's explanation of vat processes reflects both the cosmo-politan borrowings of the West Indian system in the Western Hemisphere and an openness to adopt practices of "the East" if their utility could be proved. Thus he described the manufacturing process as beginning with "the leaves of the plant, whether green or dry" being put into the steeping vat, which could be potentially made of "pottery ware, wood, or constructed of masonry." He also described the beating process followed by drying, which could take place directly under the Sun or in shade. Obviously the system presented in De Cossigny's text was an admixture of various conventions followed in the Caribbean, Spanish America, and the American mainland, not a purely French, English, or Spanish system. The syncretism of De Cossigny's text further comes out in a particularly stark way in his referring to the practice of assigning "dry" leaves of the indigo plant to the steeping vat – a practice followed typically in many regions of India – as a viable one. De Cossigny had actually experimented with "dry indigo," having received information on such a process in existence on the Coromandel Coast from an official of the French Council of Pondicherry. Having obtained color with dry indigo, De Cossigny declared, "This experiment infallibly shows that Indigo may be produced by different methods." He called for possible improvisation with the Indian system to assess whether in terms of cost and output it could be made to deliver

---

[57] De Cossigny de Palma, *Memoir Containing an Abridged Treatise of the Cultivation and Manufacture of Indigo*, p. 113.

better results than the European system. It had some merits because it did not require investment in constructing huge vats, and the difficulty of exceeding fermentation was not encountered in the Indian process, as in the European one. However, comparing the efficiency of the two systems, De Cossigny clearly advocated the European system, arguing that out of fifty bushels of the plant, the European system could yield ten to thirty pounds of indigo compared to just one pound of indigo using the Indian system. The travelers to the Coromandel Coast who had spoken highly of the Indian system and wished for European planters to embrace that system in its current form were, in his opinion, "greatly mistaken."[58]

Whatever the differences in operational practices, De Cossigny believed in the inviolability of the sequence of scientific changes involved in the production of the dye from the indigo plant. He understood and explained these changes in a certain way. The apparent "mechanical separation" of the dye from the leaves was resolvable on "chemical principles" and could be explained through such criteria. Indigo was contained in the plant in the form of a resinous extract along with a volatile alkali. The fermentation activated the alkali in the leaf and turned it into an "intermediate solvent" that facilitated the separation of the dye and other salts from the leaf. The agitation or beating occasioned the separation of "urinous salts" (meaning alkali salts) from the suspension in the vat as a result of which the dye was released and molecules of indigo started to fuse with each other and form the grain that precipitated to the bottom of the vat. No further change in composition occurred during drying since it was a process of physical change only. De Cossigny's notable contribution and advance over previous efforts lay in proposing a distinctive "theory of fermentation" on the basis of which he tried to explain the manufacturing of indigo. If anything, this conviction that a definitive set of elemental processes was the basis of vat processes represented a certain "scientific" turn in understanding. It represented a quantum move toward a reductive understanding of indigo processes.

Planters and the French Savants in Bengal

The "new" indigo manufacturing in Bengal was purportedly distinct from Indian methods that were remnants of the seventeenth-century system

---

[58] De Cossigny de Palma, *Memoir Containing an Abridged Treatise of the Cultivation and Manufacture of Indigo*, pp. 2, 130–6.

from the precolonial times. Many European planters pledged fidelity to manufacture in the "European manner." Some references point toward deliberate and specific steps by colonial officials to launch Western-style cultivation and manufacturing as indigo came to dominate private trade.[59] Prior studies have recognized concrete measures undertaken by the state to spread information about the "Western" system originally perfected in the Caribbean. These efforts included handing out samples of the West Indian dye to Bengal planters for emulation, giving them "letters of instruction" on dye manufacturing, and, in a few documented cases, awarding licenses to planters from the West Indies to settle in Bengal.[60]

But the thesis that an arriving "external" system of manufacture swiftly and completely effaced the "indigenous" system of indigo culture on the subcontinent overstates the case. The assumption of continuity in an intercontinental web of efforts to improve the agriculture, science, and technique of indigo manufacturing is more appropriate for developing an accurate historical understanding of the origin of indigo plantations in Bengal. As European planters extended their sway into the Indian subcontinent in the last quarter of the eighteenth and early nineteenth centuries, old theories continued to be tested and retested, and the planting community continually arrived at new convictions in a continuing quest for improvement. Changed agroclimatic factors and land and labor issues in a new political environment, the subject of a detailed study in the next chapter, provided a different context to the search for the natural limits to the improvement of the plant and the process. There was a simultaneous engagement with prior knowledge and practice of indigenous manufacturing of indigo that had survived in small isolated pockets on the subcontinent.

The discourse of legacy and connection to the wider world of plantation science and indigo manufacturing appears in texts and practices in late eighteenth-century colonial South Asia. The self-conscious references were especially pertinent at the time of the early "encounter" of Europeans with the Indian practices and local plant kingdom, as the

[59] H. V. Bowen's study of the company's "official" trade emphasizes the fact that products like sugar and indigo formed less than 6 percent of its sale of goods in 1810, while between 1793 and 1810, "drugs, sugar, indigo etc" accounted for 60 percent of all private trade sanctioned by the company. H. V. Bowen, *The Business of Empire: The East India Company and Imperial Britain, 1756–1833*, Cambridge: Cambridge University Press, 2006, p. 245 and note.

[60] Blair B. Kling, *The Blue Mutiny: The Indigo Disturbances in Bengal, 1859–1862*, p. 18 and note.

naturalists tried to figure out the characteristics of local germplasm and foreign planters deliberated the best cultivating practices for indigo in a new agroclimatic environment. The similarities and differences between the global and the local were articulated with a flourish. The local experience was accumulated on top of the given heritage from external sources. The "local" experience itself, it needs re-emphasizing, had been shaped by experience and scientific information flow in an arena whose geographical outlines were not confined to the territorial boundaries of South Asia or even enclosed by the formal rim of the British imperium.

Relevant actors in South Asia reached out to the important texts on indigo by Elias Monnereau, De Cossigny, and Richard Nowland. In *The Complete Indigo-Maker,* published in London in 1769, Monnereau had claimed that his purpose in writing the book was "being useful to my countrymen" and beneficial to "our colonists."[61] But in a process reflecting the wider relevance that the work had, the willing disseminators of this knowledge translated the text into English. The English rendering of the work ensured its entry into and reception in the wider Anglo-American networks. A member of the East Florida Society of London, a forum of English settlers in East Florida, took the initiative in publishing the work. Since the society had interests in the possibilities of agricultural development in the newly acquired colony in Florida, the member had dedicated the translation to the society. Monnereau's work also became a staple read for everyone interested in indigo manufacturing in London.

Indigo-Maker's larger intercontinental influence in South Asia also becomes apparent in examining the particular case of John Prinsep, an indigo pioneer in Bengal and among the colony's most successful of indigo manufacturers and traders in the later part of the eighteenth century. In 1780, Prinsep had written to the directors of the East India Company expressing his wish to produce indigo in India for the British markets.[62] The permission granted, he became one of the first private merchants to set up indigo factories in Bengal. Prinsep is especially significant to

---

[61] Elias Monnereau, *The Complete Indigo-Maker: Containing an Accurate Account of the Indigo Plant; Its Description, Culture, Preparation, and Manufacture, with Economical Rules and Necessary Directions for a Planter How to Manage a Plantation, and Employ his Negroes to the Best Advantage, To Which is Added A Treatise on the Culture of Coffee,* translated from the French of Elian Monnereau, Planter in St. Domingue, London: P. Elmsly, 1769, p. v.

[62] H. V. Bowen, *The Business of Empire: The East India Company and Imperial Britain, 1756–1833,* pp. 244–6.

our analysis because he deployed the three-vat system in Bengal and his manufactory in Bengal, as told by another pioneer, Claude Martin, was set up after the model described in the Caribbean text. After studying Monnereau's *Indigo-Maker* Prinsep had evidently wished to replicate the Caribbean manufacturing.[63] Although there are no documents to describe the actual details of Prinsep's experience, Martin attested to the several failures he suffered, offering the explanation that "indigo is not a crop that can be hurried." Indigo was indeed a picky crop, and its manufacturing process was equally capricious. The translation process involved in using the West Indian system in Bengal proved to be predictably difficult. In the first two years, Prinsep fell back upon buying indigo from other producers in Agra and Kanpur in Oudh, indigo that was possibly produced by using indigenous methods. This indigo was sold to the company for shipment to Britain in their ships.[64] Prinsep later succeeded in the enterprise and went on to make a fortune. Like many others who acquired financial and social capital in the colony, he did not remain long in India. In 1788, eight years after launching his indigo enterprise, he returned to England, where a great career awaited him as an alderman to the city of London and an influential MP from Queenborough.

Claude Martin himself was another renowned name in the early days of Indian indigo manufacture. A French military general who later capitulated and joined the East India Company's forces, he enjoyed a successful career with the latter. Born in Lyon, he first participated in the Anglo-French wars on behalf of the French Crown in Pondicherry between 1752 and 1760, before switching to the English side. A multifaceted personality, Martin also worked for a while in the English East India Company's survey operations under James Rennell in Cooch Behar and Bhutan. Finally, he accepted the position of the superintendent of armory in Oudh, which was a protectorate state of the Raj under Nawab Asaf-ud-daulah. During his time in Oudh he established his indigo factories at Najafgarh, Pundri, and Korah and went on to become one of the richest Europeans in the colony.[65]

---

[63] Rosie Llewellyn-Jones, *A Man of the Enlightenment in 18th Century India: The Letters of Claude Martin 1766–1800*, New Delhi: Permanent Black, 2003, p. 268. On John Prinsep's life in India and the connection of his descendants with affairs in colonial India, see, H. T. Prinsep, "Four Generations in India," IOR, European Manuscripts, C 97.

[64] The export of indigo from the colony was largely in the hands of private merchants until the early years of the nineteenth century. H. V. Bowen, *The Business of Empire: The East India Company and Imperial Britain, 1756–1833*, p. 245.

[65] Rosie Llewellyn-Jones, *A Very Ingenious Man: Claude Martin in Early Colonial India*, Delhi: Oxford University Press, 1999; *A Man of the Enlightenment in 18th Century*

Martin's letters and recollections touching upon his indigo operations reveal a person open to multivalent influences. He was conflicted when choosing between the West Indian system and the indigenous system. In a 1787 letter he seemingly asked his associate at the Najafgarh estate to have indigo vats constructed. His will to his successor at Najafgarh, John Queiros, contained his wish that the West Indian system of manufacturing in existence at his indigo work be continued after his death. The system had proven to be efficient and had enabled indigo to be made "cheap and good" and "very profitable." While such a claim would portray Martin as a devotee of the "three-vat system," other records suggest that toward the later part of his life he had been captivated by the indigenous method of manufacturing prevalent near Ambore in Carnatic.

As he described in an article submitted to the Asiatic Society in Bengal, on a visit to the region, Martin had witnessed a breed of indigo plant growing at a higher altitude that did not need "artificial watering" and could withstand intense heat even at the time of harvest. That this southern plant could grow in barren hill tracts offered an advantage over the common variety of *tinctoria* that Bengal planters were compelled to grow on the most fertile land available. The Carnatic plant was apparently hardy as it could survive failure of rainfall or sudden warmth in climates that typically harmed crops in Bengal. The regional manufacturing process also had enough to commend itself because it dispensed with the vats altogether, affording savings on that count. In this system, the leaves were first boiled in earthen pots and the extract poured into earthen jars embedded in "excavated ranges" in the ground. The extract was then agitated for forty-five minutes with bamboo sticks, then concocted with a mixture of red earth and water that was alkaline in nature. Left overnight to settle, the liquid at the top was disposed off and the fecula at the bottom collected and put in bags to be dried the next day.[66] Martin submitted his findings to the journal in Bengal with the express purpose of reaching out to Bengal planters in the region and proposing adoption of the "Ambore system" in the eastern province. He strongly recommended a switch to the new system since it would protect Bengal planters from investing in the "dead stock" of vats.

India: The Letters of Claude Martin 1766–1800, New Delhi: Permanent Black, 2003, pp. 268–70.

[66] "On the manufacture of Indigo at Ambore By Lieutenant Colonel Claude Martin (with) an Extract of a Treatise on the Manufacture of Indigo, by Mr. De Cossigny," London: Asiatic Researches, 1807, pp. 475–7, according to an old note of April 1791; *A Man of the Enlightenment in 18th Century India: The Letters of Claude Martin 1766–1800*, pp. 269–70.

Some of the French savants considered the specifics of the Indian system before ultimately discarding them. The rebuttals offered by savants prominently appeared in journals that were available to the planters. De Cossigny availed himself of many opportunities to get acquainted with the specifics of the Coromandel system as it existed in southern India. He owned an herbarium of plants from southern India. Indeed, De Cossigny visited southern India and Bengal in the 1780s and conferred with East India Company specialists.[67] The later edition of his book that was translated in Calcutta in 1789 contained a thorough review and critique of the Indian system. To give one example, De Cossigny praised the Indian practice of keeping dried leaves away from open air before putting them in water. This was explainable by his "theory of fermentation" from the *Memoir*. Leaving indigo exposed to air for a long time caused its slow and extended fermentation, leading to a complete loss of color. De Cossigny emphasized that the natives' treatment of crushed leaves with water constituted "a process of maceration, and not fermentation." But as long as maceration was "kept within proper bounds" and followed by "suitable agitation," the operation could be made to succeed. But extended maceration could lead to complete decomposition of indigo and failure.[68] Excerpts from De Cossigny's text that refuted Claude Martin's espousal of Indian system in Carnatic were inserted to appear side by side in a publication of *Asiatick Researches,* the most important journal for the arts and sciences in late eighteenth- and early nineteenth-century colonial India.[69]

In later experiments that were appended to his *Memoir,* De Cossigny reflected on the possibility of obtaining indigo in the absence of fermentation and agitation altogether. While the latter was theoretically possible, the processes were far from ideal. Although open to trying out all alternatives, De Cossigny left nobody in doubt about his belief in the superiority of the three-vat system.[70]

---

[67] See the reference to De Cossigny's visit to Calcutta in Rila Mukherjee, "Calcutta in the Eighteenth Century: Vignettes from Contemporary French and Scottish Travel Accounts," *Bengal Past and Present,* 110 No. 210–11 (1991): 75–91. He wrote a memoir describing his India visit that was published in French in Paris in 1799.

[68] *Memoir Containing an Abridged Treatise on the Cultivation and Manufacture of Indigo by M. De Cossigny De Palma,* pp. 1–12, 132.

[69] "On the manufacture of Indigo at Ambore by Lieutenant Colonel Claude Martin (with) an Extract of a Treatise on the Manufacture of Indigo, by Mr. De Cossigny," London: *Asiatic Researches,* 1807, pp. 475–7, according to an old note of April 1791.

[70] This part of De Cossigny's work was appended to the 1789 translation and reflected his later, separate work. "Indigo obtainable without the Fermentative and Agitation Processes," *Memoir Containing an Abridged Treatise on the Cultivation and Manufacture of Indigo By M. De Cossigny De Palma,* pp. 140–5.

William Roxburgh: The Scottish Naturalist in Calcutta

The Scottish naturalist William Roxburgh, an East India Company official, engaged full throttle with "the indigo question" at the end of the
eighteenth century. In many ways, he represented the effort to work
out the indigo processes locally in the colony. He considered all options
before him including native methods and the methods contained in normative indigo texts. He represented "local knowledge," or the knowledge
worked out in the colonial locality under colonial conditions.[71]

Trained as a botanist and physician, Roxburgh had first accepted a position in Madras as the assistant surgeon of the presidency. In Madras, he
plunged into a study of the flora on the coast of Coromandel, a pursuit
that had his undivided, full-time attention after he was put in charge of the
East India Company's garden at Samulcottah in northern Circars. He made
detailed drawings of Indian plants and studied crops of commercial value
such as sugar and pepper in southern India. He contributed to deliberations
at the Royal Society in London and routinely sent the society samples of
Asian crops. Roxburgh's botanical expertise was quickly gaining a reputation
throughout the empire. In 1793, the Bengal government invited him to take
over the superintendence of the prestigious Botanical Garden in Calcutta.
Once in Bengal, he turned his focus additionally to hemp and *lacca*, commercial crops of benefit to the region. As an East India Company employee, the
attention to agricultural commodities by Roxburgh was only predictable.

Roxburgh found himself drawn into the debates around the science
of indigo cultivation and manufacturing. His sojourn in the colony
coinciding with the rise of indigo in India in the third quarter of the
eighteenth century, he was uniquely positioned to engage the important
issues of indigo manufacturing with which the practitioners had been
engaged in the previous decades. As an accomplished naturalist he could
assess native practices of cultivation and production of indigo, scout the
"Tropics" for unique germplasms of the plant with the highest potency,
and use the empire's networks to engage the latest "theories" of indigo
science advanced by savants like De Cossigny. Indeed, he would go on to
do all these things and, in the process, establish a reputation as one of the
earliest experts on indigo in the East.[72]

---

[71] Scholars studying "colonial science" have used "colonial knowledge" as a historiographical category to highlight the proximity of newly introduced western science to the priorities of the colonial state.

[72] Roxburgh gained a reputation as an indigo expert that lasted well into the next century.
Along with De Cossigny he was considered one of the earliest experts on the plant by

Roxburgh's forays into the natural habitat of the Coromandel first led him to a new species of a tree with substantial content of the indigo-yielding principle. He had stumbled on the tree while working on accumulating a collection for an herbarium and deduced that the plant belonged to the genus *Nerium* in the Linnaean system of classification. On drying, the bruised parts of the leaves and branches of *Nerium* discharged a blue color, which made him suspect that the secretion might be indigo. He followed the observation with lengthy experiments that confirmed his suspicion. This tree that grew in the North Circars and Carnatic foothills was noticeably different from the shrub belonging to the genus *Indigofera* that the planters in Bengal commonly cultivated to produce the dye. It met the morphological characteristics of *Nerium* in the Linnaean system and since it also gave color like *indigofera tinctorium*, Roxburgh named this apparently newly discovered species *nerium tinctorium*.

The Scottish naturalist set himself to the task of popularizing the species, which he thought held the key to the improvement of the current yield of the dye. In 1790, he forwarded a description of the tree, a process for making indigo from *Nerium*, and actual samples of *Nerium* indigo to the East India Company's officials in London in an effort to convince them of the good quality of color that could be obtained from his new find.[73] The description was also made available through a contact in Calcutta to the Supreme government of the company in Bengal. On behalf of Roxburgh, his interlocutor submitted to the Bengal government that the tree, being perennial, offered the advantage of a more permanent stock over the annual shrub cultivated by Bengal planters. The letter emphasized that *Nerium* was an "unfailing resource against the devastations to which the cultivation of the more common annual different species of indigo are subject." While the varieties grown in India were "from their nature more

later chemists. He was especially valued for his work in the East, where he had had a firsthand opportunity to work with the plant species that was not available to scientists working in the temperate zone in the West. In a mid-nineteenth-century text, Edward Schunck acknowledged Roxburgh's pioneering efforts. He wrote, "Roxburgh ..., like De Cossigny, was one of the few possessing special chemical information who have examined the process of manufacturing indigo from the *Indigofera* on the spot": Edward Schunck, "On the Formation of Indigo-blue – Part 1," *London, Edinburgh, and Dublin Philosophical Magazine and Journal of Science,* vol. XI, Fourth Series, London: Taylor and Francis (University of London), August 1855, pp. 73–95, quote on p. 78.

73 This original letter of 1790 was published in 1793. William Roxburgh, "A Botanical Description, and Drawing of a new Species of Nerium (Rose-Bay) with the Process for extracting, from It's Leaves, a very beautiful Indigo. Addressed to The Honourbale Court of Directors of the East-India-Company," Alexander Dalrymple, *Oriental Repertory, Vol. 2,* London: George Briggs, 1793, pp. 39–44.

exposed to failures from an unfavorable spring or disastrous season," the *Nerium* was hardy enough to grow in barren lands and quite resistant to extreme fluctuations in climate. It grew naturally on the southern rocky hills and in the wild, its natural habitat.[74] The memorandum initially evoked a positive response from the company officials, who published Roxburgh's description of the *Nerium* plant and indigo manufacturing side by side with a description of the West Indian system for the benefit of planters. Roxburgh also sent *Nerium* seeds to the Collector in Bhagalpur and Gaya in South Bihar to be distributed among planters. More tree seeds were being grown at the company's garden to be given to European proprietors for further trial. Roxburgh had also forwarded *Nerium* seeds to the Planter's Society in St. Helena, where the hilly terrain was similar to that of the northern Circars in southern India.

Meanwhile Roxburgh's additional trials with *Nerium* further solidified his earlier belief in the higher and more predictable yield of blue dye from the new tree when using his methods. Over three seasons, from 1790 to 1793, he made seven-hundred-and-fifty pounds of indigo from the wild tree. "I now speak from experience, not conjecture," he announced, while also citing brokers in England who had affirmed that the *Nerium* indigo supplied by him had sold for the highest price of eight shillings a pound. He cited two eminent dyers in London to assert that his indigo "would be an excellent substitute for Spanish indigo," which had a high reputation for its quality in London markets. The London dyers had also informed him that quality of *Nerium* indigo was "as good, if not better, than what they had ever seen from India."[75] In order to persuade colonial state officials and the functionaries at the India House in London, Roxburgh also made political arguments in favor of *Nerium* indigo. Not only was making of dye from *Nerium* "infinitely more profitable," but the tree could be grown in nonagricultural tracts without encroaching on "one foot of land that ever has been devoted to other purposes." Thus, plantations of *Nerium* indigo could be established in the colony without dispossessing native landholders or diverting land from food crop production. The dispossession of native landholders by European planters was an issue full of political implications for the colonial state. By emphasizing

---

[74] A copy of this 1790 letter appears as a footnote in a communication from William Roxburgh to the Royal Society and was published in the society's later issue: *Transactions of the Society, Instituted at London, for the Encouragement of Arts, Manufactures, and Commerce*, vol. XXVIII, London, 1811, note, pp. 251–2.

[75] William Roxburgh, "Method of Manufacturing the Nerium Indigo," *Transactions of the Society*, vol. XXVIII, 1811, *op. cit.*, pp. 258–90, see in particular, pp. 269–72.

those realities, Roxburgh rallied political support with state officials for *Nerium* indigo works, which, he assured them, could be established in "the most laudable and politic manner."

Roxburgh's stakes in popularizing *Nerium* were superseded by an even more earnest desire to establish a new science of indigo manufacturing. Accounts left by him enable us to retrace the steps through which the science of *Nerium* manufacture was designed, was validated, and sought to be universalized. The numerous components of this science first developed during his efforts to manufacture the dye from the leaves of the *Nerium* tree. A new "hot water extraction process" used by Roxburgh involved subjecting indigo leaves submerged in water to heating or scalding instead of fermentation in cold water, the primary process in the three-vat system. The turn to hot water was partly forced upon Roxburgh because *Nerium* leaves, unlike the common variety, yielded little to no indigo with cold water. Although Roxburgh did not explicitly acknowledge his debt, a closer study of his experiments reveals the precedents of this process elsewhere as well as the likelihood that he was inspired by witnessing the use of this process by Indian peasants. The Dutch were known to use heating in manufacturing, but their use of heating did not displace the stage of fermentation. Rather, the Dutch first fermented the leaves and then heated the extract before turning it in for agitation.[76] On the other hand, Roxburgh was aware of the hot extraction method used by peasants.[77] Roxburgh informs that in parts of Coromandel, Indians made indigo exclusively with hot water and did not have any knowledge that it could also be made with cold water. "Nor is it necessary to inform them," he wrote, "for what they make [with that process] is of a very good quality." He also described the general pervasiveness of the hot water process in the rest of the subcontinent, saying, "The natives throughout the Northern provinces, or [northern] Circars, make all their Indigo by means of hot water, which I call the scalding or digesting process." At another point, while discussing the details of his manufacturing

---

[76] "The Manufacture of Indigo, in the Island of Java, from a Publication in Dutch by Mr. Dirk Goetloed, communicated to Mr. De Cossigny by Mr. Radermacher, of the Supreme Council of Batavia," *Memoir Containing an Abridged Treatise on the Cultivation and Manufacture of Indigo by M. De Cossigny De Palma*, pp. 150–1.

[77] "Process of making Indigo on the Coast near Ingeram, communicated by Mr. William Roxburgh," *Memoir Containing an Abridged Treatise on the Cultivation and Manufacture of Indigo by M. De Cossigny De Palma*, p. 157; William Roxburgh, "A Brief Account of the Result of various Experiments made with a view to throw some additional Light on the Theory of this Artificial Production," *Transactions of the Society*, vol. XXVIII, 1811, p. 288.

process, he stated that "in many parts of the Carnatic they [i.e., people on the subcontinent] also extract the colour by the same means, hot water." Thus, while he attested the antiquity of the hot water process among Indians in the Carnatic, and more broadly along the Coromandel Coast, he did not admit that he borrowed the method from them. He probably drew a distinction between his streamlined process of using hot water and the unsystematic native system that also used hot water. He criticized that the natives used the "most inconvenient" apparatus, used "an inferior agent" as a precipitant, and, in his opinion, were "very careless" in conducting their operation. In fact, he referred to their degenerate apparatus and operation to validate the hot water process. If despite such a "rude manner" the indigo produced by natives was of an excellent quality, the result "must alone be imputed to the nature of the process by which the colour ... is extracted from the plant," that is, the hot-water extraction process. The "excellent" quality of native indigo was "an indisputable proof that the scalding process is superior to fermentation for obtaining good Indigo."[78]

The operational details and the theory of Roxburgh's hot water methods were published in their full detail in a journal issue of the Royal Society in London.[79] This process was based on an eclectic foundation of theoretical and empirical assumptions. That Roxburgh would have been inspired to try the novelty used by Indian peasants seems likely given his general openness to different knowledge systems including indigenous knowledge. For instance, he also praised the native practice of using dried leaves of indigo for processing, which dispensed with the need for transporting and processing a fresh harvest all at once. Enormous savings could arise from not having to invest in permanent stock by building a number of vats to process volumes over a short season. He had regretted that his trial with dried leaves had proved a failure. Similarly, his first trials with *Nerium* had involved using a cold infusion of the locally available Jamblong bark as a precipitant, "which is what the *Hindoos* use universally on this part of the [Coromandel] Coast, to precipitate their Indigo,"[80] although after his numerous experiments he became convinced that limewater was the best precipitant.

---

[78] William Roxburgh, "A Brief Account of the Result of various Experiments made with a view to throw some additional Light on the Theory of this Artificial Production," *Transactions of the Society*, vol. XXVIII, 1811, pp. 286–90.

[79] *Transactions of the Society*, vol. XXVIII, 1811, *op. cit.*, pp. 258–90.

[80] William Roxburgh, "A Botanical Description, and Drawing of a new Species of Nerium (Rose-Bay)," 1793, first published in 1790, pp. 43–4.

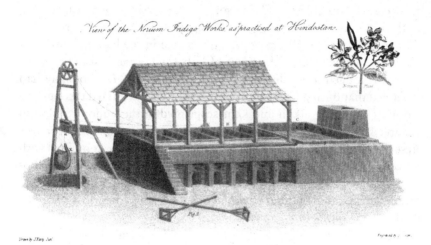

FIGURE 2. Roxburgh's experimental design using the indigenous process.
Source: Transactions of the Society Instituted at London for the Encouragement of Arts, Manufactures, and Commerce, volume 28, 1811, plate XVII, facing p. 262.

Later he began to claim that the process using hot water was equally ideal for processing the shrubs of the indigo plant (*indigofera tinctorium*) cultivated by planters in Bengal. He called the process "hot water extraction," as illustrated in Fig. 2, and was committed to universalizing this process of manufacture and its associated "science." Roxburgh scarcely hid his intent to woo Bengal planters to use his method of extraction. He stressed that the process was ideal for processing the *Indigofera* variety in Bengal. In other words, the science that he proposed was not "particular" and applicable to the *Nerium* species alone. Its potential to extract color from indigo-bearing plants of any species was universal. He continued to disseminate information on the *Nerium* tree and the "hot water extraction" process in the colony-metropolis corridor and beyond. A decade later, he claimed that he had dispersed information on the tree "from the Rajah-mundry Circar [in southern India] to various parts of the world."[81]

While substituting the stage of fermentation with hot water extraction, Roxburgh also proposed a new theory for extraction of color from

[81] William Roxburgh, "A Brief Account of the Result of various Experiments made with a view to throw some additional Light on the Theory of this Artificial Production," *Transactions of the Society*, vol. XXVIII, 1811, 272–90, see, p. 289.

the leaves. To his credit, in doing so, he countered the overbearing theoretical premise advanced by De Cossigny that lay at the core of the West Indian system: that color already existed in the leaves. This belief had been sustained since the times of Labat, who had insisted that the color was lodged in the leaves and ribs of the plant. De Cossigny had explained that during fermentation this color was extracted through the action of a volatile alkali and kept suspended in the liquor. Roxburgh disagreed. Instead, he suggested that only the "base" of the color was present in the leaves. During agitation the vegetable base combined with the color principle absorbed from the atmosphere. He made an elaborate effort to disprove experimentally that any alkali was discharged during fermentation in order to counter De Cossigny's theory of a volatile alkali.[82]

The rest of the steps in Roxburgh's system bore similarity to the West Indian system. At the completion of scalding, the extract went through the process of "agitation," just as in the three-vat system. The mix was suffused with a precipitant that catalyzed the process of grain formation. Like the French scholars De Cossigny, Monnereau, and De Beauvais-Raseau, who were at best ambivalent about the use of precipitants, Roxburgh struggled to explain the contribution of precipitants to the process. He suggested that the role of precipitant was probably to rid color particles of fixed air and to cause sedimentation. He had experimented for a long time with a variety of precipitants ranging from limewater and wood ashes to stale urine. He still could only conjecture about the importance of precipitants to the manufacturing process. It seemed to him that their contribution was important though not critical. If the operation had been conducted in an optimal fashion, one could probably dispense with having to use a precipitant. In making this suggestion, he echoed De Cossigny, who had also remarked that if the plant was not of very good quality, or if there had been an excess of fermentation or agitation, "the precipitant becomes absolutely necessary." Roxburgh may have been aware of De Cossigny's observation. But he also referred to the wider use of precipitants by Indians and elsewhere on the American mainland.[83]

[82] William Roxburgh, "A Brief Account of the Result of various Experiments made with a view to throw some additional Light on the Theory of this Artificial Production," *Transactions of the Society*, vol. XXVIII, 1811, *op. cit.*, pp. 273–7.

[83] William Roxburgh, "A Brief Account of the Result of various Experiments made with a view to throw some additional Light on the Theory of this Artificial Production," *Transactions of the Society*, vol. XXVIII, 1811, *op. cit.*, p. 282.

In the next reservoir, the supernatant water was simply run off while the precipitate was allowed to settle to the bottom and collected. Roxburgh embraced De Cossigny's counsel for washing the fecula with warm water and a small amount of acid of vitriol. He also experimented with other acids but in the end found De Cossigny's choice most appropriate on the basis of efficiency and low cost. The purpose of such a treatment was to dissolve any of the impurities and unwanted salts remaining in the dye. It reduced the amount of the dye but also enhanced its quality so it sold for a higher price. The process of purification was followed by drying. Roxburgh recommended slow drying of indigo in the shade since faster drying made the dye friable, reducing its value in the market.[84]

Roxburgh's efforts to universalize the "hot water extraction" method for large-scale manufacture of indigo in colonial India proved unsuccessful. Bengal planters remained loyal to the practice of steeping leaves in cold water rather than adopting the practice of scalding. There are indications that in some cases, the Roxburgh method was tried in areas that had *Nerium* plantations. In a communication of January 8, 1845, the planter G. J. Fischer mentioned his indigo factory based on *Nerium* and on the deployment of hot water methods at Salem in southern India. He also referred to the long history of the process at the Salem indigo factory. In 1823, one Mr. Heath had started the factory there "after the Doctor's plan," referring to the method recommended by William Roxburgh. He ran the factory until 1830, when, responding to low prices of indigo, he switched to the production of chrome dye made out of iron deposits. He shifted the boilers from the indigo factory to the new outworks for chrome that was set up at adjoining Porto Novo. Fischer took over this factory in 1833, bought boilers from the government's gunpowder factory, and set the factory running again. As late as 1842, this *Nerium* factory was in operation and giving "tolerable remuneration." Fischer wrote that the plantation and factory based on *Nerium* could only become viable if one had access to rent-free lands to grow *Nerium* trees and the time to wait for their slow maturation after sowing. But he also mentioned that southern India offered many acres of land with free mature *Nerium* plants, awaiting those with capital and enterprise to exploit them.[85] In

[84] William Roxburgh, "Method of Manufacturing the Nerium Indigo," *Transactions of the Society*, vol. XXVIII, 1811, *op. cit.*, pp. 267–8.
[85] "Correspondence and Selections, Further Particulars Regarding the Nerium Indigo," Extract of a Letter from G. J. Fischer, dated Salem, January 8, 1845, to Dr. Robert Wight, of Coimbatore, *Journal of the Agricultural and Horticultural Society of India* IV Part 1 (Jan–Dec 1845): 129–31.

the mid-1840s, some Bengal planters were debating whether it would be viable to replace the *indigofera tinctorium* variety with *Nerium* and the hot extraction method. Opinion was divided over the investment in tracts to sow *Nerium* afresh and over the cost of boilers and firewood to heat an immense quantity of water.[86] But there is no reference to the use of the hot extraction method by European planters in Bengal. Roxburgh's method had not been "universalized" in its application to large-scale plantations, although natives continued to use scalding and heating in their small-scale operations.

The system of hot water extraction certainly retained its presence in the domain of science and philosophical discussions as a credible system of indigo extraction. The fundamental process of scalding and heating indigo leaves from which Roxburgh had carved out his "hot extraction" method remained alive in native practices; the West Indian system could not dislodge it. The knowledge world of large-scale plantations remained part of a resilient information system that included a cache of viable principles offering numerous possibilities. This scientific ecumene was marked by countervailing tendencies. Thus Claude Martin's advocacy of the native system, even if not in completely unambiguous terms, coexisted with Roxburgh's quest for hybridism and the efforts by the likes of De Cossigny and Richard Nowland to make the art of indigo manufacturing conform to Western patterns. Attention to this composite information system enables a fuller understanding of the knowledge flows through which the plantation system was embedded into the land structures and broader agrarian context of colonial Bengal in the late eighteenth and early nineteenth centuries.

---

[86] For a favorable opinion on the potential of *Nerium*, see "Further Particulars Regarding the Manufacture of Indigo from *Nerium tinctorum*, Communicated by C B Taylor," *Journal of the Agricultural and Horticultural Society of India* V Part 1 (Jan–Dec 1846): 77–8; for an adversarial position to the former, see, "Result of Trials Given to Various Seeds at Chandamaree Factory, Rungpore, Communicated by H Rehling," *Journal of the Agricultural and Horticultural Society of India* IV Part 1 (Jan–Dec 1845): 27–30.

2

The Course of Colonial Modernity: Negotiating the Landscape in Bengal

The colonial modernity of indigo plantations in nineteenth-century South Asia was unleashed through the deployment of commodity production and practical application of science and inhered with a vision of agrarian improvement. On such criteria, it shared commonalities with the visions espoused by the colonial state. These compatible visions made the planters and the colonial state partners in stabilizing the plantations in the colony. The same ideas also came to pervade visions of agricultural improvement embraced by the colonial state as it gradually defined a role for itself in the development of Indian agriculture. In the early parts of the nineteenth century achieving excellence in the two domains of indigo culture and colonial agriculture seemed to many to intersect with each other. At a public meeting in Calcutta in 1829 the Bengali reformer and modernizer Raja Rammohun Roy, for example, claimed that peasants living in the indigo districts in Bengal were better clothed than other agriculturists because of the extra income from sowing indigo. He was pointing toward what appeared an amicable relationship of indigo production with colonial agriculture and improvement of Indian society. This claim echoed comments about improvement by Dwarkanath Tagore, a major indigo planter and Bengali entrepreneur, who similarly observed that "the cultivation of Indigo and residence of Europeans ... have considerably benefitted the country and the community at large." Further amplifying such resonance between indigo and improvement, the governor-general, William Bentinck, a backer of plantations, boldly asserted that "as a general truth it may be stated that every [indigo] factory is in its degree the center of a circle of improvement raising the persons employed in it and inhabitants of the immediate vicinity above the general level."[1] Such resonance was still appearing in colonial reports and publications later in the century. One important report in 1893 celebrated the system of agriculture on the indigo tracts, claiming that "the cultivation of indigo has been greatly improved by the European planter and the native growers have to some extent followed the example set them."[2] This compatibility among visions – between the realms of commoditized indigo production and peasant agriculture, between Indian and colonial interests, and between the programs of "private" European planters and the "public" colonial state – illustrates that modernity in colonial South Asia hinged on actors and ideas that moved across the worlds of indigo culture and colonial agriculture.

This chapter turns attention to the science that came to be centered in these visions of agricultural improvement. Through the nineteenth century, the nature of this science was itself in the process of taking a modern form. As it did, it transmuted from its moorings in natural history to a new base in the realm of experimental sciences. This shift was punctuated by the assumption of a central role by chemistry and the rising authority of agricultural experts. As planters, state officials, nationalists, foreigners, and naturalists attempted to alter the agricultural landscape in the colony they also dabbled in a science that was changing.

For close to a century colonial South Asia was a theater where modern agricultural science in its worldly dimension evolved on the plantations and in state imaginaries and action. This science was amenable to influence from various quarters. Knowledge and practice in the colony were influenced not only by local agrarian relations, landscape, and imperatives of colonial power, but also by modernist trends within agricultural science. In that regard, this chapter scrutinizes the overlap between

---

[1] *Report of the Proceedings at a General Meeting of the Inhabitants of Calcutta, on the 15th of December, 1829. Extracted from the Bengal Hurkaru,* etc., London: T Bretell, 1830, pp. 4–5. British Library, 08223.h.29(1.); for Bentinck's note, see Parliamentary Papers, viii, 1831–2, General Appendix 5, no. 46, cited in Chittabrata Palit, *Tensions in Bengal Rural Society: Landlords, Planters and Colonial Rule,* New Delhi: Orient Longman, 1998 (first published in 1975), p. 90. These elite references, however, belie the reality of peasant experience. Any unambiguous claims on behalf of indigo and modernity must wrestle with the grim reality of peasant exploitation on the indigo tracts and a history of insurrections against planters. Historians have reached a consensus over the era after which the Europeans evidently coerced Indian peasants to grow indigo. Most take 1825 to be the turning point. Thus, for example, Sugata Bose maintains that "in the first quarter of the nineteenth century indigo was not quite the kind of forced cultivation that it became after 1825." Sugata Bose, *The New Cambridge History of India, Peasant Labour and Colonial Capital: Rural Bengal Since 1770,* Cambridge: Cambridge University Press, 1993, p. 47.

[2] J. A. Voelcker, *Report on the Improvement of Indian Agriculture,* London: Eyre and Spottiswoode, 1893, pp. 222, 236, 257–66.

TABLE I. *Indigo production in Bengal and adjoining territories between 1795–6 and 1831–2*

Year	Maunds	Year	Maunds	Year	Maunds
1795–6	62,500	1808–9	94,539	1821–2	92,848
1796–7	32,300	1809–10	43,012	1822–3	112,606
1797–8	54,600	1810–11	73,407	1823–4	80,315
1798–9	23,800	1811–12	69,654	1824–5	110,227
1799–1800	35,540	1812–13	73,883	1825–6	156,548
1800–1	39,900	1813–14	74,585	1826–7	79,678
1801–2	38,500	1814–15	102,662	1827–8	151,699
1802–3	29,800	1815–6	114,481	1828–9	98,009
1803–4	54,048	1816–7	83,000	1829–30	132,946
1804–5	64,803	1817–18	72,000	1830–1	129,117
1805–6	85,380	1818–19	75,000	1831–2	121,000
1806–7	51,244	1819–20	106,843		
1807–8	103,950	1820–1	76,254		

Note: 1 *maund* = 33.868 kilograms.
Source: John Phipps, *A Series of Treatises on the Principal Products of Bengal*, 1832, p. 35.

colonial agriculture and commerce, on the one hand, and an evolving indigo science, on the other hand, to scrutinize the knowledge foundation of indigo culture. It illustrates two aspects with particular emphasis: first, that the colonial framework was important for the development of this knowledge; second, that this knowledge that transitioned in significant ways from its earlier craft basis and its foundations in the wisdom of "old hands" to its "modern" form was open to influence from forces of spatially dispersed origins.

Imperial Britain and Colonial Bengal: Rise of South Asia as the New Indigo Supplier

At the beginning of the nineteenth century European planters had expanded the cultivation of indigo extensively on the Indian subcontinent and created different zones of indigo production. The primary roots of the plantations lay in Lower Bengal, although their tentacles had also spread into north Bihar and, in a minor way, farther west into Oudh, Agra, or the Doab region generally. The port city of Calcutta in Bengal was the common outlet for all of this indigo received from northern and eastern India. The account of the indigo industry provided by John Phipps, a chronicler of trade and commerce, illustrates the rise and expansion of the industry in Bengal from the end of the eighteenth century to the early decades of the nineteenth century. Table I contains official statistics of

TABLE 2. *Indigo imported into England from Bengal and other parts of the world between 1783 and 1800*

Year	From Bengal (pounds)	From Other Regions (pounds)	Year	From Bengal (pounds)	From Other Regions (pounds)
1783	93,047	1,121,506	1792	581,827	1,285,927
1784	237,230	1,259,149	1793	881,554	1,015,148
1785	154,291	1,540,774	1794	1,394,620	1,464,874
1786	253,345	1,725,712	1795	2,862,684	1,412,165
1787	363,046	1,517,284	1796	3,897,120	651,550
1788	622,691	1,474,220	1797	1,754,223	390,967
1789	371,469	1,599,749	1798	3,862,188	171,218
1790	531,619	1,309,196	1790	2,429,377	549,943
1791	465,198	1,145,595	1800	2,674,317	1,004,642

Source: Adapted from Phipps, *A Series of Treatises on the Principal Products of Bengal*, 1832, p. 49.

commercial indigo manufactured in Bengal and its adjoining territories, attesting the expansion of the industry in the colony.[3]

Phipps also affirmed the new stature of South Asia as the leading exporter of indigo to the world by the first decade of the nineteenth century. In terms of quantity exported Bengal was now the leading source of the blue dye. Central America still supplied fine quality "Guatemala" indigo, but in terms of volume these regions were generally past their period of prior dominance. Signifying such a trend Phipps compares the quantity of Bengal dye to a "very inconsiderable" supply of Spanish indigo into England. Many other varieties, including Java indigo from the Dutch colonies, Manila indigo, and Madras indigo from southern India, also had a presence in the international market in London. But while the import of the first two was "very trifling," according to Phipps, the Madras indigo was of a decidedly inferior quality that sold at half the price of Bengal indigo. No other indigo came close to challenging the dominance of Bengal's indigo. The statistics in Table 2 regarding transactions at the London market reflect the larger global trade in indigo. London was, after all, not only a supplier to the textile belt in the Yorkshire and Lancashire regions of England, the world's most distinguished consumers, but also

[3] John Phipps, *A Series of Treatises on the Principal Products of Bengal No. 1, Indigo*, Calcutta: Baptist Mission Press, 1832, p. 35.

an exporter to the rest of the Western world and parts of Asia. These figures suggest that statistically Bengal indigo had already overtaken the combined export of every other type of indigo by 1795. Phipps elaborated to illustrate this dominance by India in the 1830s, saying, "At the present time, this country supplies all the Indigo, with the exception of an inconsiderable quantity, required for consumption in Europe, America, Persia, and Arabia."[4]

Profit and the Colony's Improvement: "The Fostering Hand of the Company"

John Phipps was an aficionado of the enterprise and science of indigo production and by extension of the colonial modernity that produced it. Working at the Master Attendant's office in Calcutta, he was a shipping and commerce insider with links to colonial commerce. His ties to the East India Company (EIC) were bolstered by his ideological investment in the political economy of the EIC, which sanctioned the organization of production and trade of agricultural commodities for profit to its officials and shareholders. [5]

Phipps clearly saw the relevance of the institutional space created by the action of the colonial state – "the fostering hand of the Company" – which facilitated indigo commerce in the era of pursuit of imperial free trade policies in the colony.[6] The political dominance of "free traders" in the metropolis had led to the passage of the Charter Act of 1813, which ended the EIC's monopoly in Indian trade and its career as an exclusionary, mercantilist institution. At the same time it initiated the era of EIC in the colony as an instrument of the implementation of imperial Britain's free trade policies worldwide. Speculation in indigo henceforth was no longer solely guided by the remittance requirements of EIC or the financial interests of its officials trading in their private capacity

---

[4] John Phipps, *A Series of Treatises on the Principal Products of Bengal, No. 1, Indigo*, Calcutta: Baptist Mission Press, 1832, pp. 13, 19.

[5] Phipps wrote a series of compendia on agricultural commodities as he was admittedly enamored of the subtleties of science, technique, and "enterprise" that undergirded their manufacture "as applicable to modern times." His specific allusion to the science and manufacture of indigo was also meant to underline his advocacy of this modernity. *Cf.* John Phipps, *A Series of Treatises on the Principal Products of Bengal No. 1, Indigo*, pp. ii–v.

[6] John Phipps, *A Series of Treatises on the Principal Products of Bengal, No. 1, Indigo*, p. 9.

and other licensee residents in the colony. Just a year after Phipps's treatise was published, through the Charter Act of 1833, the free traders
opened up Bengal to an even freer inflow of external capital, tying the
colonial industry directly to the actual existing international demand for
indigo. The land settlement policies followed by the company in the early
years had cautiously avoided disrupting native systems of ownership by
restricting Europeans from purchasing land owned by Indian peasants.
The colonialists were wary that Indian peasants and landlords should
not be left totally at the discretion of rapacious capitalists. In order to
keep a tab on the process of land transfers, Regulation 38 of 1793 and
another regulation of 1795 restricted land acquisition by Europeans.
But a regulation in 1829 reversed the earlier policy by giving planters
the legal right to lease land. In succession, the Charter Act of 1833 and
another regulation of 1837 sanctioned unrestricted right of acquisition,
residence, and holding of tenures by the British subjects.[7] The contract
laws regulating relationships between the planters and the Indian growers of indigo were tempered to make them more favorable to the former.
The governor-general, Lord Amherst (1823–8), enacted a regulation that
made peasants liable to pay back monetary advances along with penalties
through a summary suit if they were convicted for violating the terms of
contracts. In doing so Amherst had decided to ignore warnings issued by
the previous governor-general, Lord Minto, that such a law would "open
the door to the exercise of the greatest tyranny, oppression and exaction
[by the planters]." Against the grain of previous circumspection yet again,
an enactment in 1830 went on to make violation of contracts by peasants
punishable by imprisonment.[8]

Phipps was oblivious to any violence toward Indians in the measures
mentioned. Instead, he focused on science and enterprise alone, boldly
lauding "the great progress" of the indigo industry, which he thought
was "attributable to the freedom from restrictions." The state's drive
to facilitate entrepreneurship had caused, in his opinion, "the circumstances … of European skills, industry, and capital" to flourish in the

---

[7] See a good summation of these issues in Amiya Rao and B. G. Rao, *The Blue Devil: Indigo and Colonial Bengal*, Delhi: Oxford University Press, 1992, pp. 30, 61–4.

[8] Discussing Governor General William Bentinck's executive order of 1830, Amiya Rao and B. G. Rao rightly suggest that it was "unjustly severe on one party – the weaker one," and that it did not take into account the fact that contracts were in a way thrust upon the peasants and did not represent an agreement between two consenting parties. Amiya Rao and B. G. Rao, *The Blue Devil: Indigo and Colonial Bengal*, pp. 48–60.

colony. The exclusive focus of his treatise was precisely on the science and trade of indigo manufacturing, which he celebrated within an Adam Smithean ideological framework of commercial enterprise unfettered by statist restrictions, and as an active agent of the global British Empire. This singular focus on "enterprise" made him blind to the exploitative aspects of the industry toward the Indian peasantry or local capital. As an advocate of colonial capitalism, he visualized the Adam Smithean free-trade world as a sanitized, apolitical enterprise that was beneficial to all in one way or the other. He argued that in the absence of the indigo industry vast tracts of land in India would, after all, lie fallow, detrimentally to the interests of both indigenous landlords and the colonial state. Therefore, he argued that the plantation "adds considerably to the value of India, while conferring benefit on Great Britain," finding indigo estates benign and in the process minimizing the political conflicts inherent to the industry.[9]

The nurturing hand of the company was also evident in the government's efforts to shore up the faltering agency houses that financed the indigo factories in the early decades of the nineteenth century.[10] The trading houses controlled diverse aspects of indigo trade, manufacturing, shipping, insurance, and banking by meeting all of the capital needs. The entire set of indigo operations including fixed stocks, leases for land, advances to peasants, cartage, shipping, and insurance were financed by the agency houses based in Calcutta.[11] However, despite their overall

---

[9] John Phipps, *A Series of Treatises on the Principal Products of Bengal, No. 1, Indigo*, p. 24.

[10] The agency houses were the key institutional face of the prevailing credit system, which enabled imperial financiers in Britain to conduct their operations smoothly in India and in the East generally. Stanley Chapman has explained how these business houses represented the surplus capital of manufacturing and merchant interests based in Manchester, London, Glasgow, and Bristol. Stanley Chapman, *Merchant Enterprise in Britain: From the Industrial Revolution to World War I*, Cambridge: Cambridge University Press, 1992, pp. 107–28.

[11] Amales Tripathi recounts the overall dominance of commerce and manufacture by the agency houses in Bengal in such terms: "They controlled country trade, financed silk, indigo, sugar, and opium, ran three banks and four marine insurance companies, speculated in public securities, and negotiated bills on foreign securities." Amales Tripathi, *Trade and Finance in Bengal Presidency, 1793–1833*, Calcutta: Oxford University Press, 1979 (first published in 1956), p. 11; according to Benoy Chowdhury, around 1829–30, of a total of Rupees 20 million invested in indigo, six companies (Alexander & Co., Palmer & Co., Fergusson & Co., Colvin & Co., Cruttenden & Co., and Mackintosh & Co.) contributed Rs. 16 million. He also cites Bentinck to the effect that in 1829, of the total production of 149,285 *maunds* of indigo produced in Bengal, those very companies accounted for 108,603 *maunds*. Cf. Benoy Chowdhury, *Growth of Commercial Agriculture in Bengal, 1757–1900*, Calcutta: India Studies, 1964, p. 83 and notes.

dominance of colonial finance, these institutions were fallible as institutions of finance and had a history of repeated failure on the subcontinent. The agency houses entered into a long-term trend of decline and collapse between 1826 and 1833. But every time any particular house seemed on the verge of collapse, the East India Company stepped forward to salvage it by advancing loans and generally protected them all with the use of monetary and fiscal tools. The government's willingness to prop up agency houses in order ultimately to safeguard the indigo industry was predicated on the financial reality that indigo was far too important a medium of remittance to the East India Company to be allowed to die. The company's administrators also wished to protect the immensely profitable commerce in indigo and the investments that its own officials had made in the agency houses. Thus in the 1820s and 1830s, some argued that any large-scale failure of indigo and its financial empire would not just harm speculators, but be "of most serious public consequence."[12] Such assertions rested on the fact that indigo manufacturing was the primary colonial industry tied to the financial interests of colonial traders, European planters, and a handful of Indian planters and the needs of the metropolitan textile industries. But these assertions also verged on bolder claims that colonialists held up to Indians and themselves to legitimize colonial rule by insisting that the indigo industry was not only tied to profits by sectional interests but was, after all, for the improvement of the entire colony.

Landscape, Land Rights, and the Bid for Improvement

The effort by the Europeans to establish plantations in Bengal was inextricable from the context of landscape and monsoonal climate, on the one hand, and the nature of preexisting peasant rights, on the other hand, in the colonial locality. The expatriate planters had not arrived to colonize an uninhabited terrain. Peasants with a well-entrenched system of land rights had practiced agriculture in the region for centuries. The mutual relationship between land and land rights nourished an existing form of peasant agriculture in eastern and northern India.

---

[12] See the citations in a Court document of 1826 and William Bentinck's assertions in 1833 when he opposed the views of members of the Supreme Council to advocate advancing loans to agency houses. Benoy Chowdhury, *Growth of Commercial Agriculture in Bengal*, pp. 95–6, 104–5.

The very use of the term "plantations" to describe the cultivation of indigo in Bengal is misleading. It suggests that sowing of indigo transpired on large-scale holdings under individual or joint proprietorship. In reality the major part of new indigo cultivation took place on peasant holdings. Late eighteenth-century European immigrants had to find the means to insert indigo cultivation into a preexisting system of agriculture. The planters employed an elaborate system of contracts to facilitate the growing of indigo by the peasants and its procurement at planters' factories. The precise forms of cultivation, remuneration, and delivery were determined by a complex interplay between Europeans' financial and political power and the momentum of local agriculture. If in some ways the financial and political power of European planters held sway, in others the system of peasant agriculture induced mutations in the form of rising plantations. In the end Bengal's "plantations" adopted a unique physical appearance and organization tempered and influenced as much by the force of peasant agriculture as by colonial facilitation.

Riparian Plantations

The planters' testimonies affirm that in the early period most indigo plantations were nestled on the lowlands in Bengal close to the major river systems prone to annual flooding that remained under water for months at a time. Indeed some of these lands were in the middle of rivers as floodwater receded, throwing up new islands (Fig. 3) of cultivable land that could only be serviced by planters' boats, or *bhowalea* (Fig. 4). Such changes to land were often the cause of litigations and disputes among peasants over who exactly owned the land. Indeed, the planter George Ballard observed of the indigo tracts that "nine-tenths probably of the land bearing this crop is more or less under water by the end of July."[13] Another planter, N. Alexander, also implied that the "*chur* of the Ganges," in other words, its riverbeds, formed the core area of indigo cultivation in Bengal in the early period. In Lower Bengal, the primary indigo fields were submerged

---

[13] G. Ballard, "On the Culture of Indigo in Bengal, Read 10th June, 1829," *Transactions of the Agricultural and Horticultural Society of India* II (1836): 14–24, quote on p. 14. See also Phipps, *A Series of Treatises on the Principal Products of Bengal*, pp. 75–6. The extended submergence under floodwater would often change the contours of the land so completely as to make them unrecognizable on emergence after the floods.

FIGURE 3. Sowing of indigo, or *chitani*, in Bengal on river islands.
Source: W. M. Reid, *The Culture and Manufacture of Indigo with Description of a Planter's Life and Resources*. Calcutta: Thacker, Spink and Co., 1887, facing p. 74.

FIGURE 4. The planter's boat (*bhowalea*).
Source: W. M.Reid, *The Culture and Manufacture of Indigo with Description of a Planter's Life and Resources*. Calcutta: Thacker, Spink and Co., 1887, facing p. 75.

under water for several months. Alexander could look back at the expansion of indigo in the province and see that the year 1823 emerged as some sort of a cutoff point after which the indigo plantations seemed to have started to expand beyond the lowlands to encroach into the highlands.

This expansion had been spurred primarily by the rise in demand for export of indigo.[14]

The planters' early recourse to the peasants' low alluvial lands was to some extent enforced by the preferences of the current occupier-cultivators and the cycle of flooding that seemed to suit indigo while interfering with the cycle of cultivation of staples. The floodwater did not recede until October, by which time it was too late to sow paddy, the peasant staple. Paddy sowings in the region were known to take place generally in the month of May. Under no circumstances could the process be delayed beyond August. Thus the annual cover of valuable fertile alluvial soil left over land was of no use to the rice peasantry. The local labor situation also freed these lowlands for indigo. As Ballard argued, alluvial lands were "surplus" lands in the sense that the peasant families' labor was exhausted in the cultivation of rice on the highlands and thus their routine agriculture rarely left much labor or time at their disposal to cultivate the lowlands properly. On the other hand, the swampy lands left behind after the floods could be strewn with indigo seeds with a minimal application of labor. Indigo cultivation on swampy lands without any requirement of plowing and digging also left untouched the animal power of the peasants, which they jealously guarded. Thus the peasants did not hesitate to take cash advances from the Europeans for the purpose of sowing these "extra" lands with indigo in the early phase of expansion of plantations.

The adequacy of floodplains for indigo was also validated by the early science of soil fertility. Indigo producers had noted that the inundated areas in Bengal did not lose their productivity for indigo year after year, as compared to the highlands, where any such intensive cropping had to be alternated with years of inactivity to revive the fertility of the soil. The annual inundation from the Ganges seemed the likely source of rejuvenation. But to confirm these beliefs the planters mobilized the science of soil testing. The "fertilizing principle of the Ganges' inundation" was put to scientific analysis. Alexander reports that Henry Piddington, a scientist in the colony with more than trivial interest in indigo, had tested silt

[14] Alexander had been driven to write an account of the history of expansion in indigo cultivation in Bengal as a riposte to those who held planters responsible for a reckless expansion in its cultivation leading to a glut in the market. Alexander argued that the expansion had occurred because of rising demand for the commodity. N. Alexander, "Cultivation of Indigo, Read 13th August, 1829," *Transactions of the Agricultural and Horticultural Society of India* II (1836): 31–41.

from the Ganges collected from two different sites, one near Sukhsagar and another from the vicinity of Krishnagar, both in Bengal. These assays confirmed that while the silt contained a minimal amount of vegetable matter, it was very rich in calcareous content. The highland soils were, on the contrary, found to be high in vegetable content and low in calcium and phosphates. Piddington thus concluded that "the calcareous matter was, perhaps, the great agent" of indigo's productivity.[15] Thus, the early European settlers seemingly followed the course dictated by the line of least resistance in favoring the flood lands. But the science too suggested that it made sense to stick with lowlands.

In the northern Bihar districts of Tirhut, to the west of Lower Bengal, indigo was cultivated on both highlands and inundated lowlands. But only some of the lowlands in Tirhut were actual riverbeds. As Ballard noted, the location of his factory "on the banks of the Ganges and great Gunduck," the two prominent rivers of Tirhut, was a rarity.[16] Most indigo lands in Bihar were also inundated, though not by flooding from rivers but by high rainfall and a rising water table. Ballard refers to the numerous "*jheels* and *nullahs*" or small lakes and streams and the fact that the indigo tracts were subjected to flooding from these water bodies. Phipps also mentioned the importance of lakes, "which are common in Tirhoot; these being excavations formed by the overflowing water during the floods in the rains."[17] Colin M. Fischer classified the soil of Bihar into three categories: *bangar,* the low-lying, annually inundated, hard clay soil that was suitable for winter rice; *bhit,* the fertile, sandy loam that lay above the level of annual inundation and was fit for raising valuable cash crops; and *goenra,* a special upland soil that lay near the village and had been tended carefully over the generations by the farmers and was very fertile. Indigo occupied the low-lying areas in Bihar from its early establishment in the 1790s up to the late 1830s and only after that began to expand into the upland areas. Thus indigo displaced rice lands in the early period in Tirhut. In the later period as it expanded into zones that were given to profitable cash crops, it compromised the interests not so

[15] N. Alexander, "Cultivation of Indigo, Read 13th August, 1829," *Transactions of the Agricultural and Horticultural Society of India* II (1836): 34–5.

[16] The Ganges flowed to the south of Tirhut and separated it from southern Bihar. Most of the Gangetic floodplains lay to the south of Tirhut and only a very limited portion lay in Tirhut proper. The floodplains of Gandak, the major river system traversing Tirhut, were under a variety of crops other than indigo.

[17] G. Ballard, "On the Culture of Indigo in Bengal," p. 19; Phipps, *A Series of Treatises on the Principal Products of Bengal*, p. 99.

much of the peasantry as of the *banians* (local traders) who had monopolized cultivation and trade in remunerative cash crops.[18]

Raiyati *and* Nij: *Indigo in Bengal and Tirhut*

A number of factors minimized direct cultivation of indigo by the planters under their own demesne lands, or *nij*. The eastern districts in Bengal were covered with forests and were also sparsely populated. The inhabited lands, on the other hand, did not have a large cache of cultivable wastelands or redundant lands that could be easily brought into cultivation with incentives. Addressing the available courses for expansion of indigo, Alexander reminded, "No man, who has been connected with Indigo factories, is ignorant how difficult it is to procure new land." Alexander here was of course giving vent to the difficulties in the way of land acquisition from the planters' perspective. What should not be disregarded here is the fact that the planters overcame all these difficulties to obtain lands when they thought they must. One possible course available to the planters was to obtain leases on the landholdings already under cultivation of other crops. In Bengal most of these leases were registered in the names of planters' dependents since leasing of land by planters was not permissible under company laws until 1833. But these were the lands on which the natives grew profitable "dry" crops, the crops that could not withstand flooding. To be able to obtain leases on them, a planter had to be, Alexander reminds us, "influential" and had to have the means to pay high charges to compensate the peasants. Ballard writes that direct cultivation was "an expensive plan" that was profitable only when the price of the indigo dye was at its highest point in the international market. It was not the preferred method of growing indigo, and thus the district had witnessed an abandoning of *nij* by those few planters who took to it initially. In his estimation the cost of indigo to the planter by giving contracts to the peasants was 25 percent cheaper than through *nij*.[19]

---

[18] As Fischer says, "Whether through conscious preference or market pressure, the N. Bihar planters between 1820 and 1840 changed from lowland to highland cultivation." Colin M. Fischer, "Indigo Plantations and Agrarian Society in North Bihar in the Nineteenth and Early Twentieth Centuries," Unpublished Ph.D. dissertation, University of Cambridge, 1976, p. 76; see the description of soils on p. 2. Fischer argues that the expression of discontent against indigo was spearheaded not so much by the peasants as by the traders, whose business interests were harmed by the displacement of other cash crops by indigo.

[19] N. Alexander, "Cultivation of Indigo," p. 35 and notes; G. Ballard, "On the Culture of Indigo in Bengal," p. 20.

The turn toward *nij* in Tirhut, on the other hand, was fairly advanced by the third decade of the nineteenth century. Phipps wrote in 1832 that half of the indigo tracts in Tirhut were under the *nij* form of cultivation.[20] The explanation offered by Sugata Bose and, following him, by Jacques Pouchepadass for the rise of demesne cultivation in Bihar after mid-nineteenth century seems to be generally true for the opening decades as well. Outright purchase of land by the planters was almost nonexistent. To explain the absence of *nij* in Bengal and its prevalence in Bihar, Bose has pointed out that the two regions had a "different configuration of agrarian social classes." The landlords in Bihar had "immense" control over land. But at the same time they were dependent on planters for rural credit. In Bengal, in contrast, the landlords and the planters competed in advancing capital to the peasantry. This backdrop of land tenures and credit circulation explains why Bihar planters were able to obtain lease for *nij* lands from the landlords by advancing them credit. Jacques Pouchepadass has described the existence of limited perpetual leases (*mukarrari*) and the more common temporary leases of five to twenty years taken out by the planters in Tirhut for the later period. It is quite likely that these leases had their birth in the early decades. Writing in 1828, the planter George Ballard mentioned the emergence of the "new" system of *thika* in the region: "A still better system is lately gaining ground in the district; I mean that of taking villages in *ticka* or farm." He also alluded to the willful effort by the planters to make the *zamindar* (landlord) or the *jytedar*, or head *assamee* (the dominant cultivating class of *jotedars* in eastern India), become part of the contract through which leases were obtained in order to ensure a tighter control over the peasantry in the leased area.[21]

The Primary and Secondary Zones of Indigo Culture: The Influence of Peasant Agriculture

Bengal and Tirhut

The precise terms of the indigo contracts deployed by the European planters varied from region to region partly because they were modified to meet the needs of specific labor procurement under distinct soil and

---

[20] Phipps, *A Series of Treatises on the Principal Products of Bengal*, pp. 76, 97.
[21] Bose, *The New Cambridge History of India, Peasant Labour and Colonial Capital: Rural Bengal Since 1770*, p. 76; Jacques Pouchepadass, *Champaran and Gandhi: Planters, Peasants and Gandhian Politics*, Delhi: Oxford University Press, 1999, pp. 37–40; Ballard, "On the Culture of Indigo in Bengal," p. 17.

weather conditions, but also because they reflected the determination by class-relations as to what was possible and what was not possible for the planters to impose.[22] A monetary advance that was offered at no interest enticed many impoverished Bengali peasants with "the advantage of a little ready money" and thus induced them into contracts to commit land to indigo. In late October, soon after the floodwater had receded, the peasant sowed the seeds on the swampy beds through the method of "broadcast" or simple scattering, called *chittani* in local parlance. Such land required no prior preparation and demanded no animal power to be diverted from the peasants' own lands. The effort simply required scattering of seed received by the peasants from planters at one-third of its market price. Only a few peasants who had land on a higher elevation could afford to keep the plant after cutting, allowing it to flower again and develop seeds and were thus a little more advantageously placed than other peasants. Most lands were submerged before the seedpods would arrive. There was also a risk that the majority of plants from October sowing might not survive the ensuing winter. But such was the precarious nature of the crop that the planters wished to have this round of sowing to serve as a backup in case the second round of sowing in March did not do very well by itself. Planters engaged natives' labor by advancing them two rupees for every *bigha* that the grower planted. The effort needed for weeding and the general upkeep of plantations was minimal. The harvest often had to be rushed because of an impending swell of floodwaters from the river and all manufacturing had to come to an end by the middle to end of August. The peasant was obliged to return indigo to the factory and was paid at the rate of a rupee for every six to eight bundles. The remuneration did not usually compensate him for his costs, however. As even the planter Ballard acknowledged, unless the peasant could grow his own seeds he had no chance of returning the advance and thus remained bound to grow indigo year after year.

The Tirhut districts did not practice the double sowing of October and March used in Bengal. However, a much more extended period of preparation of land was required of the peasants leading up to spring. The preparation of land started in September and was targeted at retaining moisture in the soil. At least four rounds of plowing were followed

[22] George Ballard has provided the most comprehensive description of the different types of indigo contracts and the cultivation patterns in Bengal, Tirhut, and Oudh. G. Ballard, "On the Culture of Indigo in Bengal, Read 10th June, 1829," *Transactions of the Agricultural and Horticultural Society of India* II (1836): 14–24.

by digging and cleaning of weeds, and then the plot was subjected to another four rounds of plowing. A team of peasants with bullocks thereafter leveled the field. If rains suddenly began, they would destroy much of the preparation up to then, necessitating more rounds of plowing and harrowing to get rid of the hard crust of soil as water on the fields dried up. In between the peasants also weeded the plot with the help of family labor and crude implements that made the entire operation extremely labor-intensive. The land was sowed in February or March. Broadcasting followed by harrowing was commonly used to make the seed descend into the soil. Unexpected rains immediately after sowing could necessitate more harrowing. Heavy rains would require the planters to resow the land all over again.

In order to have the operations completed the planters in Tirhut signed a covenant called the *noviskaun* with peasants, mandating that they complete each step in the process from preparing the field to harvesting and delivering indigo. Having a validity of five years the contract provided a strict control over the cultivation process to the planter. The planters staggered their payment over the entire duration of cultivation ranging from sowing of seed to the delivery of harvest. Illustrating the strict control enforced on growers by the planters, George Ballard wrote, "[The peasant] engages to give you such land as you may select, prepare it according to instructions from the factory, sow and weed it as often as he is required, cut the plant and load the hackeries [*sic*] at his own cost, and in every other respect conform to the orders of the planter or his aumlah [clerk]."[23] The grower was exempted from paying for seeds, cartage, and the cost of seed drills. But heavy penalties were enforced in case of nonfulfillment of any of the terms specified in the contract.

Highland Agriculture and the Distinct System of Indigo Culture in Doab

The culture of indigo in the Doab started as these territories were brought under control in the first half of the nineteenth century, but these were not the primary zones of indigo plantations and were not even very desirable from the perspective of the planters. Oudh and the northwestern provinces, the latter also referred as "Upper India" by Ballard in order to distinguish it from eastern parts, were both relatively prosperous regions where Indian peasants grew crops that required high-cost inputs and returned high profits. These were the areas of "highland agriculture"

[23] G. Ballard, "On the Culture of Indigo in Bengal," p. 17.

where local landlords engaged in irrigated agriculture of staples and high-value food cash crops like wheat and barley, or nonfood cash crops like sugarcane and cotton. If indigo were to be extended into these areas, the planters would have to embrace the high-input agriculture of the region. The indigo cultivation in these areas was more prominent only in certain stretches of years when the price of indigo was extremely high in the market, enabling planters to afford the high input costs and still make a profit. In that sense indigo culture of these regions was a spillover from Bengal and Tirhut, the traditional belt of indigo culture in the colony.

Ballard's bitter complaints that the indigo system in Oudh was "ruinous" or was full of "evils" stemmed partly from the high cost of inputs including labor and rent, and the problems that the local terrain and climate in Doab posed to the planter-entrepreneur. The light and sandy soil of Oudh was less adapted to the cultivation of indigo than the clayey soil of Bihar or the alluvium in Bengal. In order to induce the peasants to give up their cultivation of wheat, barley, cotton, and sugarcane, the planters had to entice them with higher remuneration. The planters also had to battle the elements, in particular the deficient rains and colder conditions. Rains occurred later in the year and were also scarcer than in Bengal and Tirhut. The winters were colder, and thus the *khoonti* crops left to be cultivated the next year faced very frosty conditions during the winter and did not always survive them. Thus conditions overall were adverse for indigo culture, usually practiced under conditions of low rents and prices in Bengal.

The primary method of procurement in Oudh involved cash advances to local landholders under two main types of contracts – the *kush kurrea* and the *bighowty*.[24] In contrast to Bengal, the advances were made to relatively more prosperous agriculturists, and not to impoverished peasants. The monetary advance was a necessity because of what Ballard called "rack-renting" and the "onerous assessment" of cultivators by the native ruling state. The charge of overassessment resonates with the building rhetoric of colonialists about "misgovernance" under the native nawabi rule in the region of Oudh, which would soon become a ground for the company's annexation of Oudh.[25] In addition to the task of enticing the landlords to sow indigo, the planters had to negotiate the challenges

---

[24] There was some *nij* cultivation in Oudh, but its extent was negligible. "A mere trifle" is how Ballard describes the prevalence of *nij* in Oudh. G. Ballard, "On the Culture of Indigo in Bengal," p. 21.

[25] The nawabi rule in Oudh was propped by the British under the subsidiary treaty of 1801.

of climatic factors. *Kush kurrea* was the favored system of the two. It involved advancing of one rupee for every five *maunds* of *jumowah* crop, three *maunds* of *assaroo* crop, and seven *maunds* of *khoonti* crop. The first was an irrigated crop sown in May and irrigated every week utilizing expensive labor until the rains approached at the end of June. But if for some reason *jumowah* failed, a later *assaroo* crop was sowed after the onset of rains. The *assaroo* did not mature until late September, and waiting that long meant that this harvest ran into problems of weaker fermentation in manufacturing during the harsh winter, which reduced the yield. One solution attempted by the planters was to leave the *assaroo* plant in the field to give a *khoonti* crop the next year. But leaving *khoonti* in the field required extra manpower to protect the crop over the extended period. It also involved taking the risk that the crop might not be able to survive the frost in the cold season. The *bighowty* system of contracts was a very specific one prevalent in the Meerut and Moradabad districts. It involved a different regimen of payment for *jumowah, assaroo,* and *khoonti.* The planters under this system advanced nine rupees and five rupees and four *annas* (a quarter of a rupee), respectively, for one *bigha* of *jumowah* and *assaroo.* For an additional payment of eight *annas* the planter could lay a claim to the *khoonti* of both. In this system the planter had the responsibility to supply seeds.

The Oudh systems generally afforded the planter very stringent control over the cultivation process. The planter reserved the right to choose "the highest and best lands" from the cultivator. Planters generally had a preference for the land that had been under wheat, barley, and, especially, sugarcane until March or April. These were usually the best lands in peasants' possession. He closely supervised plowing, sowing, and weeding. The Oudh system also entrusted to him the right to declare any particular planted area as *naboode,* or inferior, if two competent persons adjudicated that the bad crop of indigo was due to negligence in late sowing, bad selection of soil, or lack of weeding. In such a case the planter could ask for deductions in the peasants' share at the time of the final settlement of accounts. Although Ballard complained that it was difficult to enforce *naboode,* the very existence of the convention reflects the planters' stronger position in Oudh land relations.[26]

---

[26] Ballard also attests that the contracts were certainly not profitable or beneficial to locals in any way except in terms of providing ready cash to pay off the rents. He also mentions that the preferred cash crop for the Oudh landlords was either cotton or sugarcane, not indigo.

The Gaud System and Its Demise in the "Upper Provinces"

In the zone extending from Allahabad to Aligarh in the Doab the practice of indigo culture was even more expensive than in Oudh. It was a zone of the most sought after land. Its fertility and assured returns spurred prosperous landholders in the area to dig *kucha* wells (wells with no masonry) to practice irrigated agriculture. The costs of digging wells and the labor-intensive processes that involved lifting water from the wells and managing channels to water the lands were very high. Thus traditionally the peasants in the region specialized in the cultivation of cash crops that gave a very high margin of profit as a compensation for the high cost of irrigated agriculture.

The *gaud* system of indigo culture that emerged in the Upper Provinces was unique and distinguishable from all other systems of indigo culture in the colony. The European indigo factories that had been set up in the region since 1809 directly engaged the Indian landholders not merely to cultivate indigo but also to manufacture the "fecula" or *gaud* using the processes of fermentation and beating. The system involved the purchase of the semiprocessed dye by the planters, which was then processed to a finished form at the factories. Having received the *gaud*, the planters processed the dye with the familiar "Bengal method." This manner of manufacture was a deviation from the common practice of indigo culture elsewhere in the colony wherein the planters only made the peasants grow indigo while completing the entire manufacturing process at their own plants. The longevity of *gaud* system of culture at one concern in Aligarh is illustrated in Table 3. Between 1822 and 1824 at Aligarh alone the planters purchased an average of 13,906 *maunds* of *gaud* indigo per year from the natives and produced 1,695 *maunds* of dry indigo out of them. In 1825 the total output of indigo produced with *gaud* was 29,628 *maunds*.[27]

The *gaud* system apparently did not survive the rising trend in prices for raw materials charged by native agriculturists. Thus in 1824 the European planters in the region had congregated at Fatehgarh in a meeting to deal with the question of the high price of *gaud*. They passed several resolutions announcing cartel-like common policies designed to control the market price of raw material to their advantage. The planters agreed to fix the maximum price that any planter should offer for the procurement at Rs.12 inclusive of all charges for the best *gaud*, Rs. 11 for

[27] Phipps, *A Series of Treatises on the Principal Products of Bengal, No. 1, Indigo*, pp. 103–13.

TABLE 3. *Account of the expense and profit of a concern using the* gaud *culture in Aligarh*

Year	Advance (in Rs.)	Output (in *maunds*)	Profit (in Rs.)
1809	140,610	11,543	NA
1810	137,710	11,058	NA
1811	168,967	13,601	NA
1812	146,249	7,941	NA
1813	205,775	11,251	8,842
1814	293,114	18,556	11,205
1815	221,453	13,472	8,320
1816	182,519	8,588	5,743
1817	199,180	11,540	619
1818	180,312	19,055	3,422
1819	126,766	7,735	2,799
1820	112,032	8,076	9,578
1821	275,325	15,466	59,278
1822	513,458	18,193	91,334
1823	330,778	18,964	26,224
1824	236,253	14,561	4,714
1825	179,763	10,633	2,089
1826	274,516	17,647	2,055

Source: John Phipps, *A Series of Treatises on the Principal Products of Bengal*, 1832, p. 111.

the *gaud* produced under the *kush kurrea* contract, or, under a different system of payment, Rs. 1 for four and five *maunds* of *jumowah/assaroo* and *khoontie* crops used for *gaud*, and Rs. 1 for three and two *bighas* of planted lands under *jumowah/assaroo* and *khoontie*, respectively, turned into *gaud*. It was decided that no cash advance should be made for *kush kurrea* contracts, and it was also agreed upon that any *gaud* rejected by one planter would not be purchased by another. Despite these agreements, the price of *gaud* shot up to Rs. 14 during the manufacturing season that year.[28] A pamphlet of 1827 by another planter similarly complained that the prices for the *gaud* that was Rs. 7.50 until a few years ago had gone up to "the extravagant pitch" of Rs. 17 per *maund*.[29]

Phipps reported that the hybrid system of *gaud* that arose in the Oudh subregion of the colonial locality had been abandoned by the end of the decade of 1820. The planters stopped purchasing *gaud* from natives and

[28] Phipps, *A Series of Treatises on the Principal Products of Bengal*, No. 1, Indigo, fn., 106–7.
[29] Phipps, *A Series of Treatises on the Principal Products of Bengal*, No. 1, Indigo, p. 106.

adopted the familiar system of making cash advances to the peasants in order to secure the plant itself at their factories, as in Bengal. At the time Phipps wrote his account, indigo culture, *gaud* or any other, had entirely disappeared from the western rim of the Doab, having lost out to the major indigo- growing districts of Jessore and Nadia in the Lower Province and the Tirhut district in north Bihar. Phipps stated as much, saying: "Almost all factories [in the western province] have recently been shut up; and there is no prospect of their being worked again to advantage – especially as the Lower Provinces and Tirhoot can furnish all the Indigo from Calcutta."[30]

Phipps blamed the high price charged by natives for the death of the *gaud* system. He seems to have accepted the arguments of planters in the western province who accused the natives of either "knavishness" or greed, as apparent in their propensity to charge a high price for supplying the raw material. But his conclusion in this regard seems barely to skim the surface of price data while neglecting the fact that even after abandoning *gaud* manufacturing the planters could not continue with the culture of indigo in the western province. In all fairness to Phipps, he only witnessed the early features of long-term trends in indigo culture in nineteenth-century South Asia. He reported that natives in the Aligarh region had taken to growing indigo seeds. Indeed the northwestern province emerged as the predominant supplier of indigo seeds to Bengal and Tirhut for the rest of the century. The demise of indigo in the province was also not the last episode in the history of indigo culture there. Time and again indigo culture would be taken up in the western provinces during favorable market conditions of very high prices. But even at the best of times this zone played only a subordinate role to the major cultivation areas in Bengal and Bihar. In the overall scheme of things indigo culture in the colony lodged itself on a stable and permanent basis only in Bengal and Tirhut, where it thrived on low-cost inputs, particularly the low labor costs prevailing in the region. The western provinces supported high-cost, irrigation-based "highland" agriculture that was unsuitable for the way indigo was produced and sold by the colonialists. And this may be the more accurate picture emerging from the analysis of long-term trends that Phipps missed.

In the previous chapter we have seen that the theory of "knowledge hegemony" with its ambition of explaining "all" developments as colonial does not stand scrutiny when we frame the late eighteenth-century

---

[30] Phipps, *A Series of Treatises on the Principal Products of Bengal, No. 1, Indigo*, p. 108.

rise of indigo on the Indian subcontinent as transpiring within an international information order. Such analysis proved useful in shedding light on aspects of multivalent genealogy of indigo plantations, not merely the "colonial construction" of plantation economy. But such wide genealogy did not preclude internal differentiation within indigo plantations as we see here. The influence of several nodes in the world was one aspect of the eclectic nature of planting craft and knowledge. After its entry into the subcontinent the regional forces in the colony introduced changes regionally. Thus the nature of plantations varied among Bengal, Tirhut, Oudh, and the western provinces on the basis of multiple factors including climate, landscape, and preexisting peasant agriculture conditions. The agrarian histories of the indigo tracts and histories of indigo peasantry have hinted at the standoff between the "native" peasant agriculture and the expansionist plantation economy in the process of the rise of indigo in Bengal and Tirhut in particular. As the nineteenth century progressed, cultivation under *nij,* or demesne, lands emerged in Tirhut with the planters buying the rights of superior landlords. Meanwhile the *raiyati* system of contract cultivation remained predominant in all other regions including Bengal, Oudh, and the western provinces. The peasants throughout the "plantations" were never dispossessed of their lands even if their rights were progressively eroded with time. The preexisting peasant agriculture had enough momentum behind it to ensure that large demesne-based plantations did not emerge in Bengal as had been the case in the rest of the world, excepting perhaps Spanish Central America.[31] Some of the debates in the field have tried to explain this interaction within the general outlines of mode of production analysis and highlighted the "superimposition" of a colonial, capitalist plantation on top of an indigenous semifeudal agriculture, a system that enabled the foreign capitalists to exploit cheap native labor. But outside such paradigms, which in a way presume the eventuality of a system of "progression" to capitalism, others such as the historian Sugata Bose have argued that the organization of plantations through peasant agriculture in late eighteenth- and early nineteenth-century South Asia represented a mechanism to tap into the unpaid labor of the peasant family. He explained the preponderance of *raiyati* plantations by suggesting that "colonial capital

---

[31] The peasants cultivated and manufactured indigo on their small units in Spanish Central America. The peasant units of production coexisted with the large-scale *obrajes* of the Spanish planters. The Spanish production had its own marks of distinction from the Bengali system. In the latter the peasants were only relied upon to cultivate the crop while manufacturing was centralized by the planters at their factories.

preferred a course in which the cost of labour was, quite simply, 'nothing.'" But he also alluded to elements of "resistance," arguing that from a labor perspective, the peasantry by its tenacious grip on land rights had "resisted" the extension of planters' demesnes, even though the peasant's success in this regard was built on the sacrifice of labor.[32] In line with Bose's arguments it would be correct to maintain that peasant agriculture had offered "resistance" of sorts that determined the consequent unique shape of plantations nestled on peasant plots in South Asia.

Michel Eugene Chevreul and the Rise of Organic Chemistry

Indigo plantations in the first quarter of the nineteenth century coincided with the rise of the field of organic chemistry, which enabled a new understanding of the science of indigo manufacture. This understanding made possible the introduction of new steps in the manufacturing process in colonial India to obtain a purer dye that won Bengal indigo a worldwide reputation for its quality. The new science developed as part of a new quest to understand the scientific basis of life-forms, of plants and animals, to be precise. Its locus lay far away from the Indian subcontinent, at the Musée d'Histoire Naturelle in Paris, where Antoine Lavoisier's two prominent disciples, Louis-Nicolas Vauquelin and Antoine François de Fourcroy, were working relentlessly toward clarifying the nature of organic substances. Following the lead of their mentor, who had earlier worked with the oxidation of organic bodies, they postulated the existence of carbon, hydrogen, oxygen, and nitrogen in plant and animal substances and explained the union of oxygen with other elements in organic bodies. Over the next few decades the field would develop its own analytical methods and experimental practices and become distinctly recognizable as the discipline of organic chemistry. It would also gather wider reputation through the early works of Justus von Liebig in Germany and other philosophers.[33]

Even in its rudimentary stage at the start of the nineteenth century it was clear that the new chemistry was driven by the ambition to understand

---

[32] *Cf.* Sugata Bose, *The New Cambridge History of India, Peasant Labour and Colonial Capital: Rural Bengal since 1770*, Cambridge: Cambridge University Press, 1993, pp. 74–9.

[33] For Liebig's contributions to organic chemistry and organic analysis, as well as for a more detailed treatment of the history of organic chemistry, see, William H. Brock, *Justus von Liebig: The Chemical Gatekeeper*, Cambridge: Cambridge University Press, 1997; Alan Rocke, *The Quiet Revolution: Hermann Kolbe and the Science of Organic Chemistry*, Berkeley: University of California Press, 1993.

the foundational unit that constituted living organisms. It grew in opposition to the dominant "vitalist" theories that assumed a fundamental difference between organic and material forms. Famous chemists such as Berzelius argued that a vital force in living organisms, and not ordinary chemical forces as in the case of material substances, produced the phenomena of life. In this argument, the organic bodies were held together by a different type of chemical affinity. But this view was being opposed by such insurgents as Berthollet, Thenard, and Chevreul. The opposition of these savants to any assumption of a basic difference in the chemical basis of material and organic forms opened the way for detailed study of organic bodies. Thus while the vitalists had stressed the inherent and foundational vital strengths that drove life, forces whose nature was assumed rather than further analyzed, the new science aspired to understand the foundational unit of life as constituted by elements and compounds. It was not willing to concede the difference between living/organic and the nonliving/inorganic, and it was particularly driven to unravel the chemical basis of life-forms.[34] The rise of organic chemistry transformed the field of agricultural science and gave rise to the field of natural products chemistry, focused specifically on valuable extracts from plants and animals.

As part of the emergence of organic chemistry, the French chemist Michel Eugène Chevreul (born 1786) made the first major contributions to theoretical and empirical understanding of indigo processes. His personal life history had taken him at the raw age of seventeen to Paris, where he joined Louis-Nicolas Vauquelin, a stalwart in the field with an interest in the analytical chemistry of vegetable and mineral matters, at the Musée d'Histoire Naturelle. Vauquelin was the assistant to Antoine François de Fourcroy, the professor of general chemistry at the museum. Thus being in touch with Vauquelin and Fourcroy, Chevreul was well placed to be part of the early forays into the study of organic substances. More specifically, both these mentors at the museum had worked on indigo and bequeathed that interest to Chevreul. Vauquelin had personally assigned Chevreul to work out analytical tests on the indigo dye. While Chevreul's most notable success was in his work on fatty acids, he would also retain his interest in natural dyes including logwood, yellow oak, and indigo. The continuing interest in natural dyes was partly the result of his involvement with the Manufacture de Royale des Gobelins, the national tapestry

[34] See the chapter "Organic Chemistry to 1825" in Albert B. Costa, *Michel Eugene Chevreul: Pioneer of Organic Chemistry*, Madison: State Historical Society of Wisconsin/Department of History, University of Wisconsin, 1962, pp. 20–38.

works of France. Chevreul remained at the museum for the rest of his life, succeeding Vauquelin when the latter retired. He additionally assumed examiner and teaching position in succession at Ecoles Polytechnique and Lycée Charlemagne and was nominated to the Academie des Sciences in 1826. His authority to speak on organic substances was unchallengeable not only nationally and in Europe but also, in the case of indigo, beyond the West, in Asia and elsewhere. As such he did not extensively conduct experiments on indigo or produce voluminous works on the subject. But his study of indigo between 1807 and 1811 was seminal and the singular, most accurate explanation of indigo that was available. He would go on to define a scientific and philosophical consensus on the constitution of indigo for several decades during a time when his primary thesis of "indigo white" would remain unchallenged.[35]

Chevreul's experiments on indigo were embedded in contemporary traditions of organic chemistry. Finding organic substances to be complex aggregates, the chemists had moved away from the initial ambitious program of synthesizing organic compounds and toward the simpler task of extraction and purification of primarily plant products.[36] Distillation and extraction with solvents such as hot and cold water and alcohol or even acids was the primary mode of analytical work. Chevreul began his query on indigo in the communion of many luminaries who were also exploring the nature of organic substances by examining derivatives formed on their treatment with acids. It was believed that most substances with nitrogen gave a "bitter substance" or "amer" when treated with nitric acid. In an early paper Chevreul mentioned the direction shown by Haussman, who "was the first" to examine the bitter principle obtained on treating indigo with nitric acid; by Welther, who treated silk with nitric acid and used the French word *amer* meaning "bitter" to refer to the product; and by many luminaries in France including Proust, Foucroy, and Vauqueline who had established the fact that *amer* resulted from treating any nitrogen-containing organic substance with nitric acid. Similarly, Chevreul referred to the works by Hatchett on treatment of vegetable compounds with sulfuric and nitric acids and to his own work on converting an extract of Brazil wood with nitric acid.[37] The purpose

---

[35] Albert B. Costa, *Michel Eugene Chevreul: Pioneer of Organic Chemistry*, Madison: State Historical Society of Wisconsin/Department of History, University of Wisconsin, 1962.

[36] Albert B. Costa, *Michel Eugene Chevreul: Pioneer of Organic Chemistry*, p. 22.

[37] "Abstract of a paper on the bitter substance formed by the action of nitric acid on indigo: by Mr. Chevreul," *Journal of Natural Philosophy, Chemistry and the Arts*, Sep/Dec 1811, pp. 351–63, published originally in *Ann. De chim.*, vol. lxxii, p. 113.

of Chevreul's experiments fundamentally was to advance the knowledge of the constituents of indigo in line with the urgings of the atomic theory of John Dalton. To be able to do this he subjected the indigo *amer* to processes that could differentiate its constituents such as resins, oxides, and gaseous substances. He also tried to find common properties that these elements might share with the *amer* obtained from other organic substances. He wished to follow up recent works by Braconnot, who had obtained an acid on treating aloes with nitric acid, which bore a similarity to the *amer* of indigo as well as to the substance that Fourcroy and Vauquelin had obtained on treating animal bodies with nitric acid. Following a suggestion from Vauquelin, he was also following a line of experiment that would make him compare the results recently obtained by Moretti, a professor of chemistry at Udina, of treating indigo with nitric acid with those obtained by him and by Vauquelin and Fourcroy.

Chevreul's trials reflect both the general direction of his effort to explore the constitution of indigo and the difficulty of this endeavor. It is true that his experiments failed to give the exact nature of the compound formed by the action of nitric acid on indigo.[38] But his lack of success in this regard was not atypical, as has been correctly emphasized by other historians such as W. V. Farrar. In a note celebrating the works of Edward Schunck, the natural products chemist, Farrar argued that the failure of organic chemists in deciphering the extremely complex nature of organic compounds in naturally occurring substances led them to focus on simpler versions of those compounds occurring among "unnatural" substances like coal gas and coal tar. The primary problem lay in the fact that these compounds in plants were too difficult to separate from each other. On the other hand, the same compounds that were also present in coal gas and coal tar offered themselves to easier isolation and analysis.[39] Chemists working on organic substances were uniformly having difficulty revealing the precise composition of natural products.

But the outcome of efforts by Chevreul and other organic chemists was not without some success. The same analytical experiments that failed to reveal indigo's constitution precisely nevertheless shed new light on the

[38] Chevreul admits in the same paper that the nature of *amer* formed by the union of indigo and nitric acid was "yet unknown." "Abstract of a paper on the bitter substance formed by the action of nitric acid on indigo: by Mr. Chevreul," p. 363.

[39] Farrar goes on to argue that this course had already been by and large adopted by organic chemistry from the 1820s. By the end of the nineteenth century organic chemistry had been redefined as "the chemistry of carbon compounds." W. V. Farrar, "Edward Schunck FRS: A Pioneer of Natural Products Chemistry," *Notes and Records of the Royal Society of London* 31 No. 2 (January 1977): 273–96.

characteristics and properties of indigo as a whole. These findings opened up new ways to separate the latter from the plant. The consequences of these insights were substantial for the manufacturing of indigo and thus for the future of the industry. Thus what was lost in terms of the knowledge of internal structure of indigo was more than offset by improved information on its behavior with other substances such as water, acids, and lime, and on its response when subjected to processes like boiling with water and oxidation, the type of information that would streamline and alter the way indigo was manufactured. Indeed the more important contribution of Chevreul might have been in exploring the "general characteristics" of indigo rather than its "chemical composition."[40]

Cheverul's analytical work sought to clarify the nature of the changes indigo underwent during oxidation. He subjected indigo to the action of oxidizing processes to procure indigo at maximum oxidation and minimum oxidation, noting the presence of the former by the bright purple color it obtained. He also referred to the green powder received as a precipitate – probably his "green matter" that turned blue on contact with air. These observations formed the basis of his theory of indigo-white. According to this theory indigo existed in a reduced or white state in the leaves. It was then turned into a fully formed indigo through oxidation during the manufacturing process. William Roxburgh had been the first in the colony to suggest that indigo was "oxidized" during the beating process in vats.[41] It is quite likely that Roxburgh had access to newer works being produced in the French and English metropolises on the matter and that his assertions on the Indian subcontinent were based on the new theoretical paradigms being accepted universally. Chevreul's work on the oxidation of indigo could be linked to the long tradition of organic chemists working on the oxidation of organic substances going back to Lavoisier in the late eighteenth century. It would seem that at least a decade earlier many across the English Channel believed that indigo obtained its blue color on oxidation.[42] In a way Chevreul's work

[40] The methods of analysis at this time were primarily quantitative, often weighing the products of combustion and solution to make inferences about constitutive elements.

[41] William Roxburgh, "A Brief Account of the Result of various Experiments made with a view to throw some additional Light on the Theory of this Artificial Production," *Transactions of the Society*, vol. XXVIII, 1811, *op. cit.*, pp. 273–7.

[42] A contributor in an important journal in London seemed to argue against what he thought was a common belief expressed by the likes of Berthollet, Bancroft, and Hauffemen that indigo turned blue as a result of gaining oxygen. *Cf.* "Experiments on Indigo – by a Correspondent," *Journal of Natural Philosophy, Chemistry and the Arts* 3 (1799–1800): 477–82.

on oxidation did little more than confirm what many already believed to be the case. But what is striking actually is the fact that these assertions were part of his larger theory that explained the role of oxidation in practical steps of indigo manufacturing.

The analytical works on indigo by Chevreul enriching knowledge of its characteristics would become the more operative part of his contributions. In line with similar works of his times dealing with extraction and purification of natural products, he turned his attention to processes likely to aid purification. In the French metropolis he had access to indigo from various sources of origin in the world including Guatemala indigo, Bengal indigo, and Java indigo. He found Bengal indigo to be the purest, containing the maximum percentage of color, and Java the least pure. Those inferences must have been of value to those in the trade for purposes of setting the price on indigo from various sources. But even more important was the demonstration by Chevreul of processes through which purity could be obtained. In the scientific paper being alluded to here he summed up his work as dealing with purification of indigo, emphasizing that indigo sold in the market was in a state of concoction with several other substances and that he would "proceed to examine the nature of these substances, and the methods of separating them." He revealed the possibility of purifying indigo through the "wet" and "dry" methods in the laboratory. Among the wet methods was one involving boiling indigo with water for up to twelve hours at the high temperature of 90° to 100° Fahrenheit. He followed up through other purification experiments involving dissolution of red resins by treatment with hot alcohol and with acid. He also conducted distillation experiments with indigo to obtain it in the purest possible form. He noted the action of alkalis and sulfuric acid on indigo.

Despite clinging to certain aspects of vitalism, Chevreul's study of indigo chemistry was groundbreaking. It is true that Chevreul's discussions of chemical composition betrayed lack of finesse that works by later organic chemists would assume. He talked of indigo's constitutive parts in broad terms like "green matter," "yellow matter," and "red matter." He continued to use the concept of "affinity" – a residual effect from the previous vitalists – to explain the relationship among the constituents.[43] But

---

[43] On one occasion he attributed the insolubility in water of indigo's green matter to the extreme affinity that red elements bore toward it and spoke of the dominant effect of yellow extractive matter on green, because of which the properties of the latter could not become apparent. He even explained the propensity of indigo to sublimate with reference to the weaker union of dilatable constituents to the "most fixed" constituents in indigo. "Chemical Experiments on Indigo: by M. Chevreul," *A Journal of Natural Philosophy, Chemistry and the Arts* 32 (May/August 1812): 211–16.

Chevreul's experiments and the stamp of authority put by his persona on this science achieved enough traction to be able to change the nature of indigo manufacturing materially in the early nineteenth century. Distinct features of this new science were apparent on the plantations in Bengal, the world's newest zone for the production of the blue dye.[44]

Chevreul's Influence on the Indigo Tracts of Bengal

Chevreul's science was assimilated into agricultural practices on the indigo plantations in Bengal. In other words, organic chemistry from the French metropolis took root in the material practices of colonial South Asia. There is no singular, specific instance of a Chevreul emissary ever traveling to colonial India or of his entire body of indigo works being translated purposefully. Rather his counsel fitfully seeped into the colony as the global scientific paradigm of indigo changed and the planting world reached out for this useful information.

Acknowledging that deep knowledge of indigo existed among the indigo interests in Calcutta and in Bengal generally, John Phipps had written in his treatise that he was offering his composition on indigo to "the scrutiny of a discerning community." The statement should not be passed off as a mere mark of humility of the writer. There exists overwhelming evidence that the indigo planters, merchants, shippers, and investors kept themselves acquainted with the changing science that might potentially impact indigo. Exchange of texts and translations of major works as well as attendance at public meetings of savants were routine. The local discussion of these issues at meetings of organizations like the Agricultural and Horticultural Society in Calcutta (established 1820) ensured their wider dissemination and made possible the practical use of the information in the colony. Underscoring the availability of texts and the convention of their consultation by indigo manufacturers, Phipps remarked that the information contained in his new treatise might as well be available in a scattered manner in the personal libraries of the individual indigo planters: "This compilation contains nothing but what may be found on the book-shelves, or desks, of almost every merchant connected with the India trade [in indigo]."[45] The virtue of Phipps's new text rested, as he claimed, in offering all that information concisely in one text.

---

[44] See, particularly, "Chemical Experiments on Indigo: by M. Chevreul," *A Journal of Natural Philosophy, Chemistry and the Arts* 32, (May/August 1812): 211–16.

[45] John Phipps, *A Series of Treatises on the Principal Products of Bengal, No. 1, Indigo,* 1832, pp. iii–iv.

The description of commercial indigo provided in Phipps's *Treatise* bore the mark of Chevreul's characterizations of the commodity. That Phipps presented it by the authority of a citation of the English chemist Andrew Ure further illuminates the modes through which Chevreul's counsels on indigo reached the colony. Chevreul's observations were widely debated across the channel and were also widely accepted. Ure seems to have borrowed from Chevreul's analytical experiments on the behavior of the several components of indigo like indigo-red, brown, and yellow and in maintaining that indigo on combustion gave out phosphate of lime.[46] The value of interpreters of Chevreul like Andrew Ure and others cannot be minimized in charting the passage of knowledge on indigo. Similarly, the increasing availability of professional chemists like Ure in the opening decades in Britain and France, often called "technical chemists" or "analytical chemists," would have paved the way for the recovery of useful, applicable information out of the theorists' analytical works. And, on the other end of this information chain, texts like Phipps's completed the process of the flow of Chevreul's theory of indigo into the colony.

The case of Pierre Paul-Darrac and other French planters based in Bengal suggests other possible connections among Bengal, the French metropolis, and other French colonies. Paul-Darrac was born in southern France in 1771 and became an appointee of the French Ministry of Navy and the Colonies to the position of the chief of the French settlement at Dacca in Bengal. This happened in 1816. What transpired in the following years has larger implication for the story of transmission of knowledge on indigo between British and French settlements worldwide. Paul-Darrac was not only commissioned to write a comprehensive report on the process of indigo production used by the English planters in Bengal but also specifically asked to send seeds of the local germplasm for delivery to Senegal (West Africa), where the French wished to promote indigo culture. Paul-Darrac was an astute indigo planter himself who had invested in indigo plantations at four sites in Bengal – Sankipour, Chodofferpour, Ogardipe, and Cazipour – all in the Krishnagar *sadr* subdivision of Nadia district. His larger family of son, daughter, and son-in-law were all engaged with indigo plantations in Bengal. He asserted that there was "no dearth of French indigo planters" in the province. He regularly corresponded with fellow French planters in Chandernagar, a French settlement in Bengal, and with others in Howrah. The account affirms

46 John Phipps, *A Series of Treatises on the Principal Products of Bengal, No. 1, Indigo*, p. 71.

that there remained a not so insubstantial contingent of French planters in Bengal in the colony in the 1820s. According to Paul-Darrac's testimony, not only was this community connected with the metropole and other French colonies, it was also sensitized to the usefulness of science. Deputed to find a competent person in Bengal who could be sent over to Senegal to start indigo culture there, Paul-Darrac remarked that anyone from the community of the Bengali French planters would be adequate to the task and likely to be well versed in "both practical knowledge and theoretical insights as well as some understanding of agronomy and even chemistry."[47] Such a community of planters aware of the needs of the science of agronomy and chemistry could be expected to keep up with the latest contributions made by organic chemists at home. Already before the close of the eighteenth century the planting diaspora in Bengal had been beneficiary to the translations of important French texts on indigo like those by Elias Monnereau and De Cossigny that were published in the French colonies in the Caribbean. The diaspora of French planters on the subcontinent – in addition to those from the British Isles – would have adequately served the task of keeping Bengal's indigo tracts abreast of new works by organic chemists on natural products, including those specifically touching upon indigo. The accumulated wisdom of the chemists including Chevreul with regard to the purification of plant products seems to have found its clearest impression on indigo manufacturing subsequently. Some part of their work dealing with the oxidation of natural substances also left its broad mark in terms of a better understanding of the chemical basis of the process of "beating" and the consequent effort made to streamline it.

Boiling and Filtering to Obtain Purity

The world dominance of Bengal indigo hinged on the ability of European planters on the Indian subcontinent to produce a good quality dye in addition to their ability to keep the price low by extracting cheap labor of the peasantry. Price was by no means the only factor that underwrote its competitiveness. Indeed a large variety of low-cost indigo from various sources including the North -Western Provinces and Madras in the

[47] The report prepared by Pierre Paul-Darrac in 1823 for the French Ministry has been translated into English. The translation appears in two parts with the first part constituting an introduction to the report and the second part being the report itself. Willem van Schendel, *Global Blue: Indigo and Espionage in Colonial Bengal*, Dhaka: University Press Limited, 2006, pp. 34–5, notes.

colony swamped the Western markets. But the inferior quality of these dyes checked their growth. Major dyers and printers clearly preferred the "Bengal indigo," which was considered to be the best available grade because of its unsurpassed purity.

The drive to obtain purity in manufacturing started with keeping the plant clean and choosing the right type of water. The French planter in Bengal Paul-Darrac emphasized that "the manufacture of clean, pure indigo depends on the availability of clear, limpid water," but that such water was "not available in Bengal's rivers and ponds, especially during the rainy season when indigo is manufactured." The indigo factories where the actual manufacturing operations took place were generally located on high grounds in order to prevent contamination by the floods, which contained a lot of sediments and could pollute the tanks. The planters constructed two reservoirs to purify water. The first was located a few feet above the ground. A basin was created and connected with the primary water source by digging a channel. The use of paddle chain pumps that were "operated by leg power" was common either for transporting the water to the basin, in case the level of water there was too low, or for lifting the water into the first reservoir. In other cases, hand pumps were also used. The Upper Provinces used well water to supply the reservoir. Here the water was left idle for twenty-four hours, during which time the silt deposited to the floor. The second reservoir was even higher than the first and divided into compartments connected with each other by holes; thus again the water had to be raised to it by using the china pump operated manually by a large team of workers. The height of the reservoir ensured that gravity could be used in transferring the water quickly and efficiently to the fermentation vats once the operations had begun.[48] The importance of analysis of water was realized with earnestness. Henry Piddington, whom the planters in Bengal often turned to for advice on questions related to indigo culture, warned them against any carelessness on this matter. He sensitized them to the relevance of the quality of water to producing a good quality output, saying, "If good Indigo has never been made in a factory, have the water examined by some person competent to report upon it." Emphasizing the sanctity of work performed by analytical chemists like him, he added, "Beware of native methods of clarifying water by lime, alum, & c."[49] A little later, John Phipps noted

---

[48] Pierre Paul-Darrac's report, 1823, in Willem van Schendel, *Global Blue: Indigo and Espionage in Colonial Bengal*, Dhaka: University Press Limited, 2006, pp. 92–7.
[49] "On the Manufacture of Indigo. By Henry Piddington, Read 10th June, 1829," *Transactions of the Agricultural and Horticultural Society of India* II (1838): 30.

that the extra effort made by planters to cleanse their water in reservoirs was the main factor that distinguished the quality of their dyes, as compared to the dyes made by "some factories, particularly small ones" that did not have the resources to construct reservoirs and hire additional labor for the purpose. He condemned the native indigo manufacturers' neglect as they were often found using water from a neighboring tank, a dirty pool, or the river at a time when it was usually full of sediments and impurities. Such apathy by natives in Bengal, Phipps offered, was the primary reason for the impurity of their dye.[50]

Perhaps the most distinctive aspect of absorption of useful practice out of analytical experiments completed by the organic chemists was in terms of inserting the new step of boiling in the manufacturing cycle. This was done evidently with the intent of obtaining purity for Bengal indigo. That this innovation took place in the East was fortuitous. South Asia was indeed the center of indigo production in the era after the arrival of organic chemistry. Thus, writing in 1814, it was the famous savant of dyes Edward Bancroft who alluded to the fact that "some of the manufacturers of this commodity, in the East Indies, have lately purified their indigo, by taking it immediately from the small dripping [receptacle] *vats*, and boiling it in copper vessels, with water and fossil alkali, (soda)."[51] Bancroft had likely received information at an early stage of the turn toward inclusion of boiling in the manufacturing process. That is why he tells of "some of the manufacturers" who took to the practice. As slightly later sources reveal, by the 1820s and 1830s, the practice of boiling of indigo had become too widespread in South Asia to require any emphasizing and was mentioned in a matter-of-fact way.

Across the subcontinent, but particularly so in Bengal, a practice of boiling had been adopted that involved heating the indigo with water for extended periods. To the very few in Bengal who did not practice boiling, Piddington counseled, "All methods which allow fecula to remain without boiling, are bad."[52] The planters understood well the role of heating. "Boiling water dissolves the vegetable matter that was mixed with granules during fermentation," Darrac noted and advised that as this undesirable matter was separated from indigo and made to rise to the top, it should be scooped off. He also noted that some salts are separated from

[50] John Phipps, *A Series of Treatises on the Principal Products of Bengal*, 1832, pp. 121–2.
[51] Edward Bancroft, *Experimental Researches Concerning the Philosophy of Permanent Colours: And the Best Means of Producing, by Dyeing, Calico Printing and c.* vol. I, Philadelphia: Thomas Dobson, 1814, p. 138.
[52] "On the Manufacture of Indigo. By Henry Piddington," p. 38.

the granule during boiling and precipitate to the bottom. The planters had to be careful to leave them behind at the end of boiling and have the boiler cleaned afterward before putting in the next batch of indigo for boiling. Piddington noted that boiling precluded the possibility that secondary fermentation would set in later during the drying process induced by impurities.[53]

The practice of boiling became widespread despite the fact that it was accomplished at no inconsiderable cost. The boiler was added in the series of the three vats, and a channel led from the third vat, the receptacle, to the boiler, where it was finally pumped up into the boiling pan. Boilers in Bengal, as per Darrac, were made of brass turned into sheet metal with walls made out of masonry. Phipps noted the use of an iron base for the boilers in the Upper Provinces wherever boiling was practiced. It could be fourteen feet long, seven feet wide, and three feet deep. It was lit with the help of dried stalks of plants or the *seet* refuge obtained from the beating vat. The boiling continued for up to five hours and was generally ended when an "agreeable smell" started to emanate from the boiler to overpower the earlier smell from burning of impurities.

The filtering of indigo had also been elaborated and perfected so as to become a distinct step in itself in the manufacturing process. Much of the filtering was organized to take place after boiling, though in some cases the filtering happened even as the dye emerging from the vats was being led to the boiler. Filtering was emphasized as part of an overall effort at keeping impurities away: "It is essential that the utmost cleanliness be maintained in all operations involving indigo granules," Darrac said. He described the dye, once it was out of the boiler, as passing three times through filtering cloth screens before being put on the filter. The filter was composed of a brickwork basin twenty by five feet in dimension and one and a half feet deep. With buttresses the basin supported a trellis-work made out of bamboo with a finely woven cloth stretched to cover its top in turn. The process of filtering took anywhere between eight and ten hours but was considered valuable for eliminating unwanted substances like the brown matter in order to procure an indigo of high grade. Piddington added a word of caution that extended periods for filtration could be counterproductive as they darkened the color of the dye. Thus what was gained in purity was given away as the desired purple color was lost whenever the filtering process was prolonged beyond a reasonable

[53] Paul-Darrac's report, 1823, in Willem van Schendel, *Global Blue*, pp. 81–3, notes; Piddington, p. 38.

length of time.[54] Given that filtering was a slow process, he urged the planters to look for every possible way to shorten its duration.

Germany, Britain, and Agricultural Science

Significant turmoil took place in the field of agricultural science around the middle decades of the nineteenth century that had a potential to impact world agriculture including the nature of indigo manufacturing and colonial agricultural science in South Asia. To be sure, these changes originated in the force of previous discoveries and the expanding knowledge across numerous arteries in Western and central Europe. But the transformation at the midcentury point was particularly sharp and occurred widely across scientific method, theory, and focus. Establishing connections between these developments in agricultural science in the West, on the one hand, and indigo science and colonial agricultural science in South Asia, on the other hand, requires a careful calibration of the modes of transmission. It will be seen that the passage was predicated on a variety of factors including the place of the new agricultural science within British imperialism, the working of state and nonstate forces on the Indian subcontinent that put the colony in contact with the world's forces, and the internally variegated nature of the "new" agricultural science.

Many of the scientific developments of the era were connected with the initiatives of the German chemist Justus von Liebig. Liebig was a native of the German grand duchy of Hessen-Darmstadt. In 1824 he obtained a professorship at the University of Giessen. It is from there that he revolutionized the field of agricultural science with his pioneering efforts in putting plant physiology at the center of a new functional science. Giessen itself emerged as the pivot of scientific work on agricultural chemistry in particular.

The publication of Justus von Liebig's book, *Chemistry and Its Applications to Agriculture and Physiology*, in 1840, directed at English agriculturists, was a landmark event for agriculture.[55] The book's powerful

---

[54] He does not offer any scientific explanation as to why the dye turned dark.

[55] Justus von Liebig had an abiding interest in English agriculture and science. It was partly the consequence of his global vision for science as a vehicle for the uplift not only of the people of Germany, or Europe for that matter, but the entire "humanity." No nation or continent, but "mankind" was to be the avowed beneficiary of his ideas. The focus on Britain as the major agriculturist nation resulted from the conviction that any victory for his ideas there would ensure a rapid acceptance by the rest of the world. Throughout his career Liebig maintained his interest in influencing English debates on agricultural science,

contribution lay in the claim that plants derive most of their critical nutrients in the form of soluble inorganic minerals from the soil. This so-called mineral theory cast aside the need for supplementing the soil's nitrogenous contents through the conventional method of addition of field manures. It asserted that plants received their requirement of nitrogen through atmospheric ammonia that was dissolved by rainwater and was available to plants through the water content of the soil. In putting forth this version of the nitrogen cycle Liebig cast aspersions on prior farming practices that gave ultimate emphasis to reinforcing the nitrogen content of soil through application of field manures and humus or decayed vegetation. In its early versions the new theory was also quite reductionist in orientation, basing its prescriptions entirely on a theory of exhaustion and supplementing of inorganic nutrients. The plants "robbed" the soil of its critical inorganic nutrients. The essential nutrients were removed and displaced as the plants entered the food supply involving animal and human consumption and never returned to the soil. If this process were not checked, it would inevitably lead to a dip in the soil's output. On the other hand, as long as farmers initiated a regimented system of returning the nutrients to the soil they need worry about nothing else. The theory made light of the traditional practices such as addition of nitrates and other efforts that farming communities were known to invest toward upkeep of land involving soil mechanics and texture of the subsoil and topsoil.

Liebig's ideas were challenged by the prominent agriculturists in Britain at its major centers: by John Lawes and Joseph Henry Gilbert at Rothamsted and by Philip Pussey of the Royal Agricultural Society of England (RASE) in the 1850s. Strong bonds of research and exchange between RASE and Rothamsted helped form a common front of opposition to Liebig's postulates emerging from Germany. In 1847 the RASE had published the first article by John Lawes of Rothamsted in its journal in which he opposed Liebig's assertions on nitrogen.[56] In a note appended to this article, Philip Pusey, perhaps the most influential member at RASE at this time, spoke highly of scientists at Rothamsted, praising the ability

and the British people had a wide-ranging familiarity with his theories. William Brock has drawn attention to Liebig's "particularly close relationship with Great Britain" and the his engagement with Britain's "pattern of scientific education, agriculture, and medicine." *Cf.* William H. Brock, *Justus von Liebig: The Chemical Gatekeeper*, Cambridge: Cambridge University Press, 1997, pp. vii, 113.

[56] J. B. Lawes, "On Agricultural Chemistry," *Journal of the Royal Agricultural Society of England* 8 (1847): 226–60.

of Lawes and Gilbert and calling Rothamsted "the principal source of trustworthy scientific information on Agricultural Chemistry."[57] In the 1850s when Lawes and Gilbert stood up to resist Liebig's downgrading of the importance of nitrogenous fertilizers, Pusey supported them. Together they had rallied in support of the time-tested belief among English farmers that the addition of vegetable matter in a state of decomposition, humus and field manures with animal wastes included, improved productivity. They challenged Liebig's argument that as long as key inorganic minerals were supplied to the soil, the further addition of humus or any other known source of nitrogen commonly used by them was superfluous and wasteful. The stance in support of humus was supported by the actual trials at Rothamsted. Besides, such an approach was in line with the stress on "practice" in English traditions where the age-old conventions followed successfully by the English farmers suggested that the addition of field manures and nitrogen and potash fertilizers was helpful. The "union of science with practice" was a founding ideology of the RASE and the society's members, many of whom were prominent landlords themselves, ably defended the practice of English farming over projections based on Liebig's theory of minerals. The failure of Liebig's patented manure during trials helped the Rothamsted clique further discredit Liebig's thesis emphasizing formulaic, "scientific" explanation.

In the seventh edition of the book published in 1862 Liebig made but scarce concessions to the opposite camp. Liebig's mineral theory became more eclectic and acknowledged that the addition of nitrates could supplement the amount of nitrogen the plant received from dissolved ammonia. It also acknowledged the role of humus in enhancing soil texture and mechanics. But it continued to stress the primacy of inorganic nutrients for yield improvement.[58] Such a revised perspective was more flexible and could have accommodated the rationality of age-old farming practices like crop rotation. Yet mineral theory remained the foundation of Liebig's scientific belief. He also remained steadfastly committed to his opposition to humus theory and thus continued to challenge the propriety of adding decomposing vegetable matter. It would be perhaps accurate to say that there were elements of convergence and discord in the views

---

[57] E. J. Russell, *A History of Agricultural Science in Great Britain*, pp. 117–18.

[58] Mark Finlay provides the best description of the revisions carried out by Liebig to his mineral theory, the majority of which appeared in the book's seventh edition. Mark Finlay, "The Rehabilitation of an Agricultural Chemist: Justus von Liebig and the Seventh Edition," *Ambix* 38 (1991): 155–67.

of major theoreticians of agricultural science in Britain and Germany at this point. Agricultural chemistry evidently retained aspects of such internal differences as a discipline, and these diverse strands were expressed through the varied focuses of its major interlocutors up to the end of the nineteenth century.

Edward Schunck's Work on Indigo in Manchester

Justus von Liebig had himself worked on indigo as early as 1827. But he also inspired his students to work on the natural dye.[59] One of them, Edward Schunck, a specialist renowned for his work in natural product chemistry, carried forward that work in Manchester. Edward Schunck was born in the year 1820 into a prosperous family in the textile manufacturing districts of the Yorkshire region in England. His father owned factories involved in fulling, bleaching, and printing of textiles and ran an extensive business in textile merchandising spanning the British Isles and the Continent. Schunck had taken a fancy to chemistry early in his life. When he came of age he was sent by his family to Germany to pursue his interests in higher learning in chemistry. After attending the university at Berlin and then receiving a doctorate from the university at Giessen, under Justus von Liebig, Edward Schunck returned to England in 1842. Although he joined his father's chemical factory for a while, his interests lay in active research. He soon digressed to pursue a career devoted to the chemistry of natural colors. It was his mentor, Liebig, who had initiated him in the study of natural coloring matters, an interest that he continued to pursue with considerable success. In the middle decades of the nineteenth century, Edward Schunck produced meritorious works on the extraction of colors from various substances like aloe, lichens, indigo, and madder, among others. His empirical work on indigo was seminal and paradigm changing, one that transformed the theoretical understanding of the color-bearing substance in the indigo plant.[60]

Speaking of the prevalent understanding of blue dye in 1855, Schunck commented that "its properties, composition, and products of decomposition, have been so carefully examined, that it may safely be assumed that there are few organic substances whose nature is more accurately

[59] H. Buff, "Ueber Indigsäure und Indigharz," *Jahrbuch Chemie und Physik* 21 (1827): 38–59. I am indebted to Mark Finlay for pointing out this source in a personal communication.

[60] For Edward Schunck's biography and a summary of his works, see, W. V. Farrar, "Edward Schunck FRS: A Pioneer of Natural Products Chemistry," *Notes and Records of the Royal Society of London* 31 No. 2 (January 1977): 273–96.

known than that of indigo blue." Schunck was referring to the vast body of analytical work accumulated on indigo in the preceding decades. But he also brought out the irony that despite the profusion of prior works there still was no clear understanding of "the origin and mode of formation of this body [from the parent compound in the leaf to the dye granules]."[61] He was aware of these deficiencies as he now had the newer analytical tools of organic chemistry. The insights of structural chemistry enabled him to clarify the arrangement of specific elements in the original substance of leaves and pinpoint with accuracy the products of chemical changes that transpired as indigo went through the manufacturing process. Earlier formulations broadly made hypotheses about the nature of changes undergone by organic substances in the manufacture. But now Schunck could use these methods to explain with a higher degree of specificity the nature of parent compounds, their products of decomposition, and "the relation in which they stand to each other." [62]

Schunck believed this new specificity of knowledge had implications for improving indigo manufacturing. First and foremost he clarified that the color did not exist in an already formed state in the leaves, as was believed by French philosophers. Rather, a glucoside, which Schunck named "indican," was the color-giving principle present in the leaves, which was progressively oxidized during manufacturing to give the blue granules. He showed in his laboratory that indican could be oxidized by acids to yield indigo, glucose, and a few other by-products, a process that mimicked the oxidation of indican through "beating" in the manufacturing cycle used by planters in their indigo factories. Indican was inherently a very unstable compound, "of a delinquent nature," as he called it. Knowledge of the properties of how indican decomposed would shed "great light" on the manufacturing process, he said on one occasion. The experiments he conducted illustrated that there were five by-products of the decomposition of indican including indigo. Thus he implored attention to his "method of considering them [the changes]" as it showed that many of the other compounds were produced "at the expense of indigo-blue," meaning that a certain part of indican "under certain unknown circumstances" had turned itself into products other than indigo.[63]

[61] Edward Schunck, "On the Formation of Indigo-Blue – Part 1," *London, Edinburgh and Dublin Philosophical Magazine and Journal of Science* (August 1855): 73.

[62] Edward Schunck, "On the Formation of Indigo-Blue – Part 2," *London, Edinburgh and Dublin Philosophical Magazine and Journal of Science* Nos. 97 (January 1858), 98 (February 1858), 99 (March 1858): 29–45, 117–33, 183–91.

[63] Edward Schunck, "On the Formation of Indigo-Blue – Part 2," *London, Edinburgh and Dublin Philosophical Magazine and Journal of Science* 98 (February 1858): 133.

The possibilities opened up by the revolution in agricultural science and indigo science in the mid-nineteenth century were realized in specific ways in the colony. The new knowledge, as we have seen, was a product of cosmopolitan intellectual exchange across Europe. These cosmopolitan elements were retained and emphasized in selective ways on the subcontinent as the new agricultural science was subjected to the operation of colonial conditions and its forces of exclusion. In the times to come statist and nonstatist forces, nationalists, commodity producers, and others embraced and emphasized whichever aspects of the new agricultural chemistry were in line with their individual agendas and priorities.

Eugene Schrottky: Liebig's Ideas in the Colony

If "interlopers" like the French planter Pierre Paul-Darrac had the potential to keep Bengal's indigo plantations in touch with research into indigo chemistry in France in the early nineteenth century, it fell upon "drifters" like the German Eugene Schrottky, born in Dresden, in the duchy of Hessen, to play a role in establishing a link between the new and diverse ideas on agricultural science in the West and their popularization in Asia. Such migrants, or still better, cosmopolites – who preferred to traverse continents and make careers in different nations and regions – disseminated the culture and materiality of the new sciences in the colony. Details on the personal life of Schrottky are almost nonexistent. But extensive references to his work and activities tell us that he moved several times to Germany, Britain, the Indian subcontinent, and Java in Southeast Asia. He called himself a student of Justus von Liebig and may have attended Liebig's lectures after the latter moved to Munich in 1852. Or, it is also possible that he never attended any formal instruction directly from Liebig as a student but nonetheless read and educated himself on Liebig's principles of agricultural chemistry and plant physiology.[64] In whichever of

---

[64] Eugene Schrottky does not appear in the published lists of Liebig's students. As the historian William Brock rightly remarks, "it is difficult to set forth an exact definition of a 'Liebig student,' because very few were known to actually stay at Giessen to receive a doctorate while most got entry into the elite category by being at Giessen for a couple of semesters, publishing in *Annlen*, or just by acquiring basic skills of organic chemistry with him during a short stint." William Brock, *Justus von Liebig: The Chemical Gatekeeper*, p. 62. Schrottky may as well have been in touch with Jutus von Liebig for a very short period, at Munich, not Giessen. Or it is even possible that he called himself a pupil in a figurative sense of being influenced by his ideas. What is clear from his writings is that he was very well versed in Liebig's theories and dedicated his own publication to the German baron, calling himself "his devoted pupil."

those capacities he certainly was engaged in the work of "translation" of Liebig's ideas in South Asia, as he went about disseminating the German chemist's theories of plant nutrition and soil improvement. He was an envoy for the German chemist, albeit an unauthorized one.

Schrottky probably drifted to Bombay sometime in the early 1870s in search of a rewarding career, something not entirely unusual for those times. In 1874–5 he decided to write to W. G. Peddar, the municipal commissioner of Bombay, proposing the establishment of a sewage farm near Bombay. He submitted a detailed scheme for taking the Bombay sewage to two agricultural tracts, one between Thana and Kurla and the other between Thana and Kolset, altogether measuring about nine square miles or more than five thousand acres, which would enable raising a double crop of Carolina rice and Indian corn annually while also leaving a part of the land fallow for manuring. He hoped that the municipal corporation would assist with the scheme. The project was viable and would return a profit, as he contended. Liebig's critique of the "robbing of the soil," fundamentally based on the notion that agriculture extracted nutrients from the soil that were then removed from the fields as the plant products entered the consumption cycle of humans and animals, was writ large on the aforesaid plan. Schrottky emphasized the importance of recovery of nutrients from sewage by detailing the "connexion between the fertility of our fields and the residue of the consumed food." The city with its 400,000 inhabitants consumed grains, meats, and vegetables, "which supply she draws from the 'mofussil'." The excrements contained all of the "ash-constituents," or inorganic minerals, and most of the nitrogen that had passed from the mofussil's soil into the city. The soil's productivity was compromised in proportion to the rapidity with which its nutrients were displaced, and, conversely, the productivity would be restored proportionately if these nutrients were restored. The process of recovery would only be, as he emphasized in the spirit and logic of Liebigian ideas, "according to the evident design of the nature."[65] Between 1858 and 1865, Liebig himself was involved with the project to use sewage as manure in the city of London in line with his ideas of returning inorganic nutrients to the soil,[66] and Schrottky apparently took a clue from Liebig to propose the Bombay scheme.

[65] "No. 24, Extracts from a Letter on the Drainage of Bombay from Mr. E. C. Schrottky to the Municipal Commissioner of Bombay," *Report on the Sanitary Measures in India in 1874–75*, vol. VIII, London: George Edward Eyre and William Spottiswoode, 1876, pp. 195–6.

[66] William Brock, *Justus von Liebig: The Chemical Gatekeeper*, pp. 250–72.

Eugene Schrottky failed to elicit a favorable response to his proposals from colonial officials. As a matter of fact, officials in India were not strangers to the idea he was pushing. Mridula Ramanna's work on the sanitation system of Bombay shows that in 1872, the army engineer Hector Tulloch submitted a proposal to sewer the city and use sewage as fertilizer.[67] The London plan for sewage irrigation after active consideration in the 1850s and 1860s had failed to take off and colonial experts might have been cautious, not unmindful of the precedent in London. As in the case of London, there remained doubts whether the scheme could viably work in Bombay. By the authority of the department's engineer, Reinzi Walton, the colonial officials rejected the proposal, saying, "A successful result from the adoption of sewage irrigation is as uncertain now as it ever was." [68] The rejection of his proposal did not bring an end to Schrottky's endeavors in the colony.

In 1876 Eugene Schrottky published his magnum opus, *The Principles of Rational Agriculture Applied to India and Its Staple Products,* in Bombay that seemed to offer a critique of colonial agriculture and the direction of current efforts by the colonial state to mitigate its ills. In the preface of the book he expressed the hope that his counsels for improvement of Indian agriculture would be welcomed " … by all those who have the interests of this country, and the interests of its commonwealth, at heart"[69] Claims like these for the betterment of people of India were not uncommon in the colonial context. But Schrottky's claims for progress for Indian agriculture were also made as criticisms of the colonial state, and he used Liebig to articulate this critique.

In including the phrase "rational agriculture" in the title of the book, Schrottky took a cue from Liebig, who had emphasized the importance of agricultural chemistry to making farming "a rational system."[70] Schrottky also attached his own meanings to the word "rational." He distinguished the improvements taking place in Europe's agriculture as a whole since the time of the Romans as "in keeping with the general

---

[67] Mridula Ramanna, *Western Medicine and Public Health in Colonial Bombay 1845–1895,* Chennai: Orient Longman India, 2002, p. 112.

[68] *Report on the Sanitary Measures in India in 1874–75,* vol. VIII, p. 32.

[69] Eugene C Schrottky, *The Principles of Rational Agriculture Applied to India and Its Staple Products,* Bombay: Times of India Office, 1876, p. v.

[70] The phrase was actually used by Liebig in the introduction of his book, while asserting that "a rational system of agriculture cannot be formed without the application of scientific principles," and by the latter expression he implied agricultural chemistry. Justus von Liebig, *Chemistry in Its Applications to Agriculture and Physiology,* Cambridge: John Owen, 1842, 3rd American ed., p. xv.

advance of education and civilization" in Western Europe. In contrast, the eastern countries showed no evidence of any change, "and the backward state of this most important of all arts [i.e., agriculture] is prominently apparent in India. No advancement, no improvement, has been effected during several ages."[71] He also held the colonial rulers complicit in the continuing decay of Indian agriculture. The colonial rulers were equally to be blamed, who had turned "the Garden of the East" into "a comparatively barren country." The colonial government's efforts to improve Indian agriculture seemed to "have been few and far between." He expressed surprise that much recent valuable knowledge on agriculture had not been made available locally. In this light he even criticized the functioning of model farms set up by the government in Bombay that were principally involved in the introduction of exotic plants that did not "bring any benefit to the country." Setting himself up to speak on behalf of the people, he polemically asked, "Should they not, first of all, exercise their practical and scientific knowledge in effecting an improvement in the soil of the surrounding country, as well as of the different crops grown, and try to diffuse agricultural knowledge among the native cultivators?"[72]

Liebig's theories provided content to this critique of the colonial establishment and its views for the improvement of Indian agriculture. The basic argument was that topsoil in India was in a state of complete exhaustion. For generations, Schrottky maintained, the cultivators had flouted the very first principle of agriculture: "to give back to the soil what was taken from it." Using Liebig's explanation of organic and inorganic constituents in a plant's physiology, he explained that inorganic minerals critical to plants' growth had been removed from the soil harvest after harvest, and as the crop entered the human consumption cycle, they were irretrievably lost to the soil. He claimed that Indian soil had a particularly poor content of phosphoric acid. In that light, he argued that the export of bones from the country was a mistake: "It is really a matter of deep regret that India's stores of this most important manure [i.e., the bones of animals], which she herself needs so sadly, should be thus withdrawn to fertilize the soil of a foreign country."[73] Following Liebig again, he suggested two measures that could restore the lost inorganic minerals. He first proposed the use of human excrement as was the practice

[71] *Principles of Rational Agriculture*, p. 3.
[72] *Principles of Rational Agriculture*, p. 20.
[73] *Principles of Rational Agriculture*, pp. 67.

in Japan. The other was deep soil cultivation, also as practiced in Japan. De-metropolizing Britain, he praised the agricultural practices in Japan.[74] Schrottky was a firm believer in deep soil cultivation, especially for India, because he believed that the lower part of the Indian soil was still rich in mineral resources. He thus asked the government to take steps to encourage the use of deep plows by the cultivators.[75]

If, on the one hand, Schrottky laid out his thesis for agricultural science in distinction to the colonial state's program, on the other hand, he also distinguished the social agenda attached to this science from those of the Indian nationalists. He found the occasion to bring out those distinctions during a dialogue with the leading nationalist Dadabhai Naoroji in June 1876. Naoroji's theory of the "drain of wealth," charging the colonial state with one-way transfer of resources from the colony to the metropolis, had managed to establish itself as the major plank in the developing critique of colonial rule. As the historian Bipan Chandra has remarked, it was a momentous occasion when in 1876 Naoroji delivered a revised draft of his thesis at the Bombay Branch of the East India Association. Originally proposed in 1867 and subsequently elaborated in his book *Poverty and UnBritish Rule in India*, by now the thesis had been fully developed.[76] Eugene Schrottky was in attendance on June 9 and July 24 to participate in discussions on Naoroji's presentation in Bombay, in which Naoroji also participated.[77]

Schrottky took issue with Naoroji and other nationalists. Participating in the meeting in Bombay he pointed out that their demand to give employment to Indians seemed to him to be skirting the major issue of development of resources of the country, especially through agriculture.[78] He was scathing in attacking Naoroji's demand "to obtain … for Indians a fair share of the lucrative appointments" within colonial administration. Such jobs would, after all, only benefit a few thousand natives drawn from the upper classes. Naoroji, according to Schrottky, was "wrong in connecting the obtainment or non-obtainment of his object[s] in any way with the poverty of India." Eradicating Indian poverty, Schrottky insisted,

---

[74] *Principles of Rational Agriculture*, p. 161.
[75] *Principles of Rational Agriculture*, pp. 89–91.
[76] Bipan Chandra, *The Rise and Growth of Economic Nationalism in India*, New Delhi: People's Publishing, 1982, first published in 1966, pp. 636–9.
[77] "Adjourned Meeting of the Bombay Branch of the East India Association, for discussion of the papers on 'The Poverty of India' read by Mr. Dadabhai Naoroji," *Journal of the East India Association*, vol. 10, pp. 83–96, 133–60.
[78] "Adjourned Meeting of the Bombay Branch of the East India Association, for discussion of the papers on 'The Poverty of India,'" p. 91.

required improving Indian agriculture, which in turn would improve the condition of its majority agricultural population. He melded Liebig's scientific ideas with what he thought was a true representation of nationalist demands, as he argued: "I maintain, we must improve the condition of its agricultural population, we must improve its agriculture, we must enrich the exhausted surface soil by the treasures hidden in the subsoil, we must develop its vast agricultural as well as mineral resources, and then the poverty of India will be a thing of the past." Schrottky may well have misread the entire import of the nationalist economic demands as he was indeed reminded by Naoroji when the adjourned meting resumed on July 24.[79] His primary goal lay in highlighting the potential of his science for meeting "greater claims of the poorer classes."

Scrottky later won some recognition for himself in the colony as an innovator of note and holder of a series of patents on the process of indigo manufacturing. At the end of the decade of the 1870s he relocated to Bengal, where he moved from one indigo factory to another as a guest of indigo planters. In the 1870s and 1880s Schrottky registered seven patents in Bengal on relevant chemical and mechanical processes (Table 4) and sold them to one of the largest indigo concerns in Bengal and Bihar, the Bengal Indigo Manufacturing Company. At least five other prominent indigo factories purchased the right to use these patents. The secretary of Bihar Indigo Planters' Association praised the efforts Schrottky had made toward improving the processes in indigo manufacture and thought that the planters "should be most grateful to him" for his services in advancing manufacturing conventions closer to science. Still later, a manager at the Palkondah Indigo Concern in Bengal wrote in July 1898 to attest that by working with Schrottky's processes he had obtained an output of 50 pounds of the dye per vat as compared to the earlier production of only 25.5 pounds. While his contemporaries debated the effectiveness of Schrottky's patents, the evidence of his patents indicates that Schrottky had made concerted efforts to deploy Edward Schunck's clarification of chemical processes involved in indigo manufacturing to try out chemical substances that would streamline and hasten the processes of fermentation and oxidation. Despite facing scathing critiques, especially from state officials, Schrottky remained relevant to the entrepreneurs. That he was partly successful in this endeavor is proven by his longevity on the indigo tracts. He sojourned in the colony

---

[79] "Adjourned Meeting of the Bombay Branch of the East India Association, for discussion of the papers on 'The Poverty of India,'" p. 152.

TABLE 4. *Eugene Schrottky's patents registered in Calcutta*

Date of Registration	Short Description of the Patent
Sept. 20, 1877	Use of yeast from fermenting vats and other precipitates in the manufacture of indigo
Apr. 5, 1879	Use of borax and other alkaloids in the manufacture of indigo
Aug. 16, 1881	Use of oxidizing salts in the manufacture of indigo
June 13, 1882	Use of saltpeter, nitrates, and sulfates in the manufacture of indigo
May 7, 1884	Resteeping of the indigo plant and the use of a perforated base for the fermenting vat in the manufacture of indigo
Mar. 9, 1886	Improvements in the resteeping process and the yeast process in the manufacture of indigo
Aug. 12, 1887	Use of carbolic acid and antiseptics in combination with saltpeter in the manufacture of indigo

Source: Board of Trade Papers, Public Record Office, Kew, AY4/2048/100168.

well up to the early decades of the next century and continued to find takers for his innovations.[80]

The Colonial State and Agricultural Chemistry

Chemistry also found its way into the colony through the channels of British imperialism. Since the early days of Humphrey Davy's lectures to English farmers at the start of the nineteenth century, the place of chemistry had become further solidified in the visions of improvers of agriculture in England. Among the declared objects of the RASE, founded in 1838, was "the application of chemistry to the general purposes of agriculture."[81] Much of the latest research on agricultural chemistry had emerged on the Continent, but agriculturists in Britain had enthusiastically embraced this new chemistry. Discussing the context for the establishment of RASE, the historian Nicholas Goddard explained the high degree of enthusiasm for chemistry among Britons: "Chemistry was seen as having particular relevance to the practice of agriculture and this quest for knowledge [for science] was in conformity with the spirit of the times."

[80] For the response from T. R. Filgate and the Palkondah Indigo Concern, see, Public Record Office, London, AY4/2048/100168.
[81] E. J. Russell, *A History of Agricultural Science in Great Britain, 1620–1954*, London: George Allen & Unwin, 1966, pp. 110–42, quote on p. 111.

Goddard located the founding of RASE within this spirit.[82] Around the same time, the Rothamsted established its reputation for its agricultural trials on nitrogenous fertilizers. The two centers together represented the typical traditions of English agricultural chemistry.

The specific request for sending out an agricultural chemist actually emerged not in the metropolis but rather locally in official debates in the colony in the last decades of the century. This happened only once the colonial state acknowledged a direct responsibility for the "improvement" of Indian agriculture in the last quarter of the nineteenth century.[83] The Famine Commission Report of 1880 encouraged the government to promote agricultural "improvement" through the creation of agricultural departments. The report put its trust in expertise, calling for employment of persons "with a high order of technical and scientific attainments, and trained in practical agriculture, agricultural and organic chemistry, and botany." It also asked for engaging civil servants in the proposed departments who might have "some knowledge of the science and practice of agriculture, and ... habits of observation which are best acquired by a training in physical sciences."[84] In the 1880s repeated requests were made by the colonial government for sending an agricultural chemist over to India. The 1888 Simla Agricultural Conference echoed the demand for an agricultural chemist. The Secretary of State may have been persuaded to act favorably at this juncture now that the Agriculture Departments were up and running and a basic infrastructure was available that could make use of someone with advanced knowledge of chemistry. To find a suitable candidate, he consulted one of the famine commissioners, Sir James Caird, who proposed the name of the RASE consulting chemist John Augustus Voelcker, one of the leading agricultural chemists in England at the time.

Voelcker had a high pedigree of connections with the best science that Europe had to offer in agricultural chemistry. His father, also named John

[82] Nicholas Goddard, *Harvests of Change: The Royal Agricultural Society of England, 1838–1988*, London: Quiller Press, 1988.

[83] For the argument about the colonial state's assumption of a direct role for itself in the agricultural improvement of the colony, see Peter Robb, "British Rule and Indian Improvement," *Economic History Review* 34 No. 4 (November 1981): 507–23, *passim*; "Bihar, the Colonial State and Agricultural Development in India, 1880–1920," *Indian Economic and Social History Review* 25 No. 2 (1988): pp. 205–35, especially on 205–7.

[84] *Report of the Indian Famine Commission*. Part II, *Measures for Protection and Prevention*, London: George Edward Eyre and William Spottiswoode, 1880, 137–45, quote on p. 140.

Augustus Voelcker, was trained by the best mentors in chemistry, including Justus von Liebig at Giessen, and served as the consulting chemist of RASE for thirty years, responsible in this capacity for advising farmers on the use of fertilizers. The junior Voelcker received training in chemistry at University College, London, and then at Giessen. At Giessen he took chemistry courses under Professor Naumann and agriculture courses under Professor Thaer and obtained his doctoral degree with a dissertation on the naturally occurring phosphates. Liebig had left Giessen for Munich in 1852, but nonetheless the university still boasted of the highest tradition of teaching and scholarship in agricultural chemistry. That both senior and junior Voelckers studied at Giessen simply reflects the stature of Giessen as the leading international center on agricultural chemistry. Anyone who wished to build a reputation in chemistry relating to agriculture ended up in Germany. However, it is notable that despite the Giessen connection, Voelcker's views were more aligned with the philosophy at RASE and his works on fertilizer and agricultural improvement were in agreement with the Lawes-Gilbert tradition.

Melding "Science with Practice" in a Colonial Context

Voelcker's *Report on the Improvement of Indian Agriculture*, which he composed after a tour of India in 1889–91, advocated a specific role for "new chemistry" in the colony. It is clear that he primarily saw a role for the new knowledge of chemistry for purposes of inquiry and considered his investigation to be a first initiative of that nature that could open the way for ameliorative steps. It tempered expectations about any quick fixes and noted that many of those who were knowledgeable on matters of chemistry in the metropolis had been "disposed to exaggerate the value of chemical manures and chemical analysis of soil [in the colony]." The analysis of soil, for instance, could "explain phenomena" or "suggest lines of improvement" but could not "reform" agriculture "unaided." The reform had to be a comprehensive follow-up along the lines opened by his inquiry. Indeed he agreed that in India there existed "a large field open for enquiry" but countered that the colony was hardly ready for outright application of chemistry.[85]

In many ways the report by Voelcker was a colonial document that was as conformist as many other agricultural tracts of the genre. It foreclosed the option of entry of modern fertilizers into the colony by emphasizing

---

[85] J. A. Voelcker, *Report on the Improvement of Indian Agriculture*, London: Eyre and Spottiswoode, 1893, pp. 34–5.

the "distinctions" of colonial conditions. Adapting the familiar RASE notion of "practice," he emphasized that any reform "must proceed in the direction of improvement from *within*" (italics in the original).[86] The current reality of smallholdings of Indian peasants, their lack of capital, and general impoverishment in the countryside meant that nitrogenous fertilizers would find no buyers. He emphasized that there were plenty of cheap sources of fertilizers like farm manure and dung whose use should be first emphasized among natives. But, at the same time, he was ambivalent about the export of bones from the colony to Britain, hazarding that caste prejudices would probably prevent Indian peasants from employing them even if cheap methods could be found to grind them locally in mills. Besides, according to his assays, the Indian soil was uniformly rich in phosphoric acid; therefore, bones were not critical to Indian agriculture. Thus, quite contrary to the views of Schrottky, he absolved the colonialists of any culpability in exporting bones out of the subcontinent. Voelcker, rather, advised on proper methods to store and release farm manure and night soil that seemed more appropriate under "local conditions." He emphasized that "the assistance of analytical analysis" should certainly be deployed, but it "must be employed in conjunction with an intelligent acquaintance with [local] agricultural practice and with the needs and resources of the agricultural classes."[87] In some ways, Voelcker defended colonial state officials from any charge of lack of initiative on scientific agriculture. As Peter Robb has indicated, the colonial agricultural departments across the colony were guilty of showing an utter lack of initiative. In fact, Voelcker shielded the colonial apparatus of any accusations of wrongdoing in boldly asserting that the common belief that Indian agriculture was primitive and backward and that nothing had been done to remedy its problems was "erroneous." The Indian peasant was "quite as good as, and, in some respects, the superior of, the average British farmer." Also, the health of Indian crops was "wonderfully good" given the conditions under which they were grown; whatever limitations farming exhibited stemmed from "an absence of facilities for improvement" such as he had indicated: poverty and small size of holdings for which evidently the colonial state bore no responsibility.[88]

The report's preference for humus theory and the RASE version of the nitrogen cycle further provided support to a policy of minimalism in the

---

[86] *Report on the Improvement of Indian Agriculture*, p. 13.
[87] *Report on the Improvement of Indian Agriculture*, pp. 34-5.
[88] *Report on the Improvement of Indian Agriculture*, pp. 10-11.

state's efforts in development of Indian agriculture. Voelcker's partiality toward the science he favored shone through in his detecting what the Indian soil was actually deficient in and what were the best ways to eliminate that deficiency. He expressed himself clearly in favor of protecting the organic matter and nitrogenous content of soil. Humus was the primary source of nitrogen to the soil. Voelcker emphasized the importance of organic matter for improving the physical characteristics of soil and supplying additional nitrogen. Minerals like potash, soda, and phosphates and nitrates were present in plentiful quantities, and the Indian peasant did not need to make any exceptional effort in this regard. On the basis of his analyses and tests of Indian soil Voelcker asserted optimistically that there was no positive evidence that the Indian soil was exhausted and that as long as the government developed a plan to keep the cycle of organic matters rotating, there would be no exhaustion in the years to come.

In the end, in Voelcker's vision, there was no rapid and gushing embrace of Liebigian chemistry or, for that matter, the English practical chemistry for the betterment of Indian agriculture. Steeped in the RASE tradition, he nonetheless provided a prescription that was in keeping with the colonial state's agenda of slow forward movement on state's investment in Indian agriculture. The influential report became the basis of colonial policy for a few decades, and thus the impact of his particular interpretation of new chemistry and its relevance for India cannot be overestimated. His report was a palimpsest in the successful promotion of the Lawes-Gilbert-RASE model of agricultural chemistry in the colony even if the latter's promises were not sought to be realized.

Voelcker also addressed the improvement of indigo culture in colonial India. He did not have a detailed knowledge of specialized indigo manufacturing processes. He also wrote that he would rather refrain from addressing the political aspects of indigo cultivation and manufacturing that involved socioeconomic and political relations among planters, *ryots*, and *zamindars*. Rather he focused more on indigo agriculture as part of the responsibility given to him to address the prospects and problems of Indian agriculture at large. Indigo plantations were indeed the "beau ideal" of colonial agriculture. In the last decade of the nineteenth century the colonialist still considered indigo plantations as primary evidence in building a case for the Western passage of agricultural modernity into the colony. The conviction expressed by William Bentinck in 1829 that the indigo plantations were "the center of a circle of improvement" had survived the ravages of time. Still decades after Bentinck, after the popular revolt against the indigo system in Bengal, and simmering discontent against its violence among the indigo peasantry in Bihar, Voelcker went on to praise Bihar's indigo planters for their selection of seed, the use of the drill, and the application of indigo refuse, or *seet,* as manure, concluding that "the cultivation of indigo has been greatly improved by the European planter and the native growers have to some extent followed the example set them."[89]

---

[89] *Report on the Improvement of Indian Agriculture,* pp. 222, 236, 257–66.

3

Colony and the External Arena: Seeking Validation in the Market

"Calcutta was founded, so the history-books tell us, in 1690. It was only in 1757, however, the year of Plassey that she began to rise in the world." In or around 1977, in his history of J. Thomas & Company, Dipak Roy highlighted the connection between Calcutta's rising importance and the famous victory by colonialists that started the British conquest of India. Although much historiography treats the famous battle as a watershed in the establishment of colonial rule on the subcontinent, Roy's note is striking in tracing instead, from the same victory, the emergence of Calcutta as an important economic hub in the world. The obfuscation of the violent aspects of colonialism in the account can be attributed to Roy's mind-set as a postcolonial intellectual. He was, after all, also the chairman of J. Thomas and Company at the time of writing, and in his spirit of celebrating one hundred and twenty-five years of the company's foundation that his book marked he could afford to be a little unabashed about the company's remote beginnings as a colonial brokering firm. Indeed the narrative underscored how remote that era was from the author's – "in an age – before the Crimea, before the death of the East India Company, before the apotheosis of Abraham Lincoln – how remote from our own."[1] But Roy's equivocation in the recounting also reflected the basic fact that colonialism had tied the subcontinent's destiny with economic forces of the world. These wider connections could not be addressed solely through reference to the British Empire and its networks alone.

[1] Dipak Roy, *A Hundred and Twenty-Five Years: The Story of J. Thomas & Company*, Calcutta: J Thomas & Company, not dated, pp. 2, 4, British Library, T 40074.

Roy's triangulated histories of J. Thomas & Company, Calcutta, and colonial times share a common connection with indigo and reveal that indigo had economically fused the Indian subcontinent not only to imperial Britain, as has been depicted in the historiography so far, but to a broader historical arena. J. Thomas & Company had built its fortunes around indigo, and the company's identity was tied to brokering in the colonial dye. The agency's office was then housed in a building named Nil Hat, or Indigo Mart. The street, Mission Row, where its warehouse stood, bore the marks of indigo, quite visibly, as Roy described: "The colour seeped from the packed chests [of indigo] and stained the length of Mission Row a deep abiding blue."[2] J. Thomas was one of the two concerns – the other being William Moran and Company – that dominated brokerage in indigo in Calcutta in the latter half of the nineteenth century. Calcutta, too, served as the major exit point for all of indigo produced in eastern and upper India for its distribution into a global market dominated by Britain in the nineteenth century. Thus facilitation by the colonial state may have allowed the indigo industry to strike roots in Bengal, but Bengal indigo's career on the subcontinent was organically connected with a trade that was global in orientation. And J. Thomas & Company as an agency house and Calcutta as a port were partners in the trade that connected the colony with the world.

Roy's account resonates with the effort in this study to write the history of indigo by focusing on the "externalities" that were located in a broader geographical arena whose outlines were not delimited by colonialism and imperialism. Colonialism itself was an external force because it tied the colony to the manifest material interests of a distant nation. It also enabled the participation of indigo in an international market of dye consumption. This unmediated connection with an international market meant that the colonial economy was exposed to the tribulations of international trade and finance, over which the local players had little control. But in the second half of the nineteenth century the meaning of "externality" for indigo assumed a new dimension due to the unfolding global transition from natural to synthetic dyes, a development that did not originate in British imperialism. The transition was connected with the successful commercialization of a hydrocarbon-based production system and the national industrial history of Britain and France, and, most importantly, Germany. From the successful invention of the artificial mauve color in Britain in 1856–7 to the manufacture of synthetic indigo

---

[2] Dipak Roy, *A Hundred and Twenty-Five Years: The Story of J. Thomas & Company*, p. 7.

in Germany in 1897 the new science-based industry followed a dynamic of its own. British industrial aspirations for synthetic dye production were consistently nationalist in nature in the context of rivalry with other European nations and were not directly constitutive of British imperialism. Moreover, it was the "German" synthetic indigo that would destroy the colonial indigo industry on the Indian subcontinent. The interests of British expatriate indigo planters based in colonial Bengal were distinct from those of synthetic dye manufacturers in the West generally, whether German or British. All in all, the transition and the inherent conflicts between its protagonists took shape outside the immediate orbit of colonial rule.

The unfolding process of transition intersected with indigo manufacturing on the Indian subcontinent. The successful commercialization of synthetic indigo in the late nineteenth and early twentieth centuries was able to interrupt and redirect the course of colonial history of indigo production. In that context this chapter focuses on the interaction between the global and the local. On the side of the local it illustrates natural indigo production in Tirhut within the Bengal presidency in the latter half of the nineteenth century and the response of actors there to the launch of synthetic indigo in a global market. It then turns attention to the English markets in the latter's incarnation not as a metropolitan market but as a "national" market representative of changes during the phase of transition as synthetic indigo swept one national market after another as part of the transition. It thus emphasizes the relevance of the external context that complicated the colonial history of indigo in India.[3]

Indigo Moves from Lower Bengal to North Bihar or "Tirhut"

After the mid-nineteenth century the north Bihar districts in colonial Bengal emerged as a major indigo-producing tract of colonial India. The new indigo from all of these sites came to be identified as the indigo from Tirhut or even continued to be called Bengal indigo in international markets, as before. As Jacques Pouchepadass has pointed out, indigo acreage in Tirhut (taken to mean north Bihar here and as depicted in

[3] For the rise of the synthetic dye industry, see, John J. Beer, *The Emergence of the German Dye Industry*, Urbana: University of Illinois Press, 1959; Anthony S. Travis. *The Rainbow Makers: The Origins of the Synthetic Dyestuffs Industry in Western Europe*, Bethlehem, Pa.: Lehigh University Press, 1993; Carsten Reinhardt and Anthony Travis, *Heinrich Caro and the Creation of Modern Chemical Industry*, Dordrecht: Kluwer, 2000.

MAP 2. Indigo tracts of Tirhut showing the four prominent districts, 1870

Map 2) doubled between 1830 and 1875 and then expanded by another 50 percent over the next two decades. The exponential growth in Tirhut occurred mostly in the four districts of Muzaffarpur, Darbhanga, Champaran, and Saran, where the corresponding figures for increased indigo cultivation between those years were 286 percent and 69 percent, respectively. The first phase of increase in Bihar transpired in the wake of the indigo revolt in Bengal (1859–62) that caused the complete destruction of the industry in Lower Bengal. As indigo declined in the deltaic region, the planters in Bihar expanded their cultivation of indigo in north Bihar. The later increase in indigo acreage between 1875 and 1895 was primarily the outcome of expanding market demand.[4]

Tirhut emerged as the new destination of European indigo entrepreneurs, and the tracts left by expatriate planters accordingly reflect the centering of Tirhut in their writings. Minden Wilson was one of the planters who left a record of their Tirhut reminiscences. Indeed he would go on to live in Bihar for more than five decades and compose a significant volume documenting the history of factories and planters. Minden Wilson landed in Calcutta aboard an American ship, *Albatross*, in 1847, having earlier made his fortune among the plantations in Mauritius. He later wrote about the moment of his arrival in the colony, saying he had mused about the life that lay ahead of him as a country steamer took him past the anchored American ship and up the river Hughly. Farther upstream the steamer would have waded through the mighty Ganges as it would have moved toward the indigo tracts of north Bihar.[5] James Inglis, the Scottish-born cosmopolite, spent several years as a planter in Champaran. After he left for Australia in 1877, Inglis reminisced about the life he had spent as a planter in Bihar in some of his writings. In his *Sport and Work on Nepaul Frontier* he praised the beauty of the landscape and surroundings in Bihar. He claimed that "among the many beautiful and fertile provinces of India, none can, I think, much excel that of Behar for richness of soil, diversity of race, beauty of scenery, and the energy and intelligence of its inhabitants." Above all, he thought of the hunting games and sports he had enjoyed during his sojourn in Champaran and even alluded to them in a collection of poems. He was among the very early European settlers in the district. Inglis reported that only a small community of European settlers, which included twenty-five or thirty indigo planters in the district along with colonial district officials, opium traders, and a military contingent located a little distance away from the district town, lived in Champaran then. This was a rather well-knit community of expatriates who were "all like brothers," who organized recreational get-togethers, whether hunting games or sports like cricket and hockey.[6]

[4] Jacques Pouchepadass, *Champaran and Gandhi: Planters, Peasants, and Gandhian Politics*, Delhi: Oxford University Press, 1999, pp. 20–3.

[5] Minden Wilson, *History of Behar Indigo Factories, Reminiscences of Behar, Tirhoot and Its Inhabitants of the Past*, Calcutta: Calcutta Oriental Printing Company, 1908, p. 105.

[6] James Inglis, *Sport and Work on the Nepaul Frontier or Twenty Years Sporting Reminiscences of an Indigo Planter*, London: Macmillan & Co., 1878, pp. 1–6. Inglis

FIGURE 5. Planters' bungalow, India. Photograph by Oscar Mallitte, 1877. © Science Museum/ Science & Society Picture. Image No. 10318243.

In an account of the general life of a Bihar planter, another major planter, W. M. Reid, described what an imaginary "young to-be Indigo Planter" would encounter after landing in Calcutta. The Calcutta agents would immediately assign a factory to him in "somewhere, say Tirhoot." The train line from Howrah to Muzaffarpur would ferry him into the heart of Tirhut from where the *daks*, or relays of riding or driving horses or ponies, would take him to the final destination. This was the alternative route to Bihar that a European took if he decided not to take the steamer service up the river Ganges into north Bihar as Minden Wilson did. Reid was a little ambivalent about the overall merit of a career in Bihar, discouraged as he was a little by the onerous nature of work and extremities of climate at the beginning. Nevertheless in the end he would go on to live for a very long time in the colony and partake in the lavish life with others (see a typical planter's bungalow in Fig. 5). One of his descendants, D. J. Reid of Belsund factory in Muzaffarpur district, was a distinguished

wrote under the pseudonym of "Maori," perhaps a symbol of his identification with the tribe of the same name in New Zealand, where he had first migrated from Scotland at the age of nineteen.

Bihar-based planter who lived on in the colony until long after the First World War.[7]

Planters in Bihar organized the cultivation of indigo on land under their direct control on what were called the *zirat* lands. The ownership of land was usually acquired through the purchase of leases over entire villages from the landlords. Such leases gave planters access to the traditional power of the landlords in the Bihar countryside. So even if the planters did not "own" the land in any absolute legal sense, they had sufficiently strong rights over the land in their possession. This system of indigo cultivation was different from the prior practice in Bengal, where the indigo was largely grown by giving out contracts to Indian peasants. The difference was particularly relevant because it enabled the planters to launch direct efforts at improvement that had not been possible earlier in Lower Bengal.

Through leases the planters obtained superior rights of almost seigniorial dimension over tenants in the village. Once the lease had been obtained, the planter resorted to a deliberate policy of manipulating existing agrarian relations in order to secure the cultivation of indigo by his tenants. Sometimes the planters resorted to outright eviction of peasants on different pretexts or used threats of eviction to force the peasants to grow indigo on part of their land. Occasionally some planters purchased subtenant rights from their own tenants in order then to grow indigo themselves using hired labor.[8] As Inglis subtly implied, ultimate authority in the countryside seems very much to have been the goal of planters in obtaining leases: "Having, then, got[ten] the village in lease, you summon in all your tenants, show them their rent accounts, arrange with them for the punctual payment of them, and get them to agree to cultivate a certain percentage of their land in indigo for you." With the authority of a leaseholder the planter could hire labor at will, requisition carts for ferrying indigo to the factories at harvesting season, and even fix a low purchase price for indigo grown by his tenants.[9]

[7] W. M. Reid, *The Culture and Manufacture of Indigo with a Description of a Planter's Life and Resources*, Calcutta: Thacker Spink and Company, 1887; for process of early arrival, pp. 1–7; for onerous work, pp. 12–20.

[8] Sugata Bose, *The New Cambridge History of India, Peasant Labour and Colonial Capital: Rural Bengal since 1770*, Cambridge: Cambridge University Press, 1993, p. 76.

[9] James Inglis, *Sport and Work on the Nepaul Frontier*, pp. 7–13, quote on p. 10; an amusing, imaginary conversation about the purchase of a lease between the manager of a plantation and a *zamindar* is reported in W. M. Reid, *The Culture and Manufacture of Indigo*, pp. 29–31.

The *zirats* were the focus of planters' concentrated efforts and larger resources for land improvement via the use of efficient equipment and better farming methods. It is no surprise that many commentators have spoken highly of the state of agriculture on the plantations. According to Inglis, the *zirats* in Bihar were farmed "in the most thorough manner" with the use of the expensive, imported English Howard's plow and application of manure. Inglis also claimed that "a fine, clean stretch of Zeraat in Tirhoot or Chumparun, will compare most favourably with any field in the highest farming districts of England or Scotland." Referring to the detailed plowing, hoeing, and cleaning of land with a large retinue of Dangar tribesmen and low castes, he praised such efforts, which made *zirats* "look as clean as a nobleman's garden."[10] He certainly would have spoken for a majority of plantations if not all. It is relevant to point out that Voelcker, the important English chemist who had been appointed by the colonial government to suggest measures for the improvement of Indian agriculture, also made indigo planters the beau ideal of colonial agriculture and praised many of their practices. Voelcker supported the assertion of Inglis on the use of expensive equipment by planters. The planters tactically fused adjoining areas within the village to acquire large contiguous *zirats* under their direct control that were suitable for the use of large machinery. Voelcker understood why large planters were the ones who could buy expensive machinery, explaining that "only where there are large areas to be cultivated, time being thus a matter of importance, and the economy of quick labour and improvement having room to show itself, so that the question of first cost becomes relatively of no consequence." Voelcker similarly explained why the Indian peasant did not adopt the more efficient Hindustan plow on his lands: because he "could not go to the expense of adopting it on his small plot."[11] Reid called seed drills "the offspring of indigo" in Bihar, crediting the planters for the introduction of the contrivance in the region. The planters embraced the seed drill for purposes of deep sowing as well as for efficiently sowing large tracts. A large number of seed drills were made and tested at indigo factories with the purposes of use on *zerats* and distribution to the peasants growing indigo for the planter.[12]

The Bihar planters were also pioneers in the colony in the systematic application of manures and other fertilizers. No such practice was

[10] James Inglis, *Sport and Work on the Nepaul Frontier*, pp. 8, 19.
[11] Voelcker, *Report on the Improvement of Indian Agriculture*, p. 222.
[12] Reid, *The Culture and Manufacture of Indigo*, pp. 42–8.

FIGURE 6. *Seet*, or indigo refuse, being spread in Tirhut agricultural fields as manure.
Source: From W. M. Reid, *The Culture and Manufacture of Indigo with Description of a Planter's Life and Resources*. Calcutta: Thacker, Spink and Co., 1887, facing p. 53.

known to exist anywhere else until then. The annual deposit of fertile alluvium on deltaic floodplains in Bengal had earlier made the use of any additional fertilizer unnecessary. But as the planters moved to the highlands of Bihar they noticed that their land became exhausted from plantings after only a few years. Under these new circumstances they embraced the use of manures and fertilizers. In some cases a portion of the plantation was set aside each year to be treated with *seet*, the refuse from indigo manufacturing. Heaps of *seet* were systematically placed at regular intervals in straight lines on the field. The planters dug trenches and deposited *seet* in furrows, after which it was covered with soil in an excruciatingly labor-intensive process that involved multiple steps. Reid thought that fertilizers were "both necessary and imperative" and that the *seet* was the ideal fertilizer for indigo fields (see the operation in Fig. 6).[13] Voelcker also referred to some planters as collecting, grinding, and broadcasting bones to supplement the phosphate content of their land. Some were apparently open to trying inorganic fertilizers. Voelcker, for example, wrote, "I have been with a planter whose belief in the sulphate of lime (gypsum) and other sulphates was almost unbounded, also

[13] Reid, *The Culture and Manufacture of Indigo*, pp. 53–4.

with another who thought that nitre was what was wanted for indigo, whilst many more whom I met ridiculed equally either idea."[14]

Mechanization of and Tinkering with the Manufacturing Process with Chemicals

In north Bihar factories in particular the manufacturing installations were usually vaster and of a better quality than their prior incarnation in Lower Bengal.[15] The entire apparatus from wells, reservoirs, and steeping vat to the beating vat, boiler, press machine, and drying house were sturdier and of a better quality than before. The new beating vats were grander and a significant improvement on their previous version. Formerly the beating vats were only a little different from the steeping vat, just longer and shallower. But modern beating ranges were composed of beating wheels that were operated by machine. They were no longer, as J. Bridges Lee emphasized, "quite so simple in construction." The surface area of the beating vat now measured many times the combined open face area of all the steeping vats leading into it. It was divided by longitudinal walls into three separate channels that were connected at both ends. The central channel housed the beating wheel containing rectangular paddles as spokes. The wheel struck the water at a right angle and was placed parallel to the longer side and closer to one end of the range. The wheel itself was usually connected via an axle to another beating wheel on the other side of the factory in a corresponding beating range, and both were driven by a common machine placed inside the factory building. Bridges Lee fully understood the scientific basis of beating in facilitating "extensive contact" of the liquor with oxygen in the air. Thus, he wrote that while the older method in which workers stood knee deep in the vat and tossed the liquid up in the air was a simple and fairly effective method, the beating wheel method was superior because it did away with the use of expensive labor or supervision of a large number of workers involved in beating (see Figs. 7, 8, and 9).

[14] Voelcker also praised the effort at selection of plants and use of the Persian wheel for irrigation by the planters. Their neglect of rotation was in his opinion a minor anomaly because indigo plants had bacteria in their nodules that fixed atmospheric nitrogen in the soil directly and ensured that the land would not be deprived of its most important ingredient despite continuously being put under indigo. Voelcker, *Report on the Improvement of Indian Agriculture*, pp. 236, 259, 260–6; Reid, *The Culture and Manufacture of Indigo*, p. 55.

[15] Jacques Pouchepadass has noted this in passing while making the point that large capital was sunk by the planters as fixed stock in the factories. Pouchepadass, *Champaran and Gandhi*, pp. 46–7.

FIGURE 7. Indigo factory, India, Photograph by Oscar Mallitte, 1877. © Science Museum/ Science & Society Picture. Image No. 10318250.

FIGURE 8. Indigo beaters, India, Photograph by Oscar Mallitte, 1877. © Science Museum/ Science & Society Picture. Image No. 10318262.

FIGURE 9. Indigo drying house, India, Photograph by Oscar Mallitte, 1877. © Science Museum/ Science & Society Picture. Image No. 10318258.

On the basis of a better understanding of the scientific principles underlying the process of beating more improvements in apparatus were being considered and implemented. To prolong the contact of the fermented liquid with air without "the useless concussion, and equally useless friction" involved in the beating wheels, Bridges Lee had patented a "shower bath method" of oxidation. The system involved pumping liquid to an overhead reservoir. A series of gutters ran down from the reservoir toward the oxidizing vat. The gutters were provided with tubes ending in a perforated metal head that created a shower bath over the oxidation vat. Not only did the new method cause the liquor to "traverse a considerable thickness of air," it improved economy by reducing oxidation to thirty to sixty minutes, considerably less than the previous two to three hours.[16]

The use of steam-based machinery was a distinguishing new aspect of manufacturing on the Tirhut plantations. Reid reported that the use of steam power "for beating and pumping purposes" was common at most factories.[17] Steam pumps were used to lift water from natural sources like streams and lakes to the reservoirs, where water was stored for a while to let unwanted impurities sediment to the bottom so that purer water entered the steeper. Most significantly, the beating wheels were now rotated with steam power in most factories. On leaving the vats the dye was transported to the boiler through a steam-operated ejector mechanism. The boiling function itself was accomplished with a steam jet instead of the previous practice of the use of fire supplied heat in pans, which was difficult to cut off swiftly. Last, toward the close of the nineteenth century, steam-operated compressors for pressing the indigo cake also made their entry into the indigo plantations. All of these innovations gave indigo manufacturing a clear stamp of mechanization enabled by the use of steam power at almost every step.[18]

The planters debated the efficacy of chemicals for various operations, and their potential was variously assessed for different parts of the manufacturing operation. For instance, there was general agreement on the use of salts to purify water used in manufacturing. No one doubted that the quality of water was of primary importance in securing a good product. Bridges Lee underscored that "the really essential requisites for the manufacture of good indigo are, good plant and good water." The planters ideally needed soft water that would be generally available from a running stream. But sometimes the only water readily available was the hard water from a spring or well. Bridges Lee warned that "a prudent man" must have his specimens of water tested by an analytical chemist. He even prodded the planters to learn the basic art of water analysis while holidaying in Europe. Since the rudiments of physics and chemistry were taught in nearly all good schools in England, he counseled that young would-be planters, before they set off for India, should

[16] J. Bridges Lee, *Indigo Manufacture*, Lahore, January 1892, pp. 86–114.
[17] Reid, *The Culture and Manufacture of Indigo*, p. 49.
[18] J. Bridges Lee, *Indigo Manufacture*, pp. 114–16.

become well versed in the rudimentary methods of analysis of soil, water, and manure. The planters in the colony could alternatively pool their resources to invite an analytical chemist from home to visit their station and give advice on how to conduct chemical tests. This would protect the planters against the "quack remedies" offered all too readily by "adventurous charlatans" and "chance adventurers who dub themselves chemists." He suggested the use of the inexpensive oxalate of ammonia and potassium permanganate for softening water.[19]

Most planters as yet resisted the idea of adding chemicals to vats during actual manufacturing, although some evidently tried it. William Reid admitted that he was "an old hand" with a "rooted objection to any 'doctoring' stuff."[20] J. Bridges Lee confirmed that many planters strongly objected to adding anything to the steeping vat.[21] The opposition to application of reagents does not reflect any innate conservatism or anti-science disposition of those individual planters or of the class of planters as a whole. Such reservation was not at all unreasonable in the absence of any clear proof of efficacy through demonstrated, consistent results. In the absence of such validation the planters were predictably skeptical. The science of manufacturing was still imperfectly understood. In such a situation claims and counterclaims about the role of chemicals were frequently heard. Entrepreneurs like Eugene Schrottky who registered patents for the use of chemicals were no doubt praised by some planters. But most planters slammed them as "charlatans" because their claims of success were hardly replicable on a consistent basis.

Colonial Agricultural Indigo and Textile Manufacturing: The Coming of Synthetic Dyes

Colonial trade closely tied the destiny of agricultural indigo produced in Tirhut with economic currents in the industrial systems of textile production in Britain and other nations around the world. The relevance of other nations relative to Britain rose as textile production spread to newly

---

[19] J. Bridges Lee, *Indigo Manufacture*, pp. 52–60, 74–6.
[20] Reid, *The Culture and Manufacture of Indigo*, p. 1.
[21] Bridges Lee himself tried adding sugar for inhibiting the process of oxidation in the fermentation vat. But he found that even though sugar was useful in slowing oxidation, it interfered with the subsequent processes and was thus not an appropriate additive. On balance, he decided to use the simple device of reducing the surface area of steeping liquid to inhibit oxidation instead of trying chemicals. J. Bridges Lee, *Indigo Manufacture*, pp. 17, 42.

TABLE 5. *Export of indigo from Tirhut to various destinations*
(in hundreds of pounds)

Year	Britain	United States	France	Austria	Germany	Japan
1873–4	47,169	5,144	9,287	5,491		
1874–5	33,570	2,331	8,078	5,487		
1875–6	51,524	3,912	16,178	6,411		
1876–7	40,833	6,157	11,961	5,813		
1877–8	44,030	9,826	29,982	6,618		
1878–9	35,539	10,773	14,045	7,125		
1879–80	20,897	12,194	7,770	4,911		
1880–1	51,232	NA	NA	NA		
1881–2	44,394	NA	NA	NA		
1882–3	46,908	NA	NA	NA		
1883–4	57,916	NA	NA	NA		
1884–5	50,402	NA	NA	NA		
1885–6	31,439	18,654	10,862	NA		
1886–7	31,146	25,750	12,868	NA		
1887–8	30,434	19,258	15,454	10,858		
1888–9	30,436	22,671	9,677	10,700	8,073	
1889–90	35,012	21,206	9,439	11,119	9,546	
1890–1	32,443	12,220	6,610	7,344	6,974	
1891–2	31,748	20,251	14,354	10,116	11,775	
1892–3	17,084	16,417	9,393	7,428	3,998	
1893–4	26,192	11,900	8,729	10,250	11,756	
1894–5	31,248	23,601	13,377	10,920	14,026	
1895–6	33,130	17,148	17,021	12,629	14,648	
1896–7	41,849	20,445	10,604	10,563	10,974	
1897–8	19,076	20,102	7,026	9,347	6,297	
1898–9	22,972	17,922	8,350	9,732	7,785	
1899–1900	13,215	10,405	9,497	6,755	5,494	5,597

Source: Jacques Pouchepadass, *Champaran and Gandhi*, appendix, Table H.

industrializing areas. The widening circle of recipients of Indian indigo in the second half of the nineteenth century depicted in Table 5 affirms that changing position. The connectedness with a growing comity of nations was made possible by both imperial and extraimperial forces. English brokers, shippers, and bankers based in places like Manchester, London, Glasgow, and Bristol had a major role in this dispersal of Indian indigo as they turned London into an entrepôt market that served the function of reexporting indigo originating in Tirhut to other places in Europe, America, and Asia. The agency houses of English extraction based in Calcutta alternately exerted little or no control over indigo that was directly shipped out of India to other nations. But more relevant to the point, colonial indigo's fortunes now lay ultimately in the hands of widely dispersed consuming classes across many territories. In a way colonial indigo was now taken outside the purview of influence of singular developments of national import within the British market. Only a generalized, momentous change in consumption pattern sweeping across all of these dyeing and printing houses would have the ability to impact materially the labyrinthine trade network of colonial indigo with textile manufacturing in the world.

Historians have argued that a force of precisely that dimension was slowly brewing in one key sector of textile manufacturing, that of dyestuffs, which had the potential to choke the colonial industry of natural indigo dye. John J. Beer and Anthony Travis, who have studied the dyeing and printing industry in the West, maintain that this sector was awaiting a major technological change in the mid-nineteenth century. It has been pointed out that a number of technical improvements had enhanced the productive capacity of textile manufacturing in key European nations. The mills were producing textiles cheaper and in large volumes for the mass market. Greater availability of clothing had generated a growing "hunger for different colors."[22] The textile districts in England went through rapid mechanization after 1840. The deployment of steam power and cylinder printing took Britain's calico printing to a level four times higher than that of France, its most serious rival in textile production. In the second half of the nineteenth century, the Mulhouse and Rouen regions of France, and then central and southern Europe, also began to adopt the same faster and more efficient printing techniques that the British manufacturers had earlier embraced. Together these nations accounted for a good proportion of all of the major textile manufacturing. Distinct from printing, innovations in dyeing appeared most noticeably in France. Travis has argued that the overall disparity in the pace of innovations between textile manufacturing and dyestuffs manufacturing created an opening where industrial chemistry was embraced for the development of more efficient factory-based dye production.[23]

[22] John Joseph Beer, *The Emergence of the German Dye Industry*, Urbana: University of Illinois Press, 1959, p. 3.
[23] Anthony Travis, *The Rainbow Makers: The Origins of the Synthetic Dyestuffs Industry in Western Europe*, Bethlehem, Pa.: Lehigh University Press, 1993, pp. 31–2; see also, Ernst Homburg, "The Role of Demand on the Emergence of the Dye Industry: The Roles of Chemists and Colourists," *Journal of the Society of Dyers and Colourists* 99 (November 1983): 325–32.

By the 1830s and 1840s some changes in the direction of "artificiality" in dye preparation and application were already under way. Major dye works had adopted the practices of preparing what were called the chemical "extracts" of natural dyes. Thus treatment of natural dyes with acids and alkalis was becoming common in the preparation of French purple from lichens, of garancine from madder, and of indigo carmine, the sulfonated version of blue dye, from plant indigo. This trend toward artificiality in the use of natural dyes occurred at the behest of a new generation of dye chemists attached to dye and print houses.[24]

But a truly radical shift in the field of dye manufacturing emerged from impulses located outside the orbit of existing dye works and their older, familiar frameworks of knowledge and operation. It was instead based on an alternate line of work by a new school of chemists working toward "the artificial formation of natural organic compounds."[25] The lineage of this work extended back to research traditions in organic chemistry that were started by Justus von Liebig at Giessen. After him, a cohort of students trained in his experimental and analytical methods worked relentlessly in this general direction. One of its offshoots eventually led to the invention of artificial colors.

The first major category of synthetic colors was developed out of the coal-tar derivative aniline. In 1843, August Wilhelm Hofmann, a pupil of Liebig, found out that the alkaline oil present in coal tar bore similarity with the oil extracted from the indigo plant. He coined the name "aniline" for this substance, which occurred across natural and artificial realms.[26] The anilines as a group of coal tar–derived substances became the cache of compounds that were developed into multiple colors using tools from organic chemistry. Hofmann had earlier moved from Germany to England to lead the Royal College of Chemistry. His student at the Royal College, William Henry Perkin in 1856 invented mauve, the first aniline color, which heralded the era of aniline colors as substitutes for

---

[24] Augusti Nieto-Galan, *Colouring Textiles: A History of Natural Dyestuffs in Industrial Europe*, Dordrecht: Kluwer, 2001, pp. 181–204.

[25] These were the words of the mauve inventor, William Henry Perkin, who characterized the scientific quest of early inventors in such terms as he looked back at the achievements of the coal tar dye industry some four decades after the invention of the first artificial dye. William H. Perkin, "Hofmann Memorial Lecture: The Origin of the Coal-Tar Colour Industry, and the Contributions of Hofmann and His Pupils," *Journal of the Chemical Society* 69, part 1 (1896): 603, cited in Anthony Travis, *Rainbow Makers*, p. 36.

[26] The name "aniline" was coined after *anil*, prevalent in Portuguese usage. The word had undergone numerous transmutations as it evolved from the root word in Sanskrit in India, *nila*, or blue. Arab merchants called it *an-nil* until the Portuguese further transformed it into *anil*.

natural dyes. Perkin demonstrated remarkable entrepreneurial skills in scaling up the product from the laboratory to the factory between 1856 and 1859. Once the initial hesitation of dyers wore off, the diffusion of purple dye was rapid among the silk dyers, most noticeably in the silk producing regions of France. Mauve opened the floodgates for the subsequent discovery and launch of other aniline colors such as Emmanuel Verguin's aniline red in France (1859), a violet by Hofmann (1863), and a black by the English printer John Lightfoot (1863). A wide variety of aniline blues also became available as part of the process.

The next generation of alizarin and azo dyes was synthesized and marketed in the 1870s. The Gewerbe Institut in Berlin became the major center for innovative work in the manufacture of dyes out of hydrocarbons. It was here that Carl Graebe and Carl Liebermann collaborated to synthesize alizarin, the red color that was until then derived from the roots of madder plant. This 1869 invention was a major victory for the synthetics as it caused the destruction of the age-old industry built around the cultivation of madder. It was a watershed development in the displacement of natural dyes and in the long-term transition from natural to synthetic dyestuffs. The production of artificial alizarin was earnestly undertaken by the Badische Anilin and Soda Fabrik (BASF) in Germany and by Perkin and Sons in England, but it was BASF that emerged as the leading supplier of alizarin in the long term. Historians have rightly singled out alizarin's important role in the emergence of Germany as the major dye producer in the world.[27] This dominance was further solidified as dye companies in Germany took the lead in the manufacture of a new class of azo dyes after 1877. It is a testimony to these industrial advances that Germany emerged as the leading producer and supplier of synthetic dyes in the seventies and was now able to meet as much as half of the world's demand.

British Dyeing and Printing Districts before 1897

Agricultural indigo's dominance in the market depended to a large extent on choices made by dyers and printers as mass consumers. The end users'

[27] Anthony Travis and Carsten Reinhardt unambiguously state that it was the synthesis of alizarin that enabled the "changeover" in the pivot of production of synthetic dyestuffs from England to Germany. Carsten Reinhardt and Anthony Travis, *Heinrich Caro and the Creation of the Modern Chemical Industry*, Dordrecht: Kluwer, 2000, p. 140. Asserting the dominance of Germany via production of alizarin, Anthony Travis states: "By 1873, dyers and printers in Manchester and Glasgow, as well as in Mulhouse, Elberfeld, Berlin, and Basle, were consumers of German-made alizarin." Cf. Anthony Travis, *Rainbow Makers*, p. 195.

preferences were not rendered irrelevant, but user priorities were introduced as market forces through the agency of middle-level consumers, who as a class determined the position of different dyes in the marketplace. In addition to being responsive to customer tastes the commercial users based their decision on work-related criteria like ease of application, predictable supply lines, and shelf price. Thus if the contest between agricultural indigo and other blues available in the market was sealed early, it was largely the result of how their respective potential was judged in the dye and print houses.

The signals emanating from southern Lancashire, the area with the densest concentration of dye and print works, serve very well as a representative case clarifying the terms on which the reigning crown of blue dyeing, plant indigo, retained its monopoly. Although indigo was an expensive dye, the stranglehold of indigo on the market was first and foremost based on its "fastness." It was universally appreciated by the public for its resistance to acid, alkali, water, and sunlight. The manufacturers appreciated its disinclination to milling, an aberration feared by printers in which the dye from one yarn was commonly known to rub off onto another in the weaves. All other blues of the presynthetic era failed to dislodge indigo from its entrenched position, be it the rather costly ultramarine blue and newer Prussian blue or the older woad, which was still available in the market.[28]

User calculations, however, began to change somewhat after cheaper synthetic blues were put on to the market even though agricultural indigo still maintained its supremacy over early versions of coal tar dyes. The manufacturers began to embrace artificial colors for dyeing blue to the extent that doing so helped implement savings without compromising the color and quality of work. The dyers especially adopted the practice of combining indigo with other cheaper artificial dyes in the growing convention of "topping" and "bottoming" in which the use of indigo was preceded or followed by dyeing with blue alternates. In deep blue dyeing, for instance, such methods could potentially dispense with the use of a large quantity of expensive indigo.[29] The anilines were extensively

---

[28] Prussian blue had been used by dyers since the early eighteenth century, while ultramarines were also explored as indigo alternatives. Augusti Nieto-Galan, *Colouring Textiles: A History of Natural Dyestuffs in Industrial Europe*, p. 186.

[29] "Topping for Indigo Blue on Cotton Yarn," *Textile Manufacturer*, April 15, 1893; "On Dyeing Cotton with Indigo combined with Other Dyestuffs, by Dr. M. Polonovsky and J. Nitzberg," *Textile Manufacturer*, December 15, 1893. The holdings of *Textile Manufacturer* are available in the Special Collections unit of Paul J. Gutman Library, Philadelphia University.

deployed for topping and bottoming in blue dyeing.[30] Many dyers also appreciated the brightness that anilines provided. In wool dyeing specifically the dyers mixed anilines freely with natural dyestuffs.[31] The alizarin blues in the 1880s performed even better than aniline blues in dyeing and printing. More than any other synthetic color it was alizarin blue that was embraced most widely as a substitute for indigo. The alizarin's improved fastness was secured by a mordant and considered "essential progress" in its value as a dye.[32] In wool dyeing inroads made by alizarin blues for obtaining deep color were noticeable. A report in a leading dyer's journal asserted that "the most pronounced indigophile will allow that, except against the light and medium shades, alizarin is proving itself a strong competitor."[33] Another report argued that although in the past the coal tar industry had claimed to furnish many substitutes for indigo, only alizarin colors had proven to be the "real substitutes" to natural indigo in actual terms.[34] The azo dyes made substantial entry into the world of cotton dyeing as substitutes for indigo. Overall it could be said that some of the artificial blues had found use as substitutes in Western markets.

The forays made by artificial alternatives, however, were neither able to bring about a depression in the price of natural indigo nor to reduce its export. The final price of indigo depended on many variables across very diverse markets on separate continents and was in any case dependent on the actions of a very large number of intermediaries who handled its distribution. India's indigo was both privately and publicly auctioned, not only in Calcutta and London, but also in many other national and international markets including Rotterdam, Le Havre, and Marseilles. Both Calcutta and London markets were also visited by continental buyers and buyers from America. These separate markets were both embedded in local economic currents and connected with each other. Perhaps the substitutes had been able to exercise a calming influence on the speculation in indigo trading and the high volatility in its prices. But the verdict of most seemed to be that agricultural indigo had successfully survived

---

[30] Anthony Travis, *Rainbow Makers*, pp. 72–3, 78–80, 131–4.
[31] A later submission to a textile journal mentions this fact. "Wool Bleaching. By Walter M. Gardner," *Textile Manufacturer*, June 15, 1893, p. 267.
[32] "Wool Bleaching. By Walter M. Gardner," *Textile Manufacturer*, June 15, 1893, p. 267.
[33] "Indigo v. Alizarin Blue," *Journal of the Society of Dyers and Colourists*, May 1896, pp. 96–7.
[34] "Substitutes for Natural Indigo in Wool Dyeing and the Artificial Indigo," *Textile Manufacturer*, April 15, 1898, pp. 153–5.

the challenge posed by substitutes. This had as much to do with indigo's superiority as a dye as with the fact that the marginal displacement of natural indigo by the hydrocarbon-based blues was more than compensated by the rising demand for natural indigo in the world. Thus, as late as the late nineteenth century, natural indigo had retained its position as the favored dye of choice for commercial consumers. The real material change would only occur with the arrival of synthetic indigo in the closing years of the century.

Launch of Synthetic Indigo and the Early Assessment of Competition

The quest for the synthesis of indigo was first launched at the Gewerbe Institut in Berlin at the behest of the eminent German scientist Adolf Baeyer (see Fig. 10) in 1865. The interest in finding synthetic substitutes for natural dyes was not an uncommon one at this date. Synthesis of colorants was a promising target for scientific achievement in this time of rising involvement of chemists with industrial production. After the successful fabrication of alizarin, indigo remained the last major holdout among natural dyes against the march of synthetics. Besides, commercial motives were available aplenty for funneling efforts toward the synthesis of indigo. Carsten Reinhardt and Anthony Travis have pointed out the important fact that between 1880 and 1896 the average annual value of natural indigo totaled 80 million German marks, which was same as the combined turnover of all European coal tar factories. This financial incentive provided a broader context for the efforts to synthesize indigo. That such an effort would receive most traction in Germany was also inevitable given that Germany begrudged more than any other European nation the monopoly established by British planters, shippers, and merchants over the supply of indigo from India.[35]

The search for a viable pathway for synthesis of indigo starting from a cheap source and involving a simple, controllable process was a long one. Baeyer's indigo research had established quite early on that indole was the source of the blue dye, and he was even successful in synthesizing indigo from it. But at this point the only source of indole was indigo itself. After advancing this far in 1865, Baeyer abandoned his indigo work for other experiments. He returned to the indigo work only in the 1870s in

---

[35] Carsten Reinhardt and Anthony Travis, *Heinrich Caro and the Creation of the Modern Chemical Industry*, pp. 187–8.

FIGURE 10. Portrait of Adolf von Baeyer, Photograph by Friedrich Mueller. Courtesy of the Chemical Heritage Foundation Collections.

partnership with Heinrich Caro of BASF. In 1880 Baeyer achieved the first synthesis of indigo from toluene. The high price of toluene was a problem in scaling up the process for industrial manufacturing. Although a patent was secured and a product in the form of propiolic acid was launched by Badische, the prospects for marketing this expensive synthetic indigo were minimal. Aside from the fact of its high cost, dyeing and printing with this indigo involved an inconvenient and indirect method of application.

But the ultimate success of synthetic indigo now seemed more like a realistic goal than ever before. The German dye companies were very encouraged with Baeyer's success. He won lucrative contracts from both Badische and Hoechst to continue his work on indigo. Heinrich Caro was even optimistic that the price of toluene could be brought down through development that might make the current brand competitive. Success continued to elude Baeyer and his associates in Germany who were trying to establish a viable pathway for the synthesis of indigo. It was finally Karl Heumann in 1890 at Zurich Polytechnic who suggested a new route for making indigo starting from two plentiful and cheap industrial sources, benzene and naphthalene. Both BASF and Hoechst immediately purchased the two processes. The first pathway was found to be unworkable because the output with the process was too deficient for industrial production. The second pathway seemed to meet all criteria for successful commercialization. But it still took another seven years for the naphthalene-based pathway to be perfected at the Badische Company. BASF was finally able to produce synthetic indigo commercially in February 1897.

The BASF synthetic indigo reached English markets that summer and immediately sparked an assessment of the nature of competition that the synthetic product might offer to the reigning product. The early report by the respected traders Schonlank, Engel and Co. appearing in the primary journal of dyers in Yorkshire noted that the synthetic's introduction had created "a profound sensation in the Indigo trade," but it still expressed doubt that the synthetic posed any immediate threat to plant indigo.[36] It averred that the synthetic was about 20 to 25 percent dearer than natural and that, in addition, it lacked the constitutional, inherent properties of natural indigo that allowed the pigment to fasten strongly on fabric. In any case, the synthetic's utility to the practical dyeing process was not proven yet. Under such circumstances it concluded that most dyers would stay with the current product in use while by and large neglecting the new substitute.

The report from Schonlank, Engel and Co. also provided a window into conflicting assessments of the prospect of competing products by separate constituents of the market. The same report also criticized London traders for offering very low prices for natural indigo in Calcutta for the season and thought that they were being "gullible" in thinking that the synthetic had fundamentally altered the balance between supply and demand. The buying price offered by London brokers in Calcutta was 5 d. to 6 d. lower

---

[36] "Indigo Report," *Journal of the Society of Dyers and Colourists*, January 1898, pp. 15–16.

than the previous year's price, which, in its opinion, was "going too far" in terms of undervaluation of natural indigo's worth. It reasoned that the planters were likely to hold on to their commodity rather than sell at such a "ruinous" price. Their prediction was based on the knowledge that the current year's output of indigo in Bihar was exceptionally small, the kind of assessment that had served the market players very well in the presynthetic era. The price of natural indigo was known to fluctuate annually on the basis of output every season, rising when production declined after a bad season since there was no alternative to plug the shortfall in supply, and falling whenever many good seasons followed in succession and caused a glut in the market.

Two years later the indigo importers Parsons and Keith wrote to the Bihar planters that the future of plant indigo was still secure. They were responding to a letter from the previous president of the Bihar Indigo Planters' Association, Sir W. B. Hudson, who was wary of the impact of BASF indigo on natural indigo trade. The compilers of the report illustrated the continuing loyalty of consumers to agricultural indigo even though they acknowledged that the new synthetic substitute was "the most serious competitor of natural indigo" to date. Alluding to the inertia in the natural's use, they maintained that the dyers would need to be convinced of "some very material advantage" before giving it up. But the fact that even in late 1899 the synthetic was "dearer to use at its present price than [natural] indigo" belied any such possibility. Although the synthetic had reportedly been tried out by a wide section of dyers all over the world, reports about its merit were mixed. To underscore prevarication by users the trade document highlighted the fact that even Austria and Germany, nations where synthetic had been generally assumed to have superseded the agricultural product, imported indigo in large volumes from Calcutta in the current season. It assured the planters that trade statistics did not indicate any falling off in the consumption of natural indigo in the English markets. In fact, in recent times the consumption of natural indigo seemed to have exceeded supply, as indicated by the low level of stocks held in London.[37]

But there were others who nursed an apprehension that the synthetic would have "a very serious effect" on the natural indigo trade. The brokers Mewburn and Ellis were among those who perceived the superiority of synthetic indigo in the criterion of usability. Synthetic enabled dyers to "dye to a particular shade." It was constant in composition and

[37] "Artificial v. Natural Indigo," *Chemical Trade Journal*, October 28, 1899, pp. 286–7.

was "more easily handled by the dyer." User trials had revealed that the fastness of synthetic was as good as if not better than natural's. It also predicted that the price of synthetic was likely to decrease progressively as had been the case with all other coal tar dyes launched previously. And since the brokers thought that the cost of production of indigo in India could not be reduced, they were convinced that the natural was likely to be forced out of the market in near future.[38]

Such assumptions about the inability of natural to compete on the basis of price with synthetic were contested by the interlocutors of natural indigo in the London market as well as by the planters themselves. The indigo entrepreneur and scientist Eugene C. Schrottky, who was vacationing in London at the time, was one such optimist. He thought that the report by Mewburn and Ellis was "wrong" in assuming that the cost of production of natural indigo could not be lowered. In fact, he was convinced that the cost would decrease adequately to allow natural to compete successfully with BASF indigo on price. As someone who had lived on the plantation tracts in colonial Bihar, he had seen the recent winds of change. While earlier the traders were lukewarm to trying innovations in their methods of cultivation and manufacture, more recently the planters had rallied to embrace science. The adoption of scientific methods was bound to have a salutary impact on the dye's price. Schrottky explained that with current processes not more than 30 percent of the dye was extracted from the plant and that a further waste of 20 percent occurred during manufacturing. Thus there was "an ample margin for improvement" in manufacturing alone through application of scientific methods. Besides, additional gains in productivity were likely to be achieved given the current sentiment in favor of adopting better cultivation practices, improved irrigation, and "selection" of seeds, all of which would reduce its final cost.[39]

The public debate over victory and survival of agricultural indigo unfolded internationally in keeping with the global nature of the indigo trade. Thus, in October 1900, a little more than three years after the launch of BASF indigo, the managing director of BASF, Heinrich Brunck,

---

[38] "Artificial v. Natural Indigo," *Chemical Trade Journal*, October 28, 1899, p. 287.

[39] "Artificial v. Natural Indigo," *Chemical Trade Journal*, October 28, 1899, p. 287. Eugene Schrottky had recently returned to England to get married. Church records show that he married in December 1898 at Christchurch, Hampshire. Schrottky was, of course, a major shareholder in the £150,000 Bengal Indigo Manufacturing Company registered in Manchester in 1889 that had several factories in Tirhut. Public Record Office, London, BT 31/4628/30398/100052.

announced the impending demise of natural indigo in colonial India. He and Adolf Baeyer had been invited to speak on the occasion of the inauguration of Hofmann House in Berlin. It was a commemorative moment, one involving celebrating one of the dye world's most distinguished innovators, August Wilhelm Hofmann, whose career spanned both Germany and Britain and who was acclaimed among eminent chemists everywhere. In keeping with the spirit of celebration, the two speakers chose to recount the history of indigo synthesis. While Baeyer deliberated primarily on the scientific side of the development of indigo, Brunck focused on the business side. Brunck attributed the successful commercialization of synthetic indigo at the Badische Company to major capital investment and "scientific labour." As much as £900,000 had been spent on the indigo project at BASF. A total of 152 patents had been registered on indigo in Germany alone to secure its proprietary use. The BASF was already producing synthetic indigo in factories in a quantity that could only be produced by cultivating more than a quarter-million acres of land under a crop of indigo. He had no doubt in his mind that the agricultural system of indigo was doomed. He predicted that the indigo plantations in colonial India would be correspondingly rendered useless as the production of synthetic rose. This change was imminent, and in that light he advised the colonial government in India to turn over the land currently under indigo to cultivation of crops for "breadstuffs and other food products," a prudent step in his opinion in a land where food shortages and famines were rampant. This way the colonial state could play a positive role in the transition and thus mitigate the effect of the demise of agricultural indigo, which was on the anvil.[40]

Brunck's comments in Germany immediately made their way into Britain and joined the cacophony of consternation over the future of indigo already under way in the British metropolis. The audacity of Brunck's counsels was hardly lost on the natural's supporters. Some responded to the arrogant suggestion from Brunck by emphasizing the claims of natural indigo for trade protection by the national government. Some made explicit calls to prevent the defeat of the imperial product by a foreign commodity. For F. Mollwo Perkin, the son of William H. Perkin, the prospect of victory for synthetic indigo was deeply troubling. Between the end of 1900 and 1901 he wrote a series of articles reflecting on the contest between natural and synthetic in *Nature*. To him the

---

[40] "The History of the Development of the Manufacture of Indigo, H. Brunck, Ph.D.," *Journal of the Society of Dyers and Colourists*, June 1901, pp. 158–63.

victory of synthetic seemed to represent a larger trend in which Britain had conceded to Germany its prior leadership in the coal tar dye industry, "an industry which received its birth in this country, but which has now taken up its abode on the continent." Perkin had a personal connection to this sense of loss as the son of the mauve manufacturer whose invention in Manchester of 1856 had initially set off the coal tar–based color industry in England.[41] Herbert Levinstein of Levinstsein Limited at Blackley, Britain's leading dye manufacturers, evoked fears of an impending German monopoly in indigo while reminding the dyers and printers of Yorkshire how the establishment of monopoly by Germans in the red dye alizarin had affected their trade. "History shows us that if the control of the indigo market should really fall into the same hands ... dyers and printers alike would ultimately find themselves at their mercy." Potentially, he warned them, they could be dictated terms by the German combinations of color makers, who were "similar in character and with similar aims to the great steel trust in America." He hoped that, "whether it be from motives of self interest or from the lofty vantage point of Imperial patriotism," every Yorkshire dyer and printer would give preference to natural indigo, "the product of our greatest dependency, India," against synthetic indigo.[42]

Many wished to draw larger lessons for uplifting national science and industry in Britain out of the building narrative of the indigo fiasco. They were more concerned about emphasizing the need for science in British manufacturing and the place of Britain in the larger comity of nations. Two prominent public intellectuals made a deliberate mention of Exposition Universelle at Paris in 1900 and the nature of manufactures from Germany on display there to make the point that Britain was slipping from its leadership in industrial manufacturing. This was not evident anywhere more than in the field of the coal tar industry, and the story of German success in indigo reflected this broader change. Science was the obvious answer if this sliding fortune were to be checked.

Ivan Levinstein, the dye manufacturer, stood for the protection of Britain's imperial and national interests. Even if synthetic offered some advantages, it had already inflicted serious injury "to native labour, to the Indian planter, to our shipping and forwarding trade, and to our

[41] F. M. Perkin, "The Present Condition of the Indigo Industry," *Nature*, November 1, 1900, pp. 7–9.

[42] Herbert Levinstein, "The Future of the Indigo Industry, with a Description of the Manufacture of Indigo from Naphthalene," *Journal of the Society of Dyers and Colourists*, June 1901, pp. 138–42.

merchants." Levinstein used the indigo issue further to back his call for taking steps to create a more effective national patent system that would effectively safeguard the interests of industries in Britain. He had long been an advocate for the reform of British patent laws. He tried to rally the anxiety over synthetic indigo to make policy makers and politicians "fully appreciate the folly of our patent laws." A report on the products of German chemical industries in Paris attributed the success of those industries to that nation's patent laws, which deliberately favored German manufacturers against all foreign competition. In contrast, as a case in point, Levinstein alluded to the inadequacy of English patent laws, which had given Germans a monopoly in the production and sale of artificial indigo in the British national markets. In his opinion the patent laws should have at least secured a mandatory manufacture of the article on the British soil even if the Germans held patents on its technology.[43]

The Paris Exhibition also became a referent in the indigo account of the eminent British chemist Raphael Meldola. Speaking at the Royal Society of Arts in April 1901, Meldola recalled that, speaking at the same society fifteen years ago, he had highlighted the unfortunate fact that the coal tar dye industry, a British invention, now based itself "on the banks of the Rhine and in other parts of the German Empire." What was said earlier of German dominance of coal tar dye manufacturing could be repeated "with increased emphasis" in 1901, judging by the range of coal tar products displayed by Germany at the Paris Exhibition. He thought it important to analyze the Britons' "national weaknesses" that had led to surrendering the lead in the dye trade to Germany. He maintained he had seen a long-term apathy to science among British manufacturers, who were known to turn to science only in adversity. The same "negligence" of science was apparent in the case of indigo planters, who were accepting science only now that synthetic indigo had begun to threaten natural's existence. Referring to the development of synthetic indigo over the past two decades in Germany, he accused the planters of having "allowed twenty years of activity on the part of the chemists [working to synthesize indigo] to pass by with apathy and indifference."

F. M. Perkin was bemused while citing a report on Badische Company, the manufacturers of BASF indigo, which had evidently employed 148 chemists and 75 engineers at their Ludwigshafen plant. He wondered whether there were that many chemists employed altogether by

[43] Ivan Levinstein, "Indigo and Patent Laws," *Journal of the Society of Dyers and Colourists,* January 1901, p. 28.

manufacturers all over Britain.[44] The same figures on employment of scientists by BASF were used to paint the indigo planters as conservatives who apparently invested relatively very meager resources in science. This was an unfair comparison involving two very different systems of production, and thus criticisms based on them were also misplaced. The point-to-point comparison on scientific resources, investments, and tools of production partly represented the rhetorical response of zealous nationalists to what they perceived as the apparent surrender of Britain's lead in indigo manufacturing to Germany. In part these also reflected a lack of awareness about indigo plantations or the actual steps taken by the planters in the far-off colony. Thus, referring to the deep antiquity of the production of agricultural indigo and an assumed changelessness in its culture, Meldola wondered whether "the methods of cultivation and extraction in India [are] very different now to what they were in the time of Pharaohs."[45] Such assessments missed the rhythm of change and innovation on the plantations, which followed its own pace and evolved with reference to its own dynamics, ones that have been discussed earlier.[46]

Meldola's criticism of indigo planters also involved a progressivist debunking of the potential and prospects of natural indigo. He agreed that it was still too early to say whether natural indigo would survive. But ultimately he also believed that the standoff between natural and synthetic indigo represented a conflict between "an industry of venerable antiquity, carried on by empirical methods" and another based on "modern science." Meldola's views were colored by his location on the side of what he considered to be "modern" science. From the latter's perspective, he conjectured, natural's prospects might be "gloomy" because "the laws of Nature may be against them." In perspectives like these that saw the transition as a modernist trend, the "conservative" planters were construed as standing in the way of the forward march of modern science.[47]

But other than the likes of Meldola, who believed that the synthetic would rightfully win through adjudication by the market, and those who believed in the desirability of the survival of natural, there were also other

[44] F. M. Perkin, "Indigo and Sugar," *Nature*, May 2, 1901, pp. 10–11.
[45] Raphael Meldola, "The Synthesis of Indigo," *Journal of the Society of Arts*, April 19, 1901, pp. 397–412.
[46] Beginning in 1898, the planters also launched a major effort involving the use of laboratory science for improving the yield and quality of natural indigo. These steps are the subject of study in the next chapter.
[47] Raphael Meldola, "The Synthesis of Indigo," *Journal of the Society of Arts*, 49 (1900–1): 397–412.

participants in this unfolding transnational public debate who were optimistic about the potential of natural indigo to ward off market competition on its own merit. The major indigo buyer based at Calcutta, Jules
Karpeles, explored the cost and profitability of synthetic's manufacturers in order to advise planters on a competitive price for their product.
Karpeles's analyses first appeared in several Indian newspapers and were
republished in England and discussed by the avowed sympathizers of the
plantation industry.[48] Karpeles calculated that the average cost of production of synthetic indigo stood at 2s. 6 d. per pound, which was equivalent to the cost of producing natural at Rs. 120 per *maund*. While it
is not clear how Karpeles had arrived at the figure for synthetic indigo,
the effort, even if it was of a speculative nature, was meant to clarify the
nature of price competition. It was aimed at giving some pointers to the
planters as to how much lower they would have to reduce their costs in
order to compete successfully with synthetic indigo.

Reports like Karpeles's also tried to raise optimism among consumers in Britain that the natural indigo industry would hold on its own.
Karpeles announced that the Indian plantations were likely to turn out
112,000 *maunds* of indigo in the coming season and called upon the
consumers to look forward to Calcutta sales. The report also included
the views of other planters expressing confidence that they could further
rationalize the production of natural indigo. One anonymous planter was
quoted as saying that he had reduced the cost of his indigo operation by
getting rid of the more expensive landholdings. He expected to produce
indigo at the rate of Rs. 100 a *maund* (equivalent to 1s. 9 d. per pound),
which was far below the "customary" cost of cultivation between Rs. 125
and Rs. 150 per *maund*. These savings were merely the result of effective
land management. The report added that other "capable" agriculturists
had held practical trials to secure 70 percent to 100 percent increase
in color output through crop rotation, use of chemical fertilizers, and
deployment of better manufacturing processes. One had reckoned that
in his 1,600-*bigha* factory he could produce 440 *maunds* of indigo at the
total cost of Rs. 34,200 over a season. This translated into a very competitive cost of Rs. 80 a *maund*. Another planter, L. J. Harrington of the
Bhagwanpur Indigo Factory in Bihar, wrote in the *London Times* that
the planters could realistically produce agricultural product at Rs. 100 a
maund. He made allusions to the engagement of chemists by the planters

---

[48] "Cost of Indigo Manufacture," *Journal of the Society of Dyers and Colourists*, March
1901, pp. 74–5, republished from *Chemical Trade Journal*, January 1901.

and expressed optimism that the experts would assist the planters in turning out their product at a price that the synthetic would find "impossible" to beat.[49] Such claims were received with measured hopefulness among those buyers and agents in London who were sympathetic to the cause of natural indigo.[50]

Synthetic Begins to Make Serious Dent in Natural's Share: The Case of the London Market

Synthetic indigo began to make inroads into the London market at the turn of the century as some of the users' initial reservations against the new product faded. The claim that some of the "impurities" in the natural dye facilitated fastness did not hold long as competitors organized comparative test trials and advertised the results, which showed fastness from the synthetic to be no different. The privileged position sought by natural on the basis of the claim that it provided a unique red sheen lost its relevance as German manufacturers were able to synthesize "indigo red" and began adding it to the synthetic product. The arguments about "uniqueness" of natural indigo were losing traction with mainstream dyers and printers except those serving niche markets. One report emphasized the growing irrelevance and seeming hollowness of arguments based on quality for the mainstream market, concluding that "the competition between natural and artificial indigo is very quickly resolving itself into a matter of price."[51]

The natural had achieved some sort of equivalence in price with the synthetic at the turn of the century, but continued parity proved elusive. The planters always complained that they had no knowledge of the price at which synthetic was offered to consumers because the manufacturers kept it a secret, and the final price paid by a dye house was flexibly worked out between the buyers and the agents of synthetic companies. There were indications that the price of synthetic indigo was declining consequent to innovations and savings in its manufacturing costs. The quest to capture markets caused further reduction in the price of synthetic indigo as more companies manufacturing indigo started trying to outbid each other. By late 1901, in addition to Badische Company, two

[49] Cited in F. M. Perkin, "The Present Condition of the Indigo Industry," *Nature*, January 24, 1901, pp. 302–3.
[50] "Cost of Indigo Manufacture," *Journal of the Society of Dyers and Colourists*, March 1901, pp. 74–5, republished from *Chemical Trade Journal*, January 1901.
[51] "Indigo Prices," *Chemical Trade Journal* 29 (July-December 1901): 187.

other German companies, MLB Hoechst and Bayer & Co., as well as their subsidiaries from France were sending synthetic indigo to the markets in England. As a trade report from late 1901 indicated, BASF strategically offered attractive prices to underbid natural indigo as well as other synthetic indigos. In London the average price of natural indigo had stood in the vicinity of 2 s. 11 d. in December 1900 and rose to 3 s. 8 d. per pound in March 1901. Amid such prices for natural, the report mentioned, Badische announced that it would make available its brand of 20 percent paste of BASF indigo at 1 s. 2 d. per pound. Such a price was to be offered only for "large contracts."[52] Given such fluidity on price matters it is not surprising that many planters were befuddled over the question of where the competition stood or at best found the competition's price to be a moving target.

Comparison of price by purchasers now involved subjecting agricultural indigo to "valuation" in order to determine the actual percentage of color contained in the dye. The claims that synthetic was indigo in its purest form translated into a reductionist assessment of natural indigo as indigo and "else." The other constituents present in natural indigo might or might not aid in the task of dyeing, and if they were assessed to be irrelevant, they did not merit any price paid for them. Agricultural indigo was acknowledged to contain moisture, mineral matter, and indigo gluten, in addition to "indigotin" or indigo and indirubin or indigo-red. Only the latter two were considered to be "active principles" useful to dyeing. While difference of opinion raged about the utility of other constituents, the market increasingly favored the practice of quantitative tests for assessing the percentage of indigo in the natural dye for settling its value.[53] The partisans on the side of artificial indigo had seemingly won the argument that buyers should pay only for the active principle and not for the "impurities." The older practice of buying "on looks" gave way to buying indigo on the basis of percentages, and administration of objective tests was common to measure the percentage of "indigo" in the commercial dye.[54] The buyer considered the percentage of color in plant indigo and the declared 20 percent color in synthetic indigo paste while comparing prices.

---

[52] "Synthetic Indigo: Competition with the Natural Dye," *Journal of the Society of Dyers and Colourists,* December 1901, pp. 301–2.

[53] See references to the different tests and procedures for valuation of commercial indigo: "Valuation of Indigo," *Textile Manufacturer,* June 15, 1895, p. 236; "The Valuation of Indigo," *Textile Manufacturer,* November 15, 1899, p. 437.

[54] "Artificial Indigo," *Textile Manufacturer,* May 15, 1893, p. 220.

But ultimately the purchase decision by consumers depended upon how much precisely the dyer or the printer could color with either of them in practice. The consumers' response in the market was based on the ultimate calculation of how much and how well he could dye or print with natural and synthetic indigo. In other words, the consumers' purchase decision was based on a consideration of both shelf price and usability. Along these lines some even proposed that practical dye tests rather than percentage determinations were the more accurate way to assess the value of indigo in the market. The manufacturers would often go beyond the issue of shelf price to consider additional work-related contingencies such as whether purity made the production process smoother and speedier and enabled flow of larger volumes. In the case of textile manufacturers who were producing for a niche market, the abiding loyalty to natural could result from customer tastes and the latter's willingness to pay any price for the "real" indigo.

The indeterminacy of debates over price and quality aside, indications were beginning to emerge from the market that natural indigo was losing ground among consumers. The trend of increasing consumption for synthetic indigo was undeniable. According to one report the import of synthetic indigo into Britain rose from 2,760 chests in 1898 to 6,850 in 1899, 11,000 chests in 1900, and 14,000 chests in 1901.[55] Synthetic indigo had been "extensively" adopted in Lancashire in print work. The dyers in Yorkshire, the slubbers in particular, had also moved to the synthetic en masse.[56] The London agents in Calcutta were probably responding to these signals as they refused to pay a higher price to the planters during Calcutta sales in 1902. A Calcutta correspondent of an Indian newspaper reported that public sales there were cancelled as a result of "differences of opinion as to what indigo is worth or for what it ought to sell." The planters were asking for an advance of thirty rupees over the previous year's price while the sellers were not willing to pay more than fifteen rupees over the last year's price. This trade report justified the stance by London buyers saying that the synthetic in London was selling for the equivalent price of 3s. and 2 s. 9 d. while the price being offered in

[55] "Natural Indigo: Its Present Position," *Journal of the Society of Dyers and Colourists,* February 1902, pp. 45–6. The article incorrectly makes the case that the indigo had still held its own against such foray by the synthetic substitute, neglecting the supply side reality that the planters were truly hard pressed in order to be able to compete with synthetic on prices and maintain their customer base.

[56] "Synthetic Indigo: Competition with the Natural Dye," *Journal of the Society of Dyers and Colourists,* December 1901, p. 302.

Calcutta amounted to 3 s. 5 d. In light of such comparisons the reticence of indigo planters was "puzzling." The planters found their quotation of a higher price justifiable because the production of indigo in the current year had lagged behind last year's and the prospect for the next crop was dismal. In the end, however, the planters came around to selling at a lower price after buyers made "an unanswerable case for not meeting planters' desires." This state of affairs represented the new reality of the postsynthetic era, in which the natural producers' bargaining position was weakening.[57]

The quarterly sales for natural indigo at public auctions in London continued to emit ominous signs through the course of 1902. The report on January sales that year pointed to the "inaction of principal consumers" that had probably turned to synthetic indigo.[58] The report on July sales announced that the price of natural indigo had declined to 2 s. 10 ½ d., which was "the lowest figure yet recorded." The report by Millward and Company, the leading importers of indigo, found it astonishing that even at this lowly price demand for natural indigo was not forthcoming and thus concluded that "apparently some consumers are giving a preference to artificial substitutes, notwithstanding the superiority of the natural dye [in their opinion]."[59] There is no denying the fact that the introduction of synthetic indigo had led to a slide in the price of natural indigo. Synthetic indigo had started to set the bottom line in the price for indigo. And to be able to compete with the synthetic product the planters were forced to sell their indigo at a much lower price. As the Calcutta correspondent of the *British Trade Journal* explained, the planters were finding it difficult to produce indigo at a profit as they tried to match synthetic's price offerings.[60] Many planters in the colony were divesting, a trend that led to a reduction in the acreage of indigo in Tirhut and its export out of India. The statistics for the years leading up to 1901–2 in Table 6 illustrate the declining quantity of Indian indigo supplied to London. They also reveal the decline in total value of Indian indigo to be much sharper than the cutback in the quantity sent to London. Decline

---

[57] "Indigo Prices," *Chemical Trade Journal* 30 (January–June 1902): 5; "Natural Indigo: Its Present Position," *Journal of the Society of Dyers and Colourists,* February 1902, pp. 45–46.

[58] "The Indigo Market," *Chemical Trade Journal* 30 (January-June 1902): 153.

[59] "London Quarterly Indigo Sales, July, 1902," *Chemical Trade Journal* 31 (July–December 1902): 43.

[60] The *British Trade Journal* report was republished as "Natural Indigo," *Journal of the Society of Arts* 51 (1903): 727–8.

TABLE 6. *Quantity and value of natural indigo into London, 1896–1902*

Year	Quantity (cwts.)	Value (rupees)
1896–7	169,523	43,707,570
1897–8	133,849	30,574,019
1898–9	135,187	29,704,781
1899–1900	111,420	26,925,107
1900–1	102,491	21,359,808
1901–2	89,750	18,522,554

Note: (1 cwt or centumweight = 112 lbs.)
Source: *Journal of the Society of Dyers and Colourists,* January 1903, p. 19.

had set in in the natural indigo trade, as evident in numerous parameters including the unwillingness of consumers to pay a higher price for natural indigo, lower prevailing prices, reduction in profitability, and divesting by planters.

Season 1902–1903: "The Crop under Review Is Smallest on Record"

The indigo industry of eastern India suffered its sharpest decline in output over the season of 1902–3. The output of 43,127 *maund* was the smallest from Bengal in memory, a low figure that that made a reviewer in the planters' primary trade journal call it the "smallest on record."[61] Reports from London appearing in the same journal also noted that "shipments to the United Kingdom look like being less than half of the previous smallest on record."[62] The official report of maritime trade later confirmed that in the 1902–3 season, the export of indigo out of Bengal had "declined seriously, both quantity and value falling by between 46 and 47 per cent."[63]

The signs of an industry under duress were evident especially in the fact that despite low acreage to begin with and a slippage in output on account of bad climatic conditions, the price in the terminal market did not rise correspondingly. The production of indigo was historically

[61] "Season 1902–1903," *Indian Planters' Gazette,* March 7, 1903, p. 373. In reality, the year's shipment amounted to the smallest figure since 1878–9. But it seemed serious enough to many in the market to evoke extreme reactions.
[62] Millward and Co. Report, *Indian Planters' Gazette,* March 7, 1903, p. 373.
[63] For the Report on the Maritime Trade of Bengal, *Indian Planters' Gazette,* June 13, 1903, p. 821.

TABLE 7. *Indigo output in Bengal, 1894–1903*

Output Year (based on crop previous year)	Total in *maunds* (1 *maund* = 33.868 kg)
1894	116,829
1895	160,534
1896	161,698
1897	158,923
1898	110,212
1899	124,474
1900	86,825
1901	111,670
1902	85,073
1903	43,127

Source: *Indian Planters' Gazette*, March 28, 1903, p. 475.

known to fluctuate wildly (Table 7), reflecting climatic variations and even cycles of boom and depression in patches, but the additional impact of competition from the synthetic was writ large on the 1902–3 season. The reports by J. Thomas and Company of Calcutta pointed out that the season's sales in Calcutta started with low prices offered by buyers, but gradually prices rose marginally as the season progressed. The high prices were obtained especially for good and middling varieties. The price of good quality indigo rose by about Rs. 5 a *maund* over last year's price at the start of the season and ended up to Rs. 15 to Rs. 20 higher. The "common" varieties were sold at par, whereas those of low quality were difficult to sell even at a discount. It was the surge in demand from Suez (Egypt) and Japan for high-quality indigo and the American demand for the middling variety a little later in the season that caused prices to remain firm in a context of extremely short supply.[64] But those assessing the market were worried that even in a situation of record low output, the prices had not risen by as much as they were expected to rise. The report found "the serious competition of the Synthetic Indigo" reflected in the lower prices for indigo paid in Calcutta.

At the terminal market in London trade in indigo again remained depressed. There was wide expectation that, given the very poor crop in the colony and the demand in Japan and Suez, very little of the crop would go to the London markets, where prices were low. True to that

---

[64] J. Thomas and Co.'s Indigo Report, *Indian Planters' Gazette*, February 14, 1903, p. 253; February 28, 1903, p. 333; March 28, 1903, p. 475; April 11, 1903, p. 536.

expectation a very small consignment of indigo went to London. This measly quantity found buyers although there were disappointing reports to the contrary of "meager buying" by the major combinations of textile manufacturers who were by and large the erstwhile biggest consumers of indigo. The monthly report of Millward and Company for January described the market as characterized by "small supplies, steady demand, and uniform price."[65] The April report again emphasized that because of very low supply, the relatively high prices for indigo were likely to prevail. Others reported that "business was restricted by the firmness of holders" who delayed selling as they expected prices to go up as a result of low supplies from Bengal.[66] According to a July report these trends of very low trade in natural indigo remained in effect until the end of season.[67]

Competition extended its somber impact on cropping and prices in the next season as well. Importers in Britain were reporting early in 1903 that climatic conditions were better, and therefore a higher output and a better quality of the dye were expected. But they also indicated that the area under crop remained small compared to the years in the past when the indigo trade was in a healthy state.[68] Both trade spokesmen and the colonial government linked the reduction in cultivation of indigo on the subcontinent to competition from the synthetic. The official report from the Statistical Department emphatically stated that "the contraction in area is entirely due to the competition of natural indigo with the synthetic product in the markets of Europe and America."[69] The cutback led to estimates of about 65,000 *maunds* of indigo in the season, far lower than the previous twenty-year average of 125,000 *maunds*. And yet the advance over last year's output was inviting estimates in the market that the price of indigo was going to be twenty rupees lower than last year's, a new low that would have further dampened the spirit of natural's producers.

---

[65] "The Indigo Market – Monthly Report," *Indian Planters' Gazette*, February 7, 1903, p. 215; "Messrs. Millward and Co's London Quarterly Indigo Sales," *Indian Planters' Gazette*, April 18, 1903, p. 568.

[66] "East India Indigo Sales" of H. W. Jewesbury and Co., *Indian Planters' Gazette*, May 16, 1903, p. 691.

[67] "East India Indigo," report by Messrs. Lewis and Peat, *Indian Planters' Gazette*, August 22, 1903, p. 253.

[68] Messrs. Millward and Company, London Quarterly Indigo Sales, April 1903, *Indian Planters' Gazette*, May 16, 1903, p. 692.

[69] J. Thomas and Co.'s Price-current, dated Calcutta, October 15, 1903, *Indian Planters' Gazette*, October 17, 1903, p. 511; "Final General Memorandum on the Indigo Crop of the Season 1903," *Indian Planters' Gazette*, November 7, 1903, p. 609.

The desertion of indigo cultivation by a good number of planters transpired in Bihar in a generalized state of sentiment of gloom about its future as low prices and weak demand continued to prevail in the traditional belt of consumption in Europe. Synthetic by now started to capture emerging markets of new textile manufacturing nations in addition to dominating the old ones. The glimmers of hope provided by countries like Russia and Japan and later Egypt and China for brief times were only transitory. These nations filled in the demand for a year or two before turning to synthetic indigo. A report thus appropriately titled "A Depressing Outlook" appearing in Bihar's key newspaper compared the rapid and continuous progress of the synthetic industry with the enormity of the task of beating synthetic in the market. It explained that the planters were being called upon to fight competition from a disadvantageous position of diminishing capital. Besides they were facing a series of bad seasons. The report wished that the planters were enthused about taking joint industrywide action and trying drastically new business strategies to overcome their difficulties.[70] The *Gazette* highlighted the case of a planter at the Serryah factory in Tirhut who had abandoned the plantations and was on his way out of India to try his luck in Canada. This report argued that under the circumstances the decision of the planter was a logical one. Indigo had become "the most thankless industry" for employment except in cases when one might be attached to a prosperous concern. The closing of one of the oldest indigo factories at Pupri similarly occasioned another prediction of doom: "The fact remains that there will be more concerns closing, and more wholesome [*sic*] ruin to planters during the next two years."[71] Those planters who were pessimistic in the extreme opted out of plantation business. But many others stayed on. One option they tried was to revamp their plantations completely.

"The Bihar Planter Has Completely Reorganised His Ways and Means of Working"

A good number of planters remained in the indigo business regardless of the sense of gloom. These remaining planters believed that the current phase of decline in the industry could be arrested. Some even thought

[70] "A Depressing Outlook," *Pioneer,* republished in *Indian Planters' Gazette,* May 23, 1903, p. 724.
[71] *Indian Planters' Gazette,* September 26, 1903, p. 417; *Indian Planters' Gazette,* November 14, 1903, p. 639.

that the synthetic manufacturers would be forced to raise their price from the current unsustainable levels, which had been deliberately kept low in order to force natural indigo out of the market. Thus all that they needed to do was to hold their own a little longer. Others believed that some demand would always exist for the natural no matter how much the consumption of synthetic expanded. Such hopefulness found traction in the early twentieth-century context of expansion of textile manufacturing in the newly industrializing nations and a steady emergence of markets that were waiting to be tapped. Besides, extricating oneself abruptly from the indigo business was not an easy task except at the cost of complete financial ruin to the individual investor. Contractual obligations to leases on land whose terms ran over several years had to be redeemed, soft loans on five- to ten-year terms given out to native landlords had to be recovered, and the fixed stock of buildings and machinery had to be sold to recoup maximum recoverable capital for the investor. These other grim financial realities steeled the planters to protect their investment in natural indigo.

The planters undertook a variety of measures. As a business question, the challenge was to keep the plantations profitable while selling natural at a price that would be competitive with synthetic indigo. High profit margins of the old days gone, the planters began to make efforts to reduce the cost of production of agricultural indigo. The planters launched a search for a higher-yielding variety of indigo and hired experts to improve yield during manufacturing. When indigo-specific efforts did not produce immediate results, the goals turned to improving plantations' profitability through better land use and through growing more remunerative crops so as to "subsidize" the culture of indigo. All of these efforts were launched early and represented rational responses within the realm of plantation agriculture.

The option of diversifying indigo plantations was considered early by Bihar planters. In a note of January 1900 the planter Rowland Hudson laid out a plan to combine the manufacture of cane sugar with indigo in order to ensure maximum utilization of his plantation lands. Hudson's plan was to use the existing storage facilities, steam engines, and boilers at his indigo factory to combine the manufacture of blue dye with cane sugar. Usually an indigo factory comprised the manufacturing unit surrounded by huge tracts of land for cultivating indigo. By alternately growing indigo and cane on different parts of the landed estate, Hudson wished to keep the manufacturing operations at the factory running year round. Cane would be planted in the month of February and taken to factories for processing the next year between January and early April, the months when indigo would not be ready for processing. Thereafter the land would be fertilized with indigo refuse obtained from the previous crop of indigo and allowed to lie fallow. Alternatively, the land could be rented out to natives to raise an autumn crop. In either case, the land would be free to be planted with a fresh crop of indigo next spring. He calculated that one acre of land under cane cultivation would yield a comfortable profit of £11.6s.8d., which would help the plantation remain on a sound financial basis. Hudson had already planted different varieties of cane on an experimental basis to ascertain which variety would grow best in Bihar. He wrote to J. Walter Leather, agricultural chemist to the Government of India, asking for his expert opinion on the feasibility of importing and growing the Barbados variety of cane from the West Indies.[72] Another indigo planter, Francis Murray of the Kurnool Indigo Concern, was also in correspondence with a company in Glasgow for transfer of sugar machinery to his tracts in Bihar.[73] A third, the influential Sir William Hudson, the president of Bihar Indigo Planters' Association, chose to write to the imperial government and the secretary of state in London for a loan of £80,000 for a potential large-scale conversion of indigo lands into a sugar-manufacturing estate.[74]

The idea of sugar manufacturing received ample support from the colonial government and its experts in Bihar, who were excited by the prospect of launching a major new industry in the region. Although colonial state officials declined any financial support to the planters for making this transition, they assisted by appointing a high-powered Sugar Committee to look into the feasibility of sugar manufacturing on the indigo tracts. The report of the Sugar Committee of February 1901 concurred with the planters that a potential transfer of indigo lands to sugar production was a sound business idea. The planters consulted by the committee had assured them that given the nature of landownership and conditions of cultivation, they could immediately turn over one-fifth

---

[72] "Note Re Sugar Growing in Behar," by Rowland Hudson, dated January 1900, Bihar State Archives, Agriculture, July 1900, File 2-I/3 of 1900.

[73] Letter from L. W. Macdonald to Francis Murray, dated February 14, 1900; letter from Francis Murray to Messrs. James Finlay and Company, dated March 23, 1900; letter from Messrs. James Finlay and Company to Francis Murray, dated March 26, 1900, Bihar State Archives, Agriculture, October 1900, File 2-I/3.

[74] William Hudson's letter to Under-Secretary of State for India, dated May 21, 1900, Bihar State Archives, Agriculture, October 1900, File 2-I/3 3-32, Nos. 9-10. The request was denied on the grounds that it was not the policy of the government to provide financial support to a specific industry.

of their land to sugar cultivation. The report also assessed the market potential of additional sugar produced by the planters. In the opinion of these experts there was much more demand for sugar than was met by existing native producers. It tried to satisfy the skeptics who feared that any additional sugar might lead to a glut in the market and consequent drop in prices. Through this expert study the government signaled to the planters its tacit support of the scheme.[75]

The extensive recommendations of the Sugar Committee regarding appropriate technology and organization for the planned sugar estates would have been useful for those planters who took to sugar manufacturing in these years. It criticized unequivocally the present methods used for making sugar by some of the planters. The mill currently in use, the *beheea mill*, gave a low daily output and also failed to extract all the juice from the cane. The present method of boiling cane juice in an open pan over a fire lit in a hole in the ground was also "primitive and inefficient in the highest degree." If a planter grew a hundred acres or more of cane and manufactured sugar using these methods, he would "certainly lose money." The planters' sugar would not be able to compete with sugar produced by small cultivators who used similar methods. Nor could it compete with the far superior imported sugar. To these planters, the committee suggested that they assemble on their plantations an improved version of a cane-crushing machine, evaporating pans, and a centrifuge. Such a complex of equipment could be set up at an expenditure of Rs. 5,325, would have a working cost of Rs. 16 daily, and would turn out two tons of gray sugar every day. At those costs, the committee thought the planters would be able to make a handsome profit by selling the gray sugar in local and north Indian markets. The machine on which this calculation was based was capable of servicing cane grown on 200 acres. Slightly larger machines for larger sugar estates could also be acquired for an additional investment.

In the long term, the committee proposed that the planters initiate large-scale production and refining of cane sugar in Bihar on the pattern being followed in the English dominions of Queensland and New South Wales. Inviting the planters to form a syndicate or a company, the committee outlined various elements of the Australian model for them to emulate. Local farmers would grow cane under arrangement with this company. The cultivators would be provided with working capital as well as supplied information on the kinds of cane varieties to be grown, nature

[75] Bihar State Archives, Agriculture, May 1901, File, 2-I/3 8–12, pp. 1–2.

of soil, and right climatic condition for various field processes. The company set up by the planters would build cane-crushing mills in specific places, and cane from nearby districts would be forwarded to these mills. The company would also set up one or two centralized refineries where the gray sugar supplied from various mills in the cane-cultivating districts of Bihar would be turned into white sugar.

Another major initiative in the way of divesting was made toward introducing the fiber-producing plant rhea (also called ramie and China grass) into Bihar. Although prior attempts to start rhea cultivation on a large scale had failed, including one at the government's Saharanpur farm in the United Provinces and on the estates of private companies in the province of Madras in the 1880s and 1890s, the planters were excited about its promise in Bihar now.[76] One of the planters, Jules Karpeles, played a particularly notable role in the rhea initiative. Karpeles was a seasoned indigo planter whose association with the indigo trade as a partner in Karpeles, Heilgers and Company went back a long time. In 1897 he founded Messrs. Jules Karpeles and Company, which became one of the largest buyers of indigo in Calcutta at the time. As a major trader of agricultural indigo and as someone who was an astute observer of the market in Calcutta and Europe, Karpeles sensed a possibility for making rhea a lucrative item of trade. He studied the stakes closely and visited many areas in connection with his plans. His reconnoitering of many areas led him to believe that rhea had a good chance to prosper as a combination crop on the indigo plantations. That he had studied the market very well became clear from a note he wrote to counter those skeptics worried that rhea in Bihar might not do well enough to survive the competition from rhea cultivators in the traditional zones of rhea such as Assam, and Purnea in east Bihar. On the contrary, he explained that the lower labor costs in Tirhut would allow for a better prospect than in the former regions.[77]

Karpeles showed the required enterprise in promoting the unique experiment of the Bihar Rhea Syndicate, which he helped set up to manufacture and trade in rhea. He purchased many indigo factories that were closing down and were available at a cheap price and put them under rhea. The venture took off in 1903 with eight major concerns contracting to grow the fiber crop. The syndicate took the responsibility to provide

[76] Bernard Coventry, "Rhea Experiments in India," *Agricultural Journal of India* 2 (1906): 1–14.
[77] *Indian Planters' Gazette*, March 21, 1903, p. 441; February 24, 1904, p. 248.

the expensive Faure machine of European make for decortications at Calcutta and to ship fiber to Europe. Initially 3,000 acres of land were contracted to grow the crop, although in the end only 2,000 acres could be deployed. But the enterprise had survived at least until the writing of a note in 1907.[78]

A similar project for rhea was launched at the Dalsingserai farm of Begg, Dunlop and Company under the supervision of the seasoned indigo planter Bernard Coventry. A new improved decorticator was installed at Dalsingserai. Bernard Coventry carried out detailed investigations to ascertain the profitability of cultivating rhea and exporting it to Europe. He calculated the average cost of manufacture for a ton of rhea at £10, and of bailing, insuring, and selling it in Europe at an additional £5. Since the average price for rhea in Europe stood at £25, planters could make a profit of £10 for every ton of rhea produced. Since a ton of rhea could be obtained on 2.5 acres of land, the average profit for the planters, if they switched to rhea, would be £4 for an acre that he considered to be an ample reward for an investor. Coventry also asserted that the prices for rhea would hold fast in the coming few years.[79]

Aside from these major endeavors many small initiatives taken by the Bihar planters to build complementarities between the cultivation of indigo and that of other major crops such as sugarcane, rhea, and tobacco, as well as country crops including ginger and turmeric on their land were silently changing the landscape of erstwhile indigo plantations. These striking changes caught the imagination of one observer, who wrote in 1903 that "the Bihar planter has completely reorganised his ways and means of working." The reporter was, however, guilty of some exaggeration when he went to the extent of claiming that indigo on the plantations had been reduced to the status of a "by-product," whose continued relevance was only to provide the valuable seet manure for the other remunerative crops that the planters were interested in growing. In reality it was the reciprocity between indigo culture and cultivation of other crops that was saving the Bihar indigo planter from utter financial ruin as his profits from the sale of indigo kept plummeting. The financial benefit of letting out seet-manured land to native growers of tobacco at

---

[78] *Indian Planters' Gazette*, September 10, 1904, p. 301.
[79] "Note on Rhea," Bernard Coventry, Bihar State Archives, Agriculture 1903, File, 2I/7; See the lieutenant governor's comment praising the efforts of Bernard Coventry for promoting the manufacture of rhea and of W. B. Hudson for sugar manufacturing on indigo tracts. "Planters' dinner at Mozafferpur, speech by Sir John Woodburn," *Englishman*, January 24, 1902, Bihar State Archives, Agriculture, August 1903, File, 2I/7, newspaper reports.

rupees 100 per acre was noted by Jules Karpeles. The value of seet for many other crops including sugarcane, rapeseed, tobacco, and rhea was significant in keeping the cost of natural indigo low in the market.[80] The value of complementarities of indigo and myriad other crops in reducing the outlay was again emphasized by another planter who preferred to call these establishments in their new configuration "factories carrying on as farms."[81]

Connecting with the Dutch Colony: The Natal-Java Variety of Indigo

The Bihar planters made an important attempt to study the efforts made by the indigo planters in Java, who were facing a similar threat to their industry from synthetic indigo. What started as a reconnaissance project developed into an elaborate exercise to introduce the higher-yielding plant from Java into Bihar. In 1899 the Bihar planters made a plea to Lieutenant Governor John Woodburn to send a government emissary to Java. Planters told Woodburn that according to information available to them, indigo planters in Java had devised very economical ways to cultivate and manufacture indigo. Planters wanted to learn about the improvements and adopt them locally in order to compete with cheaper synthetic. Woodburn, while sympathetic to the planters' demand, said he could not think of a person with an intimate knowledge of the field. He encouraged the planters to select one of their own to visit Java plantations, a person who would have the ability to understand the nuances of indigo manufacturing. Taking that advice, Begg, Dunlop and Company sent a private emissary to Java, who returned with news on the superior yield of the local variety grown there and a generous supply of its seeds for possible introduction in Bihar.[82]

The news of the Natal-Java's advantages over the Bengal cultivar spread quickly. A perennial, the Java variety was capable of giving up to four cuttings annually and contained a higher percentage of recoverable

[80] *Indian Planters' Gazette*, March 7, 1903, p. 373. On additional planters' perspectives emphasizing the value of *seet* and Karpeles's assertions, see, two letters by them: *Indian Planters' Gazette*, August 6, 1904, p. 162.

[81] "Rise and Fall in the Price of Indigo in Bihar," *Indian Planters' Gazette*, February 4, 1905, pp. 133–4.

[82] Bihar State Archives, Agriculture, October 1900, File 2-I/3 3–32, Nos. 7–8, "Notes and Orders," pp. 6–7; Letter of G. H. Sutherland of Begg, Dunlop and Company to F. A. Slacke, Revenue Secretary, Govt. of Bengal, dated August 22, 1900, Bihar State Archives, Agriculture, October 1900, File 2-I/3 3–32, Notes and Orders, p. 12.

color in the leaves. Could it be successfully cultivated in Bihar? Would the Javanese variety return the high figures of output for which it was renowned on the Indian subcontinent? The planters quickly rallied to investigate these questions. If the new variety offered as high an output as was claimed, it was a potentially huge find. The Java indigo could potentially become the first major avenue to beat the competition of cheaper synthetic. Planters' seriousness about examining all angles of this question is apparent from the fact that they sent another emissary to the Natal Province in South Africa to procure the mother germplasm from which the Javanese variety had been selected. Once again the colonial government supported planters' efforts by partly funding the South African visit of the private entrepreneur H. A. Baily to Natal in July 1902. Baily sent a message that, if the planters in Bihar were willing, a supply line could be established for sending Natal seeds on a regular basis. There were no takers for Baily's proposal. The import of seeds from another continent was an enormous enterprise that would have required employing people, establishing supply lines, monitoring quality during transit, and dispensing to a large number of customers. Instead, the planters opted in favor of acquiring Natal seeds and developing, selecting, and acclimatizing them in Bihar. Indeed they took to the task of the development of indigo seeds, including Natal and Java seeds, in true earnest.[83]

The development of Natal-Java seeds first started at Dalsingserai Station in Darbhanga from February 1902. The custodian at Dalsingserai made a concerted effort at cultivating the new seed from Java and reported initial success. If sown in September the plant could be ready for a first cutting in March and a second in July. In good seasons a third cutting could also be obtained. Alternatively, the plant could be sown in irrigated lands in spring and be ready for manufacture in July. The plant promised to give good yield for two or even three years. In terms of the possibility of additional cuttings and the perennial nature of the crop the Natal-Java plant's advantages over the local variety were self-evident. Doubts had lingered whether the perennial crop could survive the cold and frosty months of winter in north India, but trials so far seemed promising.

Other planters also rallied to the new cultivar. At a meeting of the Champaran planters in August 1903, "the merits of the Natal Indigo were considered to be fully proved."[84] These planters established a fund under

---

[83] "Natal Indigo Seeds," Bihar State Archives, Agriculture, December 1902, File 2I/6, 9–25, Nos. 1 – 19.
[84] "Natural Indigo – a Last Chance," *Indian Planters' Gazette*, February 6, 1904, p. 177.

the charge of the Motihari planter W. S. Irwin to procure seeds of the new variety from Java. Complementing the efforts on Natal-Java on the eastern rim of the indigo land at Dalsingserai, these planters thus sponsored a similar attempt in Champaran that lay in the western half of the Bihar plantations. By the next year they were reporting success as seeds had been procured in ample quantity, and initial problems with their germination had been overcome. Additional planters were being invited to join the fund and share the success of its current subscribers. The optimism about Natal-Java continued to build after reports of subsequent success in manufacturing trials. F. M. Coventry from Dalsingserai famously published the statistics of manufacture and output with Natal-Java crop that suggested that yield from the Javanese crops was as much as 50 percent higher than that from the native variety. These comparative trials were repeated over two seasons in 1903 and 1904, and the dye output was 43 percent and 60 percent higher, respectively, than with the variety from the North Western Provinces that planters used in Bengal. The trial also confirmed that the Javanese plant did not suffer in potency when used for the second year running. Explaining the statistics, Coventry attributed the lower output in the second year to the general trend of bad output that year in the entire region due to climatic factors. But the percentage increase in the dye obtained from the Natal-Java variety as compared to the North Western Provinces variety was consistently higher. Coventry moved beyond experimental trials to produce Natal-Java seeds on a commercial scale and advertised them for sale in the planters' journal the next year (Fig. 11). Some planters were definitely upbeat about the prospects of the new species and claimed that it was "fast getting a footing in Bengal." Others thought that the spread of the new variety was only being checked by the shortage in supply of seeds and that the *Indigofera tinctoria* variety currently in use would be completely substituted in the end. The planters' primary journal prominently editorialized that its editors were "fully convinced that Natal-Java Indigo is bound to be the sole variety which will be cultivated in Bihar as soon as sufficient seed can be locally grown."[85]

The planters' excitement about the potential of the Java variety was understandable, but these were still early days. The process of acclimatization for a foreign variety was going to be achieved over time as the new seed traveled to different areas and was put to cultivation by numerous factories. Landscape and pathogens would interact with the new cultivar

[85] *Indian Planters' Gazette*, April 29, 1905, p. 479.

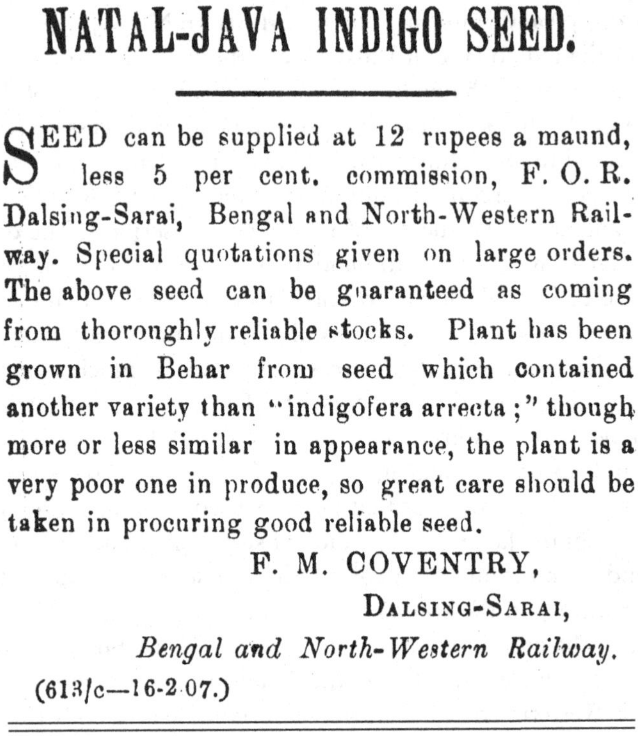

NATAL-JAVA INDIGO SEED.

SEED can be supplied at 12 rupees a maund, less 5 per cent. commission, F. O. R. Dalsing-Sarai, Bengal and North-Western Railway. Special quotations given on large orders. The above seed can be guaranteed as coming from thoroughly reliable stocks. Plant has been grown in Behar from seed which contained another variety than "indigofera arrecta;" though more or less similar in appearance, the plant is a very poor one in produce, so great care should be taken in procuring good reliable seed.

F. M. COVENTRY,

DALSING-SARAI,

Bengal and North-Western Railway.

(613/c—16-2 07.)

FIGURE 11. F. M. Coventry's advertisement for Natal-Java seed. *Source: Indian Planters' Gazette,* May 15, 1905.

over a longer trajectory, and the outcome of that engagement would only become evident gradually. But in 1905 the Java indigo seemed the major avenue by which the planters could hope to reduce the price of natural indigo by increasing yield.

Science at the Factory: Peeprah and Turcowlia

Modest trials were made at a select few factories to boost the output of the dye through application of chemical processes. These trials were different from the more detailed scientific program under way at laboratories and experiment stations elsewhere in the colony that were under the charge of scientists and agricultural experts (which will be discussed in the next chapter). In contrast, the efforts at the factories were under the direct supervision of indigo manufacturers.

The person at the center of these efforts was Eugene Schrottky, no stranger to the indigo tracts. Schrottky enjoyed a certain stature among the planting classes as a former shareholder in the Bengal Indigo Company and as the holder of several patents on indigo. He was not an establishment scientist, but was rather described by various people as "a practical indigo maker." His applications were more "on the spot" in nature, what could be categorized as "process innovations," distinct from methods resulting from detailed research and analysis. Schrottky had been more recently active in Champaran, where he assisted W. S. Irwin with the task of acclimatizing the Natal-Java seed. He had advised planters about problems with the germination of the new seed from Java. He had lived in Java earlier and was familiar with the similar problem with germination in Java as well. In all likelihood the relatively simple solution of scarifying the Natal-Java seed that spread in Bihar owed its origin to the information provided by Schrottky.

The two largest factories in Champaran invited Schrottky to test his processes during the manufacturing season in 1903 and 1904. He was first invited to conduct his trials at the Turcowlia factory of James Hill. Soon after, A. W. N. Wyatt, the planter from Peeprah, invited Schrottky to try out his processes at his factory in August 1903. Results from these trials were awaited with much anticipation as apparently many planters had pinned their hopes on his promise to deliver. The early reports appearing in the *Indian Planters' Gazette* claimed that the innovator had achieved "unqualified success" with his trials. The planters asked for "proof" of the success that Schrottky was claiming. One of them asked that his claims about the 1903 processes be substantiated by the managers at Peeprah and Turcowlia.[86] The early optimism indeed turned out to be unfounded when the results from Peeprah factory were made public. The trials after all, seemingly, had been "a failure." The results as tabulated by the factory's manager B. Norman showed that in Schrottky's trials 2,279 *maunds* of plant had returned 3 *maunds,* 30 *seers,* and 11 *chittacks* of dry indigo through processes applied between August 24 and 29. That worked out to a poor yield of 6.61 *seers* per *maund,* well below what would be normally considered a good average. The factories in Champaran were commonly known to produce as much as 8 *seers* per *maund* of indigo plant. The proprietor also reported that the quality of the indigo through Schrottky's process was so poor that he had not been able to sell it.[87]

[86] *Indian Planters' Gazette,* February 20, 1904, p. 249.
[87] *Indian Planters' Gazette,* March 19, 1904, p. 378.

Schrottky remained optimistic but his inability to furnish concrete results on a consistent basis was not inspiring the planters. He claimed that he was trying a new "synthetic process" and that the trial at Peeprah had validated his belief in those processes. He asked for more time to validate those processes during manufacturing for the *khoontie* crop (second cutting). It is likely that he was trying out a combination of processes over which he held patents and that the name "synthetic process" was merely a clever invention to draw attention in an age when the synthetic dyes were victors. The planters, in any case, did not wish to give him the luxury of detailed experimenting. As the planter Wyatt had remarked in his note, "Mr. Schrottky came over here not with a view of experimenting but in order to demonstrate to us the results obtainable from his process."[88] Schrottky's plea to the subscribers of the Natal-Java Indigo Fund to sponsor his work for the next year did not meet with a positive response. He returned to London and wrote to the editors of the *Indian Planters' Gazette* that he had given up his indigo research, only temporarily though, as developments of subsequent years showed.[89]

Lessons from Season 1903–04: "A Business Question"?

Reducing the price of natural indigo was a necessity imposed by synthetic indigo on the planters. The planters who remained in business continued their effort to reduce the price of agricultural indigo. They acted on multiple fronts to make that happen, but there were indications that they were reaching the absolute limits in the process of cheapening the agricultural indigo.

A drastic reduction in the price of synthetic during 1903–4 made the planters and the defenders of natural indigo take another hard look at market realities, and they could see for themselves now that the competition between natural and synthetic indigo had reached a different stage. Writing in June 1903, the secretary of Bihar Indigo Planters' Association, F. A. Shaw, explained that the planters were finding it impossible to reduce the cost of production of agricultural indigo any further. He was writing to make an argument in favor of tariff protection for natural indigo in the British markets and in that context was responding to some of the metropolitan commentators who held planters guilty of not doing enough to counter synthetic indigo. "Irresponsible writers on the subject never tire

---

[88] *Indian Planters' Gazette*, March 19, 1904, p. 378.
[89] *Indian Planters' Gazette*, February 27, 1904, p. 284; February 4, 1905, 134.

of advising planters to reduce the cost of production. To these kind, but ill-informed friends, I may perhaps point out that with all the desire in the world to reduce cost, it has been found impossible to go lower than present rates."[90] The price of synthetic paste of 20 percent concentration in the English markets had averaged 1 s. 1 ½ d. per pound by weight between January and December 1903. This was equivalent in price terms with natural indigo of 60 percent concentration of rupees 90 per *maund*. In 1904, the price of synthetic further fell to 8 d. per pound.[91] Natural indigo had sold at prices between rupees 110 and rupees 170 a *maund* in the English markets the same year. Evidently the planters were unable to match the 1904 price of the substitute completely. But even to sell at that price the planters had implemented a number of changes in their production system, as the sentiment expressed by Shaw reflected.

The drive to reduce costs in the natural indigo industry would continue. That was the line suggested to them by Sir Edward Law, the financial member of the Governor General's Council. Making the budget speech in the Imperial Council in 1904, he enumerated clear economic steps that he thought were required of the planters in order to prevail over synthetic. He made reduction in the outlay of plantations a central plank in his counsel. "This is a purely business question," he asserted, highlighting the relevance of both indigo-specific measures and broad measures toward effective management of plantations that alone would enable natural indigo to hold its own against synthetic indigo. In the past few years the planters had successfully matched the declining price of synthetic in the market, and he thought that there was scope for implementing further economies. These multifarious steps were already under way. Some had borne results, and there was a need to accelerate them.[92] The planters, though, seemed to be saying that their efforts on all those fronts had evidently hit a ceiling. Looking forward, it was increasingly "science" and not "business" that seemed to offer optimism to the planters.

---

[90] "The Passing of Indigo: A Plea for Protection," *Indian Planters' Gazette*, June 27, 1903, pp. 887–8, republished from *Empire Review*.
[91] For synthetic's price, see Jules Karpeles's letter to the editor of *Indian Planters' Gazette*, dated February 27, 1904, p. 283. The price had actually varied between 1 s. 5 ½ d. in January and 9 ½ d. per pound in December 1903; for natural's price in 1904, see Karpeles's letter from Paris, dated June 7, 1904, *Indian Planters' Gazette*, August 6, 1904, p. 161.
[92] "Sir Edward Law on Synthetic Indigo: A Word of Comfort to the Planter," *Indian Planters' Gazette*, March 26, 1904, pp. 410–11.

4

Local Science: Agricultural Institutions in the Age of Nationalism

The process of knowledge generation for indigo changed in a qualitative way in the years after the launch of synthetic indigo. The introduction of BASF indigo, the very first synthetic indigo in 1897, marked a definitive turn by the planters toward engagement of professional scientists. The most distinctive aspect of this shift was the proximity of the emergent process to colonial institutions. The planters took the initiative to organize experiments within laboratories and agricultural experiment stations in the colony and in the metropolis in their bid to improve the yield and quality of the natural dye, and the scientific infrastructure they created was unprecedented for the colony. But they soon turned to the colonial government for assistance to shore up their scientific experiments. The dependence on the colonial state became deeper as time passed and as profits in the natural indigo industry began to plummet.

Indigo science developed a nexus with the effort by the colonial state to create a science of agricultural development. The forays in India by indigo laboratories and experiment stations materialized in parallel with the aspiration for scientific agriculture in the colony. The growing authority of scientists lay at the foundation of these twin processes. This connection between indigo science and colonial agricultural science unfolded at the level of formation of institutions. Even as colonial officials agreed to provide assistance for indigo science, they also wished to appropriate that science for their own program of agricultural improvement in the colony. The movement of experts between indigo laboratories and government stations; the aligning of indigo science with the priorities of colonial state officials; and the involvement of metropolitan institutions with the efforts by planters and colonial state officials in Bengal created labyrinthine connections between "private" and "public" structures of science in the colony.

Self-Belief and Optimism around Laboratory Science, 1898–1901

The drive toward laboratory-based science to improve natural indigo in colonial India was facilitated by the resourcefulness of European planters. Of middle-class origin and educated in public schools in England, the typical indigo planter in Bihar had the right mind-set and the financial means, and, in the post-BASF indigo era, the motivation to exploit science. In many ways the planting class constituted an extension of the English publics in the colony and was acutely aware of important scientific developments. The more prosperous planters constituted an extremely mobile class, one that was given to frequent travels across the wide imperial networks and beyond, while most resident planters on the subcontinent moved between the colony and home as frequently as they could. On their visits to England many of them attended public meetings organized at forums like the Society of Arts in London, the Society of Dyers and Colourists at Bradford, the Society of Chemical Industry at Manchester, and similar bodies where prominent men of science and industry deliberated on important emerging philosophical debates in front of distinguished gathering. Planters heard debates in these professional organizations about the state of research on indigo, including those on the promise of chemistry. Synthetic indigo was evidently the result of painstaking efforts by German dye chemists. Could natural products chemistry, many planters wondered, do for agricultural indigo what organic chemistry had done for synthetic indigo? Evidently many of them came to believe it could. The expression of trust in science appeared at the meetings of the Bihar Indigo Planters' Association (henceforth, BIPA) in India and in the writings by planters in their trade journals and newspapers. These instances suggest the embracing of science by a cosmopolitan class that was given to mobility and exchange of information across continents rather than the disinterestedness toward science of a class of forlorn expatriates, one that was living in isolation in a colonial outpost. Many of the planters made aware in this manner of the potential of modern science canvassed support for establishing laboratories in the late nineteenth century.

The fact that planters were well organized within trade bodies facilitated coordination of their efforts to tap the best available scientific

resources for assembling laboratories. Since 1877 the BIPA had chan-
neled planters' interests institutionally and helped them build industrywide consensus on issues of common interest. As synthetic reached the
market, the BIPA promptly formed an Indigo Defence Association (IDA)
and incorporated the latter as a public limited company in London.[1] The
establishment of IDA in 1898 reflected the planters' desire to elicit pertinent information and develop the ability to intervene directly in the
key London market in the face of synthetic's threat. This company was
mandated to initiate scientific investigations, to buy patents relating to
improvement, and to pursue matters of interest with the national government. Indigo traders based in Calcutta, in addition, set up another
organization for the same cause. This second organization was named the
Indigo Improvements Syndicate (hereafter, IIS). The IIS functioned under
the overall charge of Begg, Dunlop and Company, the leading managing
agents based in Calcutta, and was given the responsibility by agents, brokers, and bankers dealing in indigo to coordinate efforts for the scientific
improvement of blue dye.

The BIPA-backed Indigo Defence Association sent the English chemist Christopher Rawson to Bihar, thus taking the first step in establishing modern indigo laboratories in the colony. Rawson was arguably the
best available expert on the chemistry of natural and synthetic indigo
outside Germany. Born in Bradford and trained at the Royal College of
Chemistry, Rawson first worked for the Royal College of Agriculture in
Cirencester before returning to Yorkshire to launch his career as an analytical chemist. It was in the dyeing districts of Yorkshire that he built his
reputation as a trade chemist. In the early 1880s this Yorkshire chemist
embarked on a detailed study of indigo and other dyestuffs and, noticeably, developed assays for checking color percentages in indigo imported
from India. In 1884, he cofounded the Society of Dyers and Colourists
and remained deeply engaged in the activities of this prominent trade
organization. Rawson continued to work along these lines into the 1890s
and enjoyed an unmatched reputation for expertise in dyeing and printing
circles. Given these credentials it is no surprise that IDA selected Rawson,
as he both was knowledgeable in theoretical aspects of indigo chemistry
and knew the dye market well.

As a trade chemist, Rawson had an appreciation of the relative merits
of natural and synthetic indigo shaped by his understanding of consumer

---

[1] Incorporation papers of Indigo Defence Association Limited, Public Record Office,
London, BT31/8154/58924/100052 (henceforth, PRO).

demand and shop floor practices in the English dye and print houses. In an early communication with the members of the Indigo Defence Association Rawson warned that the synthetic was "no chimera," alerting those few planters who vainly minimized the challenge to their "real" indigo from the German "imitation." He had reached this understanding on the basis of a tour of Yorkshire and Lancashire districts undertaken specifically to ascertain the response of commercial users to the two competing products. It was clear that calico printers preferred the synthetic because of its advantages in offering "purity of shade and uniformity of strength." According to his information the cotton dyers who had so far tried synthetic indigo believed it to be dearer than natural indigo, though not by a wide margin. To be sure, the natural indigo also had some qualities that the dyers appreciated, such as its ability to provide a unique red tinge and the fastness with which it held on to the fabric. But these finer distinctions in favor of natural indigo did not materially reduce the threat of purer synthetic indigo. The overriding message of the chemist's note to IDA was that improvement of natural indigo was a necessity and that any program for its improvement must have as its central element the objective of price reduction through higher yield and greater purity.[2]

Rawson's early explorations of the problem of indigo manufacturing in the colony confirmed his belief that he could meet the challenge of improving the commodity. Reaching India in the summer of 1898, Rawson embarked on a trip to several factories in order to observe manufacturing operations firsthand and to collect samples. His initial experiments at two indigo laboratories that were set up in succession at the Planters' Club and the Mosheri factory of the planter G. Collingridge and further trials of specific measures at numerous factories of BIPA subscribers reinforced his optimism. Having surveyed current practices and considered the potential for science to make them better, he expressed confidence about the prospect of improving indigo. In his first report to BIPA, he declared, "The yield of colouring matter from a given weight of plant can be considerably increased."[3]

In keeping with his understanding of the nature of competition Rawson's work involved trials to raise yield and increase purity. Early in 1898, he visited the Dalsingserai factory to examine the innovation introduced by

[2] Rawson's letter to the Indigo Defence Association, London, dated February 28, 1898, Bihar State Archives, Agriculture, File 2I/3, March 1901.

[3] Mr. Rawson's report No 1, dated July 14, 1898, Bihar State Archives, Agriculture, File 2I/3, March 1901.

its in-house manager, Bernard Coventry, who added slaked lime to the oxidation vat to induce sedimentation. Rawson approved of Coventry's method and worked to perfect it further. He devised a new method of oxidation by passing air currents in deep vats using air compressors and blowers, in place of the current practice of manual "beating." He claimed that his method gave 32 percent greater yield. In addition to conducting a demonstration of the refining process for improving the purity of indigo, one that could not be scaled up for commercial use, however, he also made a concrete suggestion to wash the dye with acid in boiling water to get rid of soluble impurities. Hopefulness marked Rawson's first report to IDA written from his home in Bradford early in 1900. A little later, he echoed this sentiment in a letter to the senior planter Sir W. B. Hudson. Hudson was retiring in England at that time and had contacted Rawson to ask what he thought about the crisis faced by natural indigo.[4]

As work progressed, Rawson called for expanding the scope of scientific research on indigo by engaging a bacteriologist. He believed that the steeping stage in manufacture involving fermentation had the "greatest room for improvement." There were many aspects of steeping that he only vaguely understood. The theory of Edward Schunck, the Manchester-based chemist, maintained that the relevant glucoside found in the leaves first split into indigo and glucose during steeping with the former being further reduced to "indigo white" as fermentation progressed. During beating the reduced indigo was then oxidized back into the blue dye. In some recent revisions to Schunck's work, other theorists had claimed that in the first stage of decomposition indican actually broke up into indoxyl (two fused molecules of indigo) and sugar in the steeping vat and that before entering the oxidizing vat the steeped liquor contained indoxyl not indigo-white. Rawson believed the latter postulate more closely matched his own observations. In his scientific reports he explained that the slow oxidation of steeped liquor proved the presence of indoxyl not indigo-white, as Schunck had presumed, because the latter was known to react more speedily with oxygen.

---

[4] "Mr. Rawson's Report No. 2," dated August 19, 1898; "Mr. Rawson's Report No. 3," dated September 26, 1898; Rawson's letter to Indigo Defence Association, dated July 31, 1899 and February 6, 1900; Rawson's letter to Sir W. B. Hudson, dated April 14, 1900; Rawson's letter to the Bihar Indigo Planters Association, dated August 16, 1900; Rawson's letter to BIPA, Private and Confidential, dated October 4, 1900, Bihar State Archives, Agriculture, File 2I/3, Nos. 1–7, March 1901; Christopher Rawson, *Report on the Cultivation and Manufacture of Indigo in Bengal*, Bradford: William Byles and Sons, 1899.

If the chemical composition of steeping liquor evaded certainty, the identification of the vector that brought about fermentation was even more clouded. Experts around the world disagreed whether fermentation in indigo was caused by an enzyme or a bacterium. Rawson leaned toward the camp that argued for enzymatic fermentation, although he was not absolutely certain about this.[5] The resolution of these questions required the presence of a specialist bacteriologist who could devise ways to understand and control the processes. Thus Rawson suggested to BIPA that an English bacteriologist be hired to work under his supervision at Mosheri. He suggested that any of the bacteriologists working for the many large breweries in England could be invited.[6] In a confidential report to BIPA, he apparently tried to defend himself against the assumption gaining ground among planters that he had so far paid little attention to steeping, saying, "If I have said little about the steeping vat in my reports, it is not because its importance has been overlooked but merely for the reason that I have not yet discovered anything of value to communicate."[7]

Rawson also carefully considered prior works in the colony on agricultural science and soil studies in making clear recommendations about the use of fertilizers for improving the plant in the field. In a Liebigian fashion, he analyzed mineral content in indigo ash to highlight the excessive scale and diversity of minerals extracted by the plant. He justified the application of artificial manures to the indigo field to recompense this loss. He admitted that lack of precise understanding of the constitution of indican, the color-producing principle in indigo, was a handicap. If one knew exactly the constitution of indican, the addition of fertilizers could be suggested with specificity. Nonetheless, since it was known that indigo was a nitrogenous compound, he was convinced that the application

[5] Henri ter Meulen and Hoogerwerf published a paper in Dutch in Amsterdam in 1900 in which they described a procedure for isolating the crystalline indican. Indican had probably been isolated earlier in the Dutch colonies in Java but been concealed discreetly as information of commercial value to the indigo planters. Rawson showed familiarity with the work of another Dutch chemist, Van Lookeren Bandreat, who had recently published a paper on the indigo enzyme indimulsin. Much of these results flowed out of indigo research sponsored by the Dutch colonial government and private indigo entrepreneurs at Klaten, Semarang, and elsewhere in Java beginning 1892.

[6] Letter from E. Macnaghten, General Secretary, BIPA to Secretary, Government of Bengal, Revenue Department, Bihar State Archives, Agriculture, File 2I/3, Nos. 3–32, No. 13, October 1900.

[7] Rawson's letter to BIPA, Private and Confidential, dated October 4, 1900, p. 2, Bihar State Archives, Agriculture, File 2I/3, March 1901.

of nitrogenous fertilizers would be beneficial. Even though indigo was a legume and capable of fixing nitrogen directly from the atmosphere, Rawson argued in favor of the need for fertilizers, stating that the soil in Bihar was too deficient in nitrogenous content. In support of this view he cited the report by colonial India's first agricultural chemist, J. W. Leather, and the 1893 report of J. A. Voelcker. These two chemists had earlier identified the problem of low nitrogenous content in most Indian soils. Rawson suggested the use of *seet*, or indigo refuse, for the fields to make up for this deficiency. The latter was a rich source of nitrogen and other key minerals such as potash and phosphoric acid. Its value also lay in the fact that it was readily available to the planters as refuse from the manufacturing process.[8]

During these years the other indigo body, IIS, established an agricultural station at Dalsingserai dedicated to the task of improving indigo. The setting at Dalsingserai was ideal for the establishment of the center. The subdivision comprised the floodplains of Gandak and was the site of a major indigo factory owned by Begg, Dunlop and Company. The soil and weather factors were apparently ideal for growing and manufacturing indigo. The local station manager, Bernard Coventry, an astute agriculturist with indigo experience spanning more than two decades, was well up to the task of leading agricultural trials.[9] He had previously shown considerable scientific acumen by introducing innovations to vat processes and was trusted by the planters. As he consented to participate in indigo experiments, Begg, Dunlop and Company set aside a huge plot for the station and appointed a new recruit from England, E. A. Hancock, as the station chemist to assist with the scientific work. The agricultural experiments were launched at Dalsingserai in true earnest from July 1899. A chemical laboratory furnished with all essential apparatus was added subsequently.[10]

The IIS station focused on enhancing the plant in the field as distinct from the effort at BIPA laboratories on improving the postharvest extraction processes. Coventry and Hancock put an expansive tract under

---

[8] Christopher Rawson, *Report on the Cultivation and Manufacture of Indigo in Bengal*, Bradford: William Byles and Sons, 1899, pp. 4–15.

[9] For a reference to Bernard Coventry's passage to India and early life on the plantations, see his letter from Bihar to his brother in Hampshire dating from December 1880, Hampshire Archives and Local Studies, Hants., UK, 8M59/783.

[10] E. A. Hancock, "Note on the Work of the Indigo Improvements Syndicate at Dalsingserai," Bihar State Archives, Agriculture, May 1901, File 2-I/3 1–7, Nos. 3(b)–3(c); IIS's letter to Revenue Secretary, Government of Bengal, dated January 31, 1901, Bihar State Archives, Agriculture, May 1901, File, 2-I/3 1–7, Nos. 1–2.

fertilizer experiments. Nearly one hundred and twenty plots measuring one-quarter of an acre or more each were dedicated to trials in the use of artificial fertilizers and manure and to improvement of cultivation practices. Effort was invested in other crops as well that could be potentially used as rotation crops on the plantation. The chemical laboratory nestled among the agricultural tracts helped with the task of analysis of soil and water, in checking color-yielding capacity of different indigo plants, and in ascertaining color percentages in the dye. A set of four vats was also provided for processing indigo so that experts might observe the manufacturing processes under ordinary factory conditions. But these analytical experiments were subsidiary to the biological and agricultural effort to improve indigo and other local rotation crops, unlike the work at BIPA laboratories, which was mostly aligned with the currents of metropolitan chemistry and shaped by an understanding of user priorities.

The station experts at Dalsingserai began to put together a plan for the selection of indigo plants under the overall charge of Bernard Coventry. It is likely that Coventry's methods involved not only a mass selection process whereby he selected the seeds of plants that appeared to be stronger, but also "individual selection" based on overall strengths and qualities of specific plants, and built stronger lines over generations. This was no easy task given the fact that planters in Bihar were not producers of indigo seed themselves but rather received their supply of seeds from merchants, who in turn bought them from diverse groups of farmers in the NorthWestern Provinces. Coventry first applied himself to distinguishing different types of plants according to the identifiable places of origin. He was well up to undertaking such a task. He had apparently sensed an entrepreneurial opportunity for seedsmen in the current condition of chaos in the market for seeds in Bihar. Anyone who could guarantee higher yield varieties and could be trusted by the indigo planters would make a fortune.

His long experience as an agriculturist, broad scientific training, resourcefulness, and openness to trying out the proven methods of selection widely in practice by breeders at home and elsewhere served him well. The basic process of mass selection was simple, merely involved choosing seeds of better varieties of plants over generations, and was commonly practiced by farmers. But astute agriculturists and breeders in many lands were also engaged in the more delicate "individual" selection, including Britain.[11] Selection was a commonly deployed practice

---

[11] Deborah Fitzgerald has characterized late nineteenth- and early twentieth-century breeding and selection practices as fundamentally "democratic" practices because of the

by English farmers in the last three decades of the nineteenth century. Breeders of note were already marking their presence on the English national market with some success.[12] As someone who hailed from a family of Hampshire-based squires, Coventry was apparently not unaware of the pre-Mendelian selection processes prevalent among agriculturists and seed developers in England. As a representative of English agriculture in the colony, he was well poised to try those principles for improvement of indigo cultivars. He was ably assisted in the task by Major David Prain, the director of the Calcutta Royal Botanic Garden, who was one of the most influential voices in colonial botany at the time.[13]

The Dalsingserai pioneers were galvanized by the success story of selection work on indigo in Southeast Asia. They learned that the Java variety of Southeast Asia was actually a foreign species that grew wild in the Natal Province in South Africa (*indigeofera arrecta*). The *I. arrecta* species was introduced into Southeast Asia about half a century ago. The Dutch planters in Java had in the last decade or so further perfected the Java plant with ample success. The Dutch microbiologist Van Lookeren Bandreat played a key role in this process of isolating a vigorous type that gave a high yield. Could Bengal planters achieve through selection what the Dutch planters in Java had accomplished? Could the Java plant be experimentally launched in Bihar and a critical stock of Java plants doing well in the new environment saved for producing seeds? Aspiring to make it possible, the IIS started to work with the seeds obtained from Java. The foreign variety was tested along with local varieties obtained from other parts of India by planting them side by side at both Dalsingserai and the

---

simplicity of the method, which could be easily embraced by farmers and breeders alike. She called selection, "a method so commonsensical that ordinary farmers with little scientific inclination could practice it as a matter of course," discussing the case of American corn specifically in Illinois. Deborah Fitzgerald, *The Business of Breeding: Hybrid Corn in Illinois, 1890–1940*, Ithaca, N.Y., and London: Cornell University Press, 1990, p. 9. What was true in Illinois corn agriculture was commonly true of agriculture in Europe as well.

[12] The subject of pre-Mendelian selection has been more often assumed than analyzed. Indeed most existing works have focused on the Mendelian phase for both Germany and Britain. In most of these writings the thrust lies in explaining the influence of rediscovery of Mendelism in 1900 that turned the "art" of plant breeding into a systematic, rational, controlled process. For a brief treatment of the 1870–1900 case of breeders in Britain, see, Berris Charnley and Gregory Radick, "Plant Breeding and Intellectual Property before and after the Rise of Mendelism: The Case of Britain," http://www.ipbio.org/berris.htm, accessed on June 30, 2010.

[13] David Prain, the Scottish botanist, joined the Royal Botanic Garden in 1897 and remained its superintendent until 1904. Leaving India that year, he returned to England to become the director of the Kew Garden in London.

government's Botanic Gardens at Calcutta. Under test conditions the foreign plant was giving a higher yield of color than all local varieties grown in India. The initial plan was to "select" the Java plant that would survive in Bihar's natural environment and then save those specific cultivars to raise seeds. A total of one hundred acres at Dalsingserai were sowed with indigo, as noted by Hancock, some of which undoubtedly were devoted to selection work.[14]

Indigo Entrepreneurs Reach
Out to Metropolitan Institutions, circa 1900

Indigo manufacturers' outreach to metropolitan bodies was only natural given the fact that the latter were manifestly more resourceful. Those bodies were staffed by people who had a perspective enriched by their location at the helm of an expansive empire. Key imperial figures had a panoptical view that could assess the nature of the indigo market in London from close quarters and of the other major ports in the West that traded in indigo. Scientific institutions located at the apex of a network of personnel and institutions across the empire could boast of a unique level of expertise. These resources could be potentially useful to those in the colony who stood to defend natural indigo.

Indeed the India Office in London, headed by the secretary of state for India, closely monitored developments related to Indian indigo.[15] From the very beginning this office counseled state officials in India and Bengal over the steps required to secure the future of the colonial product, readily apprising them of any commercial information that was considered pertinent to indigo. For instance, a letter from the office of the secretary of state, George Hamilton, clarified the trade situation on the Continent. It included a report submitted by Her Majesty's consul at Marseilles that spoke quite pessimistically about the future of natural indigo in France. It pointed out that while ten years ago France imported 1,400 to 1,500

[14] Notes by Major Prain, director of Botanical Gardens, Calcutta, and by F. Prange of Amsterdam in the letter from Messrs. Begg, Dunlop and Company, Honorary Secretaries, Indigo Improvements Syndicate, to Secretary, Government of India, Revenue and Agricultural Department, dated January 31, 1901, Bihar State Archives, Agriculture, May 1901, File, 2-I/3 1–7, Nos. 1–2; E. A. Hancock, "Note on the Work of the Indigo Improvements Syndicate at Dalsingserai," *op. cit.*

[15] The Indian secretary of state assisted by the council of India, based at the India House, was significant as the principal adviser to the Crown on Indian affairs. For a comprehensive description of the working of the India Office, see Arnold P. Kaminsky, *The India Office, 1880–1910*, Westport, Conn.: Greenwood Press, 1986.

chests of indigo, in 1899 only 600 chests were imported. Of those 600 chests, 130 originated in Java, 50 in Bengal, and 420 in the Coromandel (Madras). Auction of Bengal indigo had practically ceased at Marseilles, it pointed out, illustrating the dire situation that Bengal indigo was finding itself in. Only a few small buyers retained interest in Bengal indigo, whereas bulk purchasers had moved away.[16]

In this particular instance, the India House transmitted the varied perspectives of traders, buyers, and consumers about the criteria on which natural indigo was fast losing out to synthetic indigo on the Continent, information that was hardly soothing, but useful nevertheless for those who wanted to fight the artificial product in the marketplace. The Marseilles report uniquely illustrated the developments in a key international market where some traders and importers deserted Bengal indigo despite believing in its ability to dye fast. It surmised their perspective thus: "Durability has less charm. And if the [synthetic] dye will last the cloth in cotton prints, and in most woolen fabrics except in cloths for uniforms exposed to sun and rain, what advantage can the manufacturer on the continent find in a dearer though better dye, if the cheaper [synthetic dye] is more attractive to the eye, and gives so much greater profit!" The report concluded that the primary reason for the decline of natural indigo in France was its high price relative to synthetic's. It was only the minor purchasers from Japan and the Levant who continued to buy Bengal indigo at the existing price, whereas substantial "European [continental] buyers" were no longer willing to pay more. Ironically the cheaper "Coromandel indigo" from the southern coast in India, though poor in quality, still had some demand at the lower end of the market because of its low price. But there were few takers for the high-quality and expensive Bengal indigo.

The Imperial Institute of London that became operational in 1893 to act as a nodal point of distribution for scientific resources over the entire empire was also poised to play an important role in facilitating the connections of colonial indigo science with metropolitan chemistry. As Michael Worboys's study has shown, the major scientific work at the institute was performed in its Scientific and Technical Department, headed by its renowned chemist, Wyndham R. Dunstan. It was through Dunstan's initiatives that the Imperial Institute won its reputation as the

---

[16] Enclosure, "Report by Consul Gurney on the competition of Artificial Indigo with Indian Natural Indigo on the French market," Letter from India Office to Governor General of India in Council, dated July 19, 1900, PRO, AY4/2047/100168.

"Kew of chemistry." In a note Dunstan described the work at Imperial as involving scientific investigation of potential of little-known products, comparative investigations of these products with established ones, and advice on scientific policy measures relating to natural products in the empire.[17] Investigations on indigo would fall neatly within the remit of such responsibilities. Also, since 1900, the operational control of Imperial had been transferred to the Board of Trade, a body with obvious interests in commodities of value to the Empire. It routinely called upon experts at Imperial to provide input on various commodities and on general matters of commercial interest.

In early 1900, Begg, Dunlop and Company wrote to the Commercial Intelligence Department of the Board of Trade in London, asking whether they might have the facts relating to the comparative merits of natural and artificial indigo examined and clarified. They had heard different claims about the superiority of natural or synthetic indigo by partisans on both sides. Such colored claims were especially common in the early years of competition between natural and synthetic indigo. This was not at all unusual in a context where a new substitute was trying to displace a reigning product in the market. The commercial department forwarded the query from the indigo planters to the Scientific and Technical Department of the Imperial Institute for examination by its scientists.[18]

It was Wyndham R. Dunstan at Imperial who took the responsibility to study the issue in great detail. He gave due attention to all aspects of the question by inquiring with manufacturers, traders, and scientists. Dunstan was a qualified chemist himself and a "leading authority on chemistry of natural products."[19] He promptly dispatched letters of inquiry to several people, including one to the suppliers of synthetic indigo to the British markets, Zilz and Stott, asking how much synthetic indigo they were selling in England and at what price. The agents refused to disclose any information, maintaining that it was not in their trade interest to do so.[20] But Dunstan had other avenues that he

---

[17] W. R. Dunstan, "The Work of the Scientific and Technical Department of the Imperial Institute," *Nature*, No. 55, November 19, 1896, p. 62, cited in Michael Worboys, "Science and Colonial Imperialism in the Development of the Colonial Empire, 1895–1940," Unpublished D.Phil. dissertation, University of Sussex, 1979, p. 151.

[18] Letter from Board of Trade to the Imperial Institute, dated March 21, 1900, Public Record Office, AY4/2047/100168.

[19] Michael Worboys, "Science and Imperialism in the Development of the Colonial Empire, 1895–1940," p. 180, fn. 16.

[20] Letter from Zilz and Stott, dated May 17, 1900, addressed to W. R. Dunstan, Director, Scientific and Technical Department, Imperial Institute. PRO, London, AY4/2047, 100168.

could pursue. He sent another note to Heinrich Caro in Germany, the ex-manager of BASF and one who had been closely associated with BASF's indigo operations in the past, asking him to clarify whether synthetic and natural indigo were identical. He raised several other questions in the memo to get to the bottom of the truth: Did synthetic indigo dye as effectively as natural indigo? How was the synthetic manufactured? What was the estimated total output of BASF Indigo currently? Dunstan belonged to the network of British and German chemists who regularly communicated with each other against the grain of existing trade rivalry between the two countries that one might think would forbid such interactions.[21] Caro in his response emphasized that natural and synthetic indigo were absolutely identical, notwithstanding the claims of some of natural's producers to the contrary. He also countered the claim of a few that the presence of additional elements like resins and gums in natural indigo added to its fastness. On the contrary, Caro argued that the artificial product was better because it imparted brighter color to clothing. He disclosed the chemical pathway that BASF had initially used to manufacture synthetic indigo starting from naphthalene in order to illustrate the process of industrial production of a product existing in nature. Having retired some time before, he had not seen the manufacturing pathway in use currently and therefore could not confirm whether the original system was still in use. In his opinion it was plainly the "more economic production" of synthetic indigo that had enabled BASF to sell its product at a cheaper price and to challenge the supremacy of natural indigo in the market. Caro also wrote that although he did not have data on the total output of synthetic in the market, he was convinced that the production was on the rise at BASF. He also informed that both Hoechst and Basle would soon begin the production of synthetic indigo, a development that would flood the market with additional quantities of artificial indigo and further complicate the problems for natural indigo.[22]

---

[21] Anthony Travis has pointed to the personal channels of communication existing between English and German chemists in the late nineteenth and early twentieth centuries. *Cf.* Anthony Travis, "Heinrich Caro and Ivan Levinstein: Uniting the Colours of Ludwigshafen and Lancashire," in Ernst Homburg, Anthony S. Travis, and Harm G. Schröter (eds.), *The Chemical Industry in Europe, 1850–1914: Industrial Growth, Pollution, and Professionalization,* Boston: Kluwer Academic, 1998.

[22] Letter from Heinrich Caro to W. R. Dunstan, dated May 16, 1900, PRO, AY4/2047/100168.

Dunstan also wrote to two other chemists in this regard. The first was Christopher Rawson, who had presented a paper in March 1900 on the issue of the relative capabilities of natural and synthetic indigo at the Society of Arts in London. Rawson was already in the employ of BIPA and was conducting experiments in Bihar on natural indigo. He informed Dunstan that the experiments he had conducted in England and Bihar proved that natural and synthetic indigo were identical. He additionally claimed that in Britain the use of synthetic was so far limited to cotton printing and to those sectors of wool dyeing where shades of lighter blue were desired. The printers in particular expressed a strong preference for synthetic indigo.[23] The third scientist Dunstan wrote to was Hugo Muller, another coal-tar dye expert and industrial chemist resident in London at that time. Muller replied that it was perhaps too early to conclude definitively whether synthetic would replace natural indigo in the long run. Besides, the answer to this question had to be inconclusive in the absence of key data on price and production for synthetic indigo that its manufacturers never disclosed.[24]

On the basis of all the information that he had collected and analyzed, Wyndham R. Dunstan submitted a report to India House in May 1900. Dunstan refuted all claims of superiority of natural indigo over artificial indigo. He argued that the basic chemical composition of the dye extracted from coal tar and plant leaves was fundamentally the same. Thus his report called the artificial dye "in every respect identical with the natural blue coloring matter of indigo." Dunstan advised interests connected with the imperiled natural indigo industry in Bengal to focus their efforts on finding a cheaper way of producing the dye and selling it in a form liked by the consumers. While he pointed out that BASF would further lower the cost of artificial dye in the future and flood the markets with additional quantities of artificial dye, all this did not mean that the natural indigo industry was necessarily doomed. Dunstan believed that natural indigo could survive synthetic indigo's competition. But for that to happen, he called for investigations "on scientific lines" to improve the methods of cultivation of the indigo plant, the process of extraction of coloring matter from the plant, and the preparation of the final product for sale.[25]

[23] Letter from Christopher Rawson to W. R. Dunstan, dated March 27, 1900, PRO, AY4/2047/100168.
[24] Letter from Hugo Muller to W. R. Dunstan, dated April 5, 1900, PRO, AY4/2047/100168.
[25] Bihar State Archives, Agriculture, October 1900, File 2-I/3 3–32, Nos. 11–12.

Dunstan's report became a major turning point in the forward march of indigo experiments in colonial India. As a major report on indigo that was composed by a credible scientist at a metropolitan institution, it sealed the debate on questions that had earlier muddied the water and had served as stumbling blocks in the emergence of clear agendas for the way forward on indigo experiments. In addition, it spurred the enervated provincial government in Bengal to provide support to indigo experiments. Begg, Dunlop and Company thanked Dunstan for the information he had provided. The London branch of the company wrote him a letter saying how "indebted" they were for his expert opinion. They could not have overstated their case.[26]

Dunstan's report won the strong endorsement of officials at the apex of the metropolitan wing of colonial government at India House. Secretary of State George Hamilton concurred with Dunstan, stating unambiguously that "if the Indian indigo industry is to compete successfully with the Badische (BASF) dye, the process of manufacture and of production must be improved and cheapened after full scientific investigation." The office of George Hamilton not only forwarded the report to the government in India, but also recommended that in view of Dunstan's conclusions it would be worthwhile for the local government in India to give "assistance and guidance" to indigo-related research. Descending through the normal chain of command, Dunstan's report and the secretary of state's strong endorsement of the government's role in scientific experiments reached Bengal officials. The Bengal officials now turned to the indigo question in its scientific dimension with true earnestness. It would even be appropriate to say that the push from the metropolis was instrumental in priming the official apparatus in Bengal in favor of planters' indigo experiments.[27]

Defining a Government Role in Indigo Experiments

As indigo experiments progressed the planters staked a claim on the colonial state's funds for their experiments. They variously argued that support for indigo experiments was tantamount to fulfillment of a "public duty" by the government in the colony. Establishing the meaning of

---

[26] Letter from Begg, Dunlop and Company in London to W. R. Dunstan, dated June 25, 1900, PRO, AY4/2047/100168.

[27] For secretary of state's letter, dated June 21, 1900, see Bihar State Archives, Agriculture, October 1900, File 2-I/3 3–32, Nos. 11–12.

"public" in this bargain was not an easy task and involved mutual acts of accommodation between European economic interests who were pushing sectional interests and the colonial state, which pursued broader agendas of colonial governance. The planters had before them the recent instance of the government's denying a loan to a compatriot planter, Sir W. B. Hudson, who had wished to use that money for transitioning to sugar manufacturing. The government had communicated to him in no uncertain terms that it was against its policy to give loans to any particular member of the commercial community. A government committee had also rejected the claims that the precedent of financial assistance by the colonial government to a sugar syndicate in Queensland be the guide for similar assistance to indigo planters in India. Conditions were different in Australia, the committee had argued. The Crown's dominion government in Australia was trying to invite settlers into a new area and introduce a new agricultural industry. In contrast, in India, sugar manufacturing was already "very extensive" and required no such subsidy. Moreover, given the fact that sugar was commonly manufactured by both Europeans and Indians, the government could not single out the Europeans for assistance and ignore the claims of Indian producers of sugar in the region, especially in light of the fact that if promoted, European enterprise was likely to displace the sugar produced by natives who used inefficient and costly methods. The government also could not ignore the claims of other industries on the subcontinent like cotton, jute, oilseeds, and tea, while bestowing special privileges on one industry.[28]

The planters approached the government in 1900 for assistance in their scientific experiments, arguing that even if financial support for the industry were inappropriate, the government's support for indigo science was a legitimate public agenda. Indigo was, after all, the major industry of Bihar, and in that light, as BIPA's E. Macnaghten argued, state support for "scientific research" to an important industry of the region was proper. The scope of indigo research, in any case, was not going to be limited to indigo alone. Its benefits would extend to other crops grown by planters and Indian agriculturists in the region. Thus Macnaghten asserted that "the effect [of this science] would be general, and might be expected to be of enormous value to the country at large." He argued that "information and knowledge would be disseminated over the Province

[28] Report of the Committee Appointed to Inquire into the Prospects of the Cultivation of Sugar by Indigo Planters in Bihar, Bihar State Archives, Agriculture, May 1901, File, 2-I/3 8–12, p. 9.

in a way which I venture could not be obtained by other means."[29] The secretary was alluding to the engagement of Indian workers on the indigo tracts and the interspersing of indigo crop with other crops on the fields of native agriculturists, an intermingling that would ensure dispersal of agricultural knowledge resulting from scientific experiments on indigo ... The IIS representatives also made a similar claim that the seed program of the type they were developing commended itself to government support because it was in line with the work at the government's model farms in the colony and would only add to knowledge of that kind. Such experiments were valuable in a land "where it is recognized that deterioration of crops has resulted from the application of unchanged varieties of seeds to impoverished lands." In more specific terms, the seeds grown under their program would benefit the native seed growers in the North Western Provinces and Punjab as well as in Bengal, who were, after all, the major producers and sellers of indigo seeds, besides helping out the indigo industry in its current state of crisis.[30]

The planters cited the instance of state aid to science by advanced nations to argue that providing assistance for the development of agricultural science was a universally valid agenda. They also appealed to the sense of imperial vanity among state officials to persuade them somehow. The example of government assistance to agriculture in the United States was highlighted with flourish. At the turn of the century it made sense to spotlight the United States. The American agricultural landscape had been fundamentally transformed in the last three decades of the nineteenth century through the establishment of land grant institutions and experiment stations, progress that could be linked to American endeavors in extending the network of institutions. Just as an evidently enlightened and prosperous nation like the United States was providing money for agricultural research, with "whole departments being maintained by it," the colonial government in India should do the same. Planters also pointed out that the Dutch imperial government was providing assistance to indigo experiments on the island of Java where the indigo planters

---

[29] Letter from E. Macnaghten, BIPA, to Revenue Secretary, Government of Bengal, dated August 7, 1900, Bihar State Archives, Agriculture, October 1900, File 2 I/3 3–32, No. 13; Letter from E. Macnaghten, BIPA to Revenue Secretary, Government of Bengal, dated November 13, 1900, Bihar State Archives, Agriculture, October 1900, File 2 I/3 29–30, Nos. 154–5.

[30] Letter from Messrs. Begg, Dunlop and Company, IIS, to Secretary, Government of India, Revenue and Agriculture Department, dated January 31, 1901, Bihar State Archives, Agriculture, May 1901, File 2 I/3 1–7, Nos. 1–2.

were facing a similar threat to their industry.[31] Macnaghten went to the extent of writing to the secretary of agriculture in Washington, D.C., and to the British consul-general at Batavia eliciting further information on the nature of involvement of the American government and the Dutch colonial government with agricultural science.[32]

Planters' organizations showcased the substantial scale of their current scientific program for indigo improvement that they sought to expand. The BIPA first asked for support to hire a bacteriologist and returned with a much expanded plan that was drawn by its in-house chemist, Christopher Rawson, and that involved hiring an additional chemist, a bacteriologist, an agricultural chemist, a botanist, an entomologist, and four assistant analytical chemists.[33] The IIS asked for assistance to expand its seed development program running at Dalsingserai through the employment of three additional chemists whom they wished to house in new laboratories in different agroclimatic zones of Bihar. Soil conditions in Bihar were diverse, and IIS contended that the new data from plots far afield would help fine-tune the task of improving cultivation practices and the acclimatization of plant varieties. In making its case the IIS highlighted the success of current work at its center in Dalsingserai. In some trials experimenters at Dalsingserai had made the color content in leaves increase by 140 percent after they treated the crop with a particular combination of fertilizers. Some of this work by the IIS's scientists had been carried out in collaboration with experts at the government's own Royal Botanic Garden in Calcutta.[34]

The planters found state officials in Bengal open to their suggestion for support for their scientific experiments. The government found moving on assistance with scientific inquiries more acceptable than providing financial assistance that they had refused earlier. There was a remarkable change in the tone of colonial state officials that was attributable at least

[31] Letter from E. Macnaghten, BIPA, to Revenue Secretary, Government of Bengal, dated, November 13, 1900, Bihar State Archives, Agriculture, October 1900, File 2 I/3 29–30, Nos. 154–5.

[32] Letter from E. Macnaghten, to Revenue Secretary, Government of Bengal, dated December 27, 1901, Bihar State Archives, Agriculture, January 1901, File 2 I/3 28–9, Nos. 40–1.

[33] E. Macnaghten's letter to Secretary, Government of Bengal, dated August 7, 1900, Bihar State Archives, Agriculture, October 1900, File 2 I/3 3–32, Nos. 13; Christopher Rawson's letter to BIPA, dated November 12, 1900, contained in BIPA's communication with Government of Bengal, dated November 13, 1900, Bihar State Archives, Agriculture, November 1900, File 2 I/3 29–30, Nos. 154–5.

[34] Letter from Messrs. Begg, Dunlop and Company, IIS, to Secretary, Government of Bengal, Revenue Department, dated January 31, 1901, Bihar State Archives, Agriculture, May 1901, File 2 I/3 1–7, Nos. 1–2.

partly to the general building of consensus in favor of scientific investigation of natural indigo along the entire axis from the metropolis to the colony. The Bengal government passed an important resolution in favor of support to indigo science, which conceded that "the indigo planters as a body had claims on the government." They had, after all, rendered "valuable services" of note in times of "administrative stress." Nobody had forgotten the assistance by indigo planters at the time of the 1857 mutiny when the latter had organized themselves into a military corps to quell the Indian insurgency. In time of drought the planters often assisted with the distribution of relief work partly out of genuine philanthropy and partly out of self-interest in preserving the cultivating ability of agriculturists in the region. "Their disappearance," the report noted, "would be in many ways a great administrative loss." Besides, the government also noted this time that the planters had invested a large amount of capital in the industry, which seemed threatened by foreign competition. On that criterion alone, in the opinion of some officials, the expatriate community deserved the government's support.[35] Nobody wished for an important colonial industry in the region to wither away.

Bureaucratic Vision and Indigo Experiments

The decision by colonial officials to assist indigo experiments stemmed from enduring political and economic concerns to preserve a constituent of the ruling class, to protect a colonial industry, and to realize the promise of science in the aftermath of the Dunstan Report. All these aspects were evident in internal official discussions. The important state official in Bengal, a career civil servant, F. A. Slacke, appreciated very well the potential of chemistry in the task of improvement of Bengal indigo. He made an example of the role played by chemists in the Javanese indigo industry in a note, highlighting that "no factory of any size in Java is without a chemist resident on the spot." He was in favor of inviting, say, "two really good chemists" to India to work on indigo improvement under the government's watch. The outcome of such research could then be made available to all planters. In the months after the planters had approached the government for a subsidy Slacke wrote to the lieutenant governor, J. Woodburn, arguing that "looking at the very large interests at stake politically and otherwise, I do not think it would be

---

[35] "Resolution – by the Government of Bengal, Revenue Department," Bihar State Archives, Agriculture, October 1900, File 2-I/3 3–32, No. 25.

too much for the State to aid [indigo experiments] for a few years with a large grant." This constituted the clearest expression of interest within the bureaucracy in assisting indigo and indigo planters with science. He again referred to the Java example to make a case for providing government support for experiments on indigo and other crops. Inquiries should be made of the scope of the Dutch government's assistance to its planters in Java and a similar effort could then be launched in India. "If such a petty Government as the Dutch consider[s] such expenditure by the State essential, there would probably be a strong reason for getting Imperial assistance here," he remarked, playing to British imperial pride.[36] In line with such sentiments the government first sanctioned funds for the appointment of a bacteriologist to BIPA and then seriously began to consider additional support as it received further requests for funds from the two planters' bodies.

The beginning of the colonial state's involvement with indigo experiments, even at the planning stage, inevitably brought the evolving program of indigo laboratories and experiment stations in line with government norms and priorities. The scheme for experiments and trials proposed by the planters' scientists was scrutinized on grounds of economy and efficiency by the bureaucrats. The details of the programs were evaluated by the government's in-house experts. Slacke set the tone for the beginning of this turn toward government scrutiny by asking that the schemes proposed by the two planters' bodies be placed before George Watt, Reporter on Economic Products, the highest-ranking government expert on agricultural matters. Watt was to determine what was really needed, the appropriate cost that the government should bear, and the duration of the experiments.

State officials imputed their own norms of rationalization to the planters' indigo program. The officials were above all uncomfortable with what they saw as duplication and overlap in the research programs proposed separately by the BIPA and IIS and insisted that the two associations merge for purposes of research. As Slacke noted, "To give aid to both bodies would, I think, seeing that each is working on the same lines, be a waste of money."[37] The drive to force a merger was also influenced by a strong sentiment among officials that all resources in the colony be pooled in order to optimize efforts on agricultural experiments of all

---

[36] Notes and Orders, dated July 28, 1900, p. 10, Bihar State Archives, Agriculture, October 1900, File 2-I/3 3–32; Notes and Orders, dated November 17, 1900, p. 2, Bihar State Archives, Agriculture, November 1900, File 2 I/3 29–30.

[37] Notes and Orders, p. 3, Bihar State Archives, Agriculture, May 1901, File 2 I/3 1–7.

kinds including indigo. At an important conference held in the office of the Bengal revenue secretary, central and provincial officials gathered to take stock of the need for institutions and infrastructure for agricultural experiments in Bengal. The attendees thought that pooling of indigo planters' resources and the government's resources was "essential" for the purpose of establishing agricultural institutions in Bengal. Even though Slacke's conviction at this time for deferring any immediate measure for establishing public-private cooperation carried the day, the idea of such a partnership would resurface later. Slacke had argued that the time was not appropriate for such steps, given that the indigo industry required immediate assistance and forging government-industry collaboration would require many consultations at various levels over an extended period. But there was not going to be any holding back on allowing the planter bodies to follow their research programs separately. As officials made coalescing of planters' experimental programs a precondition for financial support, the two associations fell in line. A merger was more or less thrust on them, at least temporarily.

Governmental style of implementing economy also cast its influence on the program for indigo experiments. Lieutenant Governor Woodburn was taken aback by the seeming profligacy of the budget for scientific experiments initially submitted by BIPA's Christopher Rawson. He remarked in official correspondence that Rawson's letter requesting government funds "reads as if somebody had told him he had at his disposal any money he liked."[38] As was the practice, the bureaucrats always ended up asking, "Is the establishment [incurring expenses] neither more nor less than that required?" And Rawson's proposals did not pass that test. Even George Watt, more discerning on technical matters than Slacke, assailed the budget submitted by Rawson as thriftless and asked planters "not to ask for more assistance from Government than is absolutely necessary." He sarcastically wrote that "Mr. Rawson proposes a larger scientific staff to investigate indigo alone than is in the employment of the Government for all branches of the agriculture and industry of the entire empire [on the Indian subcontinent]." In demanding curtailment of the proposal, Watt made an important distinction between a research program on public funds and one that might be organized through private initiative in the industry. He admitted that he did not doubt that all of staff proposed by Rawson could be "fully occupied" in the task of indigo improvement, and in fact the size of the proposed staff by Rawson was still smaller than

---

[38] Notes and Orders, p. 2, Bihar State Archives, Agriculture, November 1900, File 2 I/3 29–30.

the "army of research[ers] that is engaged on the production of artificial indigo." But that said, the government could not entertain a demand for such a large pool of scientists for natural indigo, which was but a single industry in the colony. Officials looking after the government's finances were likely to remind the planters, if any such comparisons were drawn with the German synthetic industry, that the researchers working on synthetic indigo were paid by private industry. The government would not sanction expenditure of "public money" on indigo when several other industries were "very nearly in as necessitous a condition as indigo."[39]

Watt suggested frugality in indigo experiments if planters wanted to have any realistic chance for obtaining government aid. Indeed he believed that the scheme proposed by planters "might with advantage be cut down by fully one-half." Instead of a research chemist, an agricultural chemist, and four assistant analytical chemists, Watt recommended the hiring of just one chemist. And instead of a separate entomologist and botanist, Watt thought that the employment of one biologist would suffice. Now that the indigo research program was about to become a part of the government's own research program, he recommended its aligning with the different arms of the public apparatus of research. An entomologist had been recently employed at the Indian Museum and a mycologist's post had been sanctioned. Such specialists could fulfill the function of specialists in identifying insect pests and fungal blights that infested the indigo crops. With such experts already available there was no need for industries to hire such specialists individually.

Watt made another observation that almost seems prophetic with respect to the long-term trajectory of the evolution of indigo program. He noted a lack of balance in the indigo experiments that were currently under way. The planters' program in his opinion had put a rather disproportionate emphasis on improving postharvest extractive processes at the expense of adequate attention to the plant in the field. This could be said to be true despite the notable work on biological improvement of the plant completed at Dalsingserai. And in Watt's opinion the road to success actually lay via experiments of a broadly agricultural nature. "It is," he wrote, "a disgrace to the industry that so little should be known of the botany and agriculture of a plant upon which so much capital has been invested." In particular, Watt noticed a lack of past efforts on "selection" of plants, which to his mind was "the rational line of improvement"

---

[39] Demi-official from Dr. G. W. Watt to Secretary, BIPA, dated January 31, 1901, No. 288, Notes and Orders, pp. 16–19, Bihar State Archives, Agriculture, May 1901, File 2 I/3.

for the future. Looking forward, Watt specifically suggested a role for a biologist at newly established experimental farms who could work on the selection of indigo plants for the colony. The biologist would scout the different provinces for better seeds and engage in the work of selection at farms. This process would likely take years, but at the end of this period of sustained effort it was possible that "just as we have pedigree wheats and blight-resisting wheats, so should we possess corresponding stocks of indigo." Notably, the reference to "pedigree selection," certainly one of the most sophisticated forms of selection practiced anywhere then, showed that colonial actors considered the route of selection and hybridization of plants in the colony to be both desirable and feasible.[40] The colony's indigo experiments in the later days increasingly fell back on the agricultural and botanical path suggested by George Watt.

In more immediate terms also, Watt's vision for rationalizing indigo experimentation influenced the shape of the new joint program for indigo science that was launched in the colony beginning April 1, 1901. BIPA and IIS amalgamated and proposed a joint plan for research along the lines advocated by George Watt. The plan called for the employment of a bacteriologist, a chemist, and a biologist to add to the current staff of the chemist, Christopher Rawson, and the "expert" agriculturist, Bernard Coventry. In the unified program the manufacturing experiments were to be conducted at Peeprah under the leadership of Rawson and agricultural experiments at Dalsingserai under the supervision of Coventry. The two arms were to be centrally supervised by a research committee. As for the money part, the planters proposed a total budget of Rs. 150,000 with even contributions of 75,000 each from the planters and the government. The officials scaled down the total budget to 125,000, putting the government's contribution at 50,000 per annum and the share of the planters at 75,000 on the principle of two-fifths and three-fifths contribution from the government and planters, respectively. Planters belonging to the BIPA and IIS would contribute 50,000 and 25,000, respectively. While two-thirds of the funds would meet the expense of experiments at Peeprah, one-third would go toward Dalsingserai's trials and experiments. The officials also clarified that the government's share of research money

---

[40] It is no surprise that Watt should mention selection of wheat since substantial success in breeding vigorous strains of wheat was a crowning achievement of English pre-Mendelian breeders. Watt probably also had in the back of his mind the spectacular achievements of R. H. Lock on maize at the Peradeniya Royal Botanic Gardens in Ceylon in the early twentieth century. In Britain and in the British Empire meritorious work on selection was becoming commonplace.

would be forthcoming only after the accountant general confirmed that the planters had submitted their share to the central research committee. The stage was set for the commencement of the joint funding of experiments at Peeprah and Dalsingserai.[41]

Patronage and Indigo Science, 1902–1904

During the extremely difficult years for the indigo industry from 1902 to 1904 planters struggled to send their financial contribution to laboratories. In 1901, the very first year of the joint research program, the amalgamated association barely met its entire obligation. The funds from indigo planters and traders continued to taper off through 1902. The ability of planters to collaborate on indigo experiments was further eroded by the state of total disarray into which their organization fell. The planting community became beset with divisiveness, and the organization's functioning was compromised by the alignment of planter-members along regional lines. Planters from the western districts of Champaran, Saran, and Muzaffarpur wanted their contribution to go toward supporting experiments at Peeprah only. Another section of planters including those belonging to the Darbhanga subcommittee of the now-dissolved Research Committee wanted their subscription to fund only Dalsingserai work. The trading groups formed a group of their own. By the middle of 1902 the cooperative mechanism for the supervision of indigo experiments formally split on account of irreconcilable interests within them.[42]

In one form or another, the planters now wished to be "free riders" on indigo experiments where they would be beneficiaries of indigo science without having to contribute to its conduct. Some planters met with Lt. Governor Woodburn in Darjeeling in June 1902 to explain their position. On behalf of the BIPA, H. Hudson pointed out that the planters had been incurring heavy losses due to the low price of indigo in recent years. In such a situation BIPA was unable to raise subscriptions from

---

[41] Slacke's letter to BIPA, dated March 27, 1901, Agriculture – No. 1711, Bihar State Archives, Agriculture, May 1901, File 2 I/3, No. 4; extract from the "Proceedings of a General Committee Meeting of the Bihar Indigo Planters' Association held at Muzaffarpur on the 8th December 1900," Bihar State Archives, Agriculture, January 1901, File, 2I/3 28–9, Nos. 40–1; Bihar State Archives, Agriculture, May 1901, File, 2I/3 1–7, Nos. 3(b), 3(c), 4–8, and miscellaneous remarks in Notes and Orders.

[42] Letter of L. Hare, Commissioner, Patna, to Revenue Secretary, Bengal, dated May 12, 1902, Bihar State Archives, Agriculture, June 1902, File, 2I/3 of 10–14, Nos. 66–7; Notes and Orders, pp. 3–4, Bihar State Archives, Agriculture, December 1902, File 2I/8, 15–16.

the planters and was in turn unequipped to pay for the experiments. Many planters belonging to BIPA had completely divested from indigo and were no longer interested in supporting indigo experiments. Hudson suggested that the government put Peeprah, the new chemical laboratory of BIPA, and Dalsingserai under its control and assume overall responsibility for the experiments with the support of those few planters who were still willing to send funds. Other members of BIPA present on the occasion, J. B. Norman and L. I. Harrington, echoed the view of Hudson. Bernard Coventry was the lone voice from IIS. He expressed his wish to continue the experiments at Dalsingserai with hopes of support from the state exchequer.[43] The BIPA completely stopped funding the Peeprah laboratory on January 1, 1903, and began to wind up its operations. Coventry barely managed to keep operations running at Dalsingserai. His trials were largely funded with the help of carryover money from the previous year, and with contributions from Calcutta-based traders and a few planters around Dalsingserai. Begg, Dunlop and Company, the honorary secretaries of the IIS, suggested to the government that an exception should be made to the earlier rule of three-fifths and two-fifths contribution from planters and the government, and that the government should match whatever amount the planters could muster. Even this late maneuver to maximize inflow of resources into Dalsingserai did not prove adequate. The funding situation for Dalsingserai worsened, and in early 1903 the station had exhausted its resources and its closure seemed imminent. Only the grant of new government funds in March 1903 gave a fresh life to Dalsingserai.[44]

Peeprah

The scientists at BIPA's new laboratory at Peeprah near modern district of Gopalganj were ready to pursue a new strategy in manufacturing experiments by renewing focus on steeping. This strategy was the outcome of Rawson's belief that unwanted reactions during fermentation were leading to a loss of color. Peeprah was now well placed to conduct such experiments as a result of the appointment of the new bacteriologist, Cyril

---

[43] Revenue Secretary W. C. Macpherson's note, dated June 3, 1902, and Lt. Governor J. Woodburn's note, dated June 9, 1902, Notes and Orders, pp. 6–7, Bihar State Archives, Agriculture, June 1902, File, 2I/3 of 10–14.

[44] Letter from Begg, Dunlop and Company to Commissioner, Patna Division, dated May 6, 1902; Notes and Orders, pp. 4–5, Bihar State Archives, Agriculture, June 1902, File, 2I/3 of 10–14; Letter of A. Earle, Revenue Secretary to Commissioner, Patna, dated March 21, 1903, Bihar State Archives, Agriculture, December 1903, File, 2I/8 3.

Bergtheil. Trained initially in chemistry at the Industrieschule in Germany and at University College, London, Bergtheil had then obtained further specialized instruction in agricultural chemistry and bacteriology at the newly established agricultural college at Wye. Bergtheil was appointed under the terms of the new research program jointly sponsored by the government and the planters.[45] Under his supervision experiments on steeping became the centerpiece of scientific work at the Peeprah Indigo Station.

Early results from the effort at Peeprah provided an understanding of the relevant processes in steeping but stopped short of providing ready solutions that the planters had wanted. Planters wanted their scientists to tell them when exactly to stop steeping since they knew from experience that changing the length of fermentation resulted in difference in output. In fact, planters had often modulated the time of fermentation across different batches of steeping to optimize yield in their factory. They wondered whether any formulaic system could be devised to predict the optimal length of the steeping operation. Bergtheil explained that the conditions under which steeping was performed varied widely. The actual time of steeping would depend on the type of plant, the conditions under which it had been cultivated, the time lag between harvesting and manufacturing, the method of its placement in the vat, and so on. Given the number of variables Bergtheil found it impossible to suggest any fixed predetermined estimate for the time of steeping.[46]

Efforts were made to control the behavior of microorganisms in the steeping vat by using a combination of chemicals and antiseptics and by heating. But in each case making the process economical for commercial agriculture was proving a bottleneck, although trials were still ongoing. Bergtheil explained that the enzymes already present in the leaves and bacteria of the steeping liquor jointly acted on the leaves' glucoside, forming a body that on oxidation gave color. But the liquor contained several kinds of bacteria, some of which produced the desired change in glucoside, a few that remained nonreactive, and a few others that caused

---

[45] Cyril Berkeley, *My Autobiography* (privately published, n.d.), pp. 1–6; Cyril Bergtheil, after his immigration to Canada around the time of the world war, changed his name to Cyril Berkeley, to avoid the anti-Germen sentiments prevailing there. His earlier surname readily indicated his German Jewish ancestry, and he saw sense in dropping it for good (henceforth "*Autobiography*"). A copy of the autobiography has survived with Professor Mary Arai, granddaughter of the scientist, at the University of Calgary in Canada.

[46] "Bacteriologist's Note I" Cyril Bergtheil, dated August 9, 1902, p. 2, Bihar State Archives, Agriculture, December 1903, File 2-I/7 3.

undesirable changes leading to loss of recoverable indigotin. It was practically impossible to isolate useful bacteria while removing others. To achieve controlled steeping, therefore, Bergtheil experimented with a process in which all bacteria were killed with antiseptics to obtain a "sterile" environment. Chemical change thereafter could be brought about by the action of enzymes. However, the use of enzymes on an industrial scale was not feasible given their high cost, and thus this thread was discarded. On trials with heating, Bergtheil found that fermentation became "more speedy and complete," possibly by exterminating the wasteful bacteria or making them inactive. Three alternatives existed on the use of heat: First, steam the plant and then conduct steeping with warm water or water at ordinary temperature. Second, use boiling water to destroy both bacteria and enzymes, and then complete fermentation with specially cultured enzymes, an option that had already been ruled out. Third, use hot water at a temperature of 150°–160° F for steeping. Bergtheil told the planters that experiments were currently under way to determine the best and most economical way of using heat for fermentation.

As the year progressed the sense of hopelessness among planters with the results emerging from chemical experiments at Peeprah became noticeable. The planters were frustrated by the lack of applicability of Rawson's suggestions. His recommendations to date on the use of blowers and deep vats or the use of acid in boilers to remove impurities did not find many takers. While the yield increase from the former innovation did not justify costs, the latter ran into resistance from traders who feared that any remaining traces of acid in the dye might harm clothing dyed with it. Inspector general of Agriculture, James Mollison, sounded warning bells by prophesying that the planters were not likely to continue funding manufacturing experiments much longer. "Mr. Rawson and his assistants have accomplished striking work…. Yet very few useful results have been obtained…. The cost of applying Mr. Rawson's suggested improvements exceeds the value of increase of produce at existing market rates."[47] Later in the year Rawson presented a report on the direction of steeping experiments that also did not raise optimism. On certain days, when conditions turned out to be perfect, Rawson believed that the output from current methods of fermentation was "not more than 10 per cent below the theoretical." But at most other times, the extraction from fermentation "was far from complete." He knew that the hope for any real improvement lay

---

[47] Mollison's notes are cited in Notes and Orders, p. 2, Bihar State Archives, Agriculture, December 1903, File 2I/8 3.

with steeping. But he was getting pessimistic. "The steeping process," he said, "is receiving every possible attention but the difficulties of regulating the operation are very great."[48] The BIPA did not renew Rawson's contract and began negotiations with Bergtheil for an early termination of his contract, in the end even pleading with the government to pay Bergtheil's salary for the rest of his contract term.

Dalsingserai

It was at Dalsingserai that George Watt's vision for giving indigo experiments an additional agricultural orientation was realized. The agricultural chemist, E. A. Hancock, had left the station while Bernard Coventry, the biologist H. M. Leake, and the chemist William P. Bloxam collaborated at Dalsingserai. Although Bloxam continued his purely chemical analysis on color percentages in the dye at different stages of manufacturing, the dominant thrust of work at Dalsingserai remained focused on the plant itself. Indeed the continuing work at Dalsingserai with developing higher-yielding varieties seemed to offer hope of meeting the competition of synthetic indigo, as against the promise of incremental and often marginal increase in yield through innovations suggested by the chemists at BIPA laboratories.

The selection program on the indigo plant at Dalsingserai formed a partnership with the Royal Botanic Garden of Calcutta and underwent considerable expansion. This enlargement in scope reflected a natural growth in the program. Because a broader base of germplasm was desirable to raise the possibility of transferring a greater number of traits across species, the station continued to procure more plants. Inclusion of more plant types also stemmed from the initial lack of success in acclimatizing the Java variety. It was noticed that while the Java plant thrived in the surroundings of Calcutta and probably would fare well in all of central Bengal, its cultivation in Bihar, the primary zone of indigo agriculture in the colony now, was proving to be problematic. The director of the Royal Botanic Garden, Major David Prain, warned that "the time has not yet come for the complete abandonment of the species that is now generally grown." As a consequence experts at Dalsingserai began to focus on indigenous species as it started to seem possible that selection of indigenous varieties might also prove to be promising. In this

---

[48] "Notes on Experimental Work Done at Peeprah during the Morhan Mahai 1902," by Christopher Rawson, dated 11 August 1902, p. 4, Bihar State Archives, Agriculture, August 1903, File 2I/7.

endeavor they decided to develop pure lines of seeds of plant varieties from northwestern India that were widely in use in Bihar. These seeds were grown over an expansive zone extending from Kanpur and Delhi to Multan and Dera Ghazi Khan in West Punjab. The planters were not even sure whether the seeds claimed to be emanating from a particular subregion were actually grown there or were imported from somewhere else by the traders based in those areas. In order to move ahead on the plan for isolating pure lines of local varieties, Prain proposed sending the curator of Calcutta Herbarium, A. T. Cage, and the in-house botanist H. M. Leake to these areas on a scouting mission, a suggestion that was promptly accepted by Bengal officials. Such an expedition would not only help establish identifiable "lines" for the selection work but also choose the right candidates for that work. After their expedition Cage and Leake advised the planters strictly to avoid seeds originating from areas farthest west. They informed them that territories between Kanpur and Delhi had climate and cultivation conditions approximating Bihar's, and thus seeds grown only from those areas were appropriate for use by the planters.[49]

Dalsingserai's most notable contribution came in the way of acclimatization of the *I. arrecta* species. This was the mother stock out of which the Java variety had been developed in the Dutch Southeast Asia. The planters had earlier procured Java seed. Through H. A. Baily's visit to South Africa they had also received the "pure" *I. arrecta* breed from Africa and started a separate lines of cultivars for their selection work. Apparently the experimenters at Dalsingserai were indulging in broad-based mass selection and individual selection of indigo plants involving screening over generations. The government also ensured that the Natal seed was tried out in other provinces. Having partly financed H. A. Baily's trip, the state officials ensured the delivery of Natal seed to the provincial governments of Madras and the United Provinces of Agra and Oudh and possibly others. Coventry was also directed to share information with others on ongoing acclimatization efforts on the Natal variety.[50]

[49] Letter of Superintendent, Royal Botanic Garden to Revenue Secretary, Bengal, dated August 30, 1902, containing the report of Captain Cage; letter of L. E. B. Cobden-Ramsay, Revenue Under Secretary, Bengal, to Commissioner of Patna Division, dated September 15, 1902, Bihar State Archives, Agriculture, November 1902, File, 2I/11 1–5, Nos. 63–4; A. T. Cage's report to Superintendent of Royal Botanic Garden, Calcutta, dated November 5, 1902, Bihar State Archives, Agriculture, November 1902, File, 2I/11 1–5, Nos. 66–7.

[50] "Natal Indigo Seed," Bihar State Archives, Agriculture, December 1902, File, 2I/6 9–25, Nos. 1–19.

Bernard Coventry's report on the progress of acclimatization trials with *I. arrecta*, both Natal and Java lines, reflected the fact that the program of selection was essentially a work in progress. Coventry tried to contain optimism with regard to Natal-Java indigo. There was no doubt this new species of indigo had the potential to yield far more color than the native variety; however, Coventry noted that "its cultivation has so far been [only] a partial success." In late 1902, the trials on Natal-Java indigo, he stressed, were "still in the stage of experiment."[51] The nature of acclimatization had to be characteristically slow and one that paid dividends only with trials over several seasons because of the difference in the annual cycle of weather between Java and Bihar. In Java the plant was sown in June and harvested thrice, once each in September, December, and March. The weather conditions in Java were moderately hot and humid throughout, in contrast with conditions in Bihar, where an extremely hot and dry season was interrupted by frosty winter. The conventional time of sowing indigo in Bihar in the month of March was unsuitable for the new variety. Neither could the planters sow Natal-Java indigo after the end of rainy season because the new variety required moisture in the air after sowing. Coventry thought that sowing in June or July would be best. But that raised new questions over whether the manufacturing, which would then have to be extended well into winter, would give a good-quality output, as past experience showed that indigo manufactured during the cold season did not turn out the best quality. Coventry and his cohorts were also experimenting to discover whether the practice of transplantation, which was routine in Java, was practical in Bihar. They were also devising the best ways to control weeds for the new variety.

But the difficulty of the task of acclimatizing notwithstanding, planters and state officials alike pinned their greatest hopes on trials at Dalsingserai Station. Coventry's men seemed to provide to the planters a realistic chance of selecting a potentially higher-yielding variety. The biological manipulation of plants through selection was simply first-rate science by a team of competent experts that also made Dalsingserai the coveted station of colonial officials. The selection experiments of the nature attempted there were unprecedented for the colony. Thus planters and officials alike stood solidly behind such work in the coming years.

---

[51] Bernard Coventry's report to Begg, Dunlop and Company, dated November 13, 1902, Bihar State Archives, Agriculture, December 1902, File, 2I/6 9–25, No. 15.

Tapping into Indigo Science and American Beneficence

The indigo laboratories became embroiled in colonial contestation of another type as their paths crossed with the state-initiated effort to create an infrastructure of agricultural institutions. The Bengal officials were initially drawn to indigo experiments in their effort to save a major colonial industry. The indigo experience for the officials wherein they came in close quarters with indigo science was fortuitous in making them acutely knowledgeable about the promise of science for the development of the region's crops. Besides, the generic character of indigo science offered the possibility of transferring experts and infrastructure involved in indigo development to the task of developing a science for agricultural improvement in the colony, a prospect that provincial officials in Bengal and subsequently those at the center were quick to notice.

The increasingly widespread interest in developing colonial agricultural science was the outcome of what Peter Robb has rightly identified as the colonial state's assumption for the first time of a direct role in agricultural development in the last three decades of the nineteenth century. Robb has also indicated that in this statist vision of development "knowledge was regarded as an instrument of policy and a channel for improvement."[52] In order to create knowledge – the new "byword" that determined policy in the newly set up imperial and provincial Agricultural Departments – the officials built data on demographics, agricultural productivity, and native cultivation practices through numerous initiatives. Efforts at creating knowledge of a scientific variety specifically had begun in the 1870s. These policies were palpable both at the center and in Bengal and were broadly reflected in the openness of colonial state officials to agricultural science, their willingness to part with public funds to sponsor the laboratories of indigo industry partially, a desire to collaborate with the indigo industry on a joint research program on regional crops, and later even to absorb the indigo scientists to work in public institutions of science that were beginning to be set up in the colony.

In a significant effort of institution building, Bengal officials in 1899 initiated efforts to set up a regional agricultural center. The officials were beginning to settle on a site for this center. They discussed among themselves that the center could be set up at Pusa in Darbhanga district as

[52] Peter Robb, "Bihar, the Colonial State and Agricultural Development in India, 1880–1920," *Indian Economic and Social History Review* 25, No. 2 (1988): 205–35, quote on p. 216.

there was prospect of a large public farmland becoming avalible there. The Bengal government owned a sprawling estate at Pusa, where it had earlier run a model farm from 1875 to 1876. The land was subsequently leased out to Begg, Dunlop and Company for experiments on tobacco culture from 1877 to 1897. At the expiration of the lease, Begg, Dunlop and Company asked for permission to turn Pusa to other endeavors. The officials rejected that proposal, maintaining that the introduction of tobacco culture in the region was the raison d'etre of the government's earlier subvention and that it would rather not endow the land for any purpose other than agricultural development. A little later the government decided to take over the land itself and made plans to turn it into a cattle and dairy center. They wrote to officials in the revenue department at the center and made the case that Bengal was facing a problem of "severe" degeneration of its livestock. They argued that a cattle-breeding center at Pusa dedicated to improving the native stock would serve Bengal well.[53]

Ambitions soared subsequently, and Bengal officials began to envision a composite center of a grandiose scale unprecedented for the colony, a center that would be dedicated to research and teaching, as well as extension work. In a seeming amplification of the plan for mobilizing first-rate agricultural science, the provincial officials now wished to bring together a team of specialist scientists, start laboratories, and establish an experimental farm and a cattle-breeding facility. The center would be at the apex of existing small farms provincewide. Until now some elementary level of scientific agricultural trials were conducted at the government's experimental farms and model farms. The first seven model farms were simultaneously set up by the provincial government in 1871. All of them disappeared soon after in the following famine. But they were quickly replaced by others. Such a quick turnover rate in the establishment of experimental farms was in keeping with the temporary and transient nature of such small centers with modest goals. The government frequently established such farms on government land, on a plot from the Court of Wards, or even on a private holding that the government specifically acquired for the purpose. These centers typically conducted trials with fertilizers and seeds, introduced new crops, and served as model farms for demonstration purposes. In some cases the farms served usefully as centers for distributing seeds. The existing farms in the province

---

[53] Letter of F. A. Slacke, Secretary, Govt. of Bengal, to Secretary, Revenue and Agriculture, Govt. of India, dated January 7, 1899, National Archives of India, GOI, R&A, Proc., File No. 8 of 1899, A Proceedings, Nos. 11–14.

at the end of the nineteenth century were apparently considered too modest for the reinvigorated agenda of agricultural improvement in the province. The new plan of Bengal officials therefore envisaged marshaling the best available science by employing expert agriculturists, biologists, and chemists. These experts would work in the laboratories, while some would additionally teach at the agricultural college attached to the center. The planned institution at Pusa would also centralize the existing infrastructure for scientific research on agriculture. The head of the new center would thus also supervise work at the three experimental farms owned by the government at Dumraon, Burdwan, and Chittagong. The preexisting farms would continue to serve as outstations in various subregions, generating information and testing measures suitable to the varied agroclimatic zones.[54]

If a later-date communication from the viceroy is to be believed, the conduct of indigo experiments in Bengal had enlivened the provincial officials to the potential of agricultural science. The decision in principle by state officials to fund indigo experiments led them on a path where they were poised to scrutinize indigo science from close quarters. Over the course of consultations relating to organization of indigo experiments the officials reconnoitered the indigo project and gained awareness of the potential of agricultural laboratories. Indeed there are indications that the officials had early on come to appreciate the potential of indigo science for broader application. At the level of organization they did not see a lot of distance separating targeted indigo science from crop science in general, believing that similar if not the same institutions and experts could be deployed for scientific work with crops other than indigo. As discussed before, they were open to the idea of a single public program for agricultural science that could possibly accommodate planters' private science and meet their purposes as well.

It was in the midst of all the deliberations over indigo research that the revenue secretary, F. A. Slacke, called in parallel a meeting of the most significant imperial agricultural officials and the agricultural and education department officials of Bengal and North Western Provinces in Calcutta to formulate an agricultural science policy. Notably, the meeting jointly discussed the subject of indigo science and science for Bengal's agriculture,

[54] Letter from W. C. Macpherson, Secretary, Govt. of Bengal, to Secretary, Rev and Agriculture, Govt. of India, dated June 30, 1902, Govt. of India, Proceedings of the Dept. of Rev. and Agriculture, February 1903, "Utilization of Government Estate of Pusa in Darbhanga for Agricultural Experiments," No. 9, File No. 72 of 1902, India Office Records, P/6592.

with most participants arguing that a partnership between the planters and the government was "essential." George Watt, the economic reporter, pointed out the precedent in Ceylon, where the government and the local tea industry were collaborating on scientific research. Pooling of resources was of utmost importance. The officials acknowledged the merit of work being done in indigo laboratories. They were aware of the credentials of the scientists working there. Besides, these indigo experts knew local agricultural conditions in India, having lived and worked in the colony for a while. Their value as experts was appreciated in light of difficulties known to be encountered by experts freshly brought in from England or anywhere else, who typically took time to learn the local conditions. In fact, James Mollison, who was soon to be appointed as India's first Inspector General of Agriculture, the highest-ranking agricultural official in the colony, as well as others, called for an outright "absorption" of the scientific work done by the indigo experts Hancock and Rawson within the proposed center for agricultural science at Pusa. It seemed logical to everyone "to combine all the [scientific] establishments now at work [on indigo], and to carry on their proceedings together as one series of operations [for agricultural research in the colony]."[55]

The agricultural center proposed by Bengal officials incorporated broader goals by adopting a common approach to the improvement of indigo and "all country crops," one that would be collectively useful to European planters and Indian peasants alike. From the Bengal government's perspective the effort on indigo had to be a component of the larger goal of crop improvement and agricultural development in Bengal. The work at the new station, as a note prepared by James Mollison repeatedly emphasized, must suggest practical measures that would lie "within the reach and means of ordinary cultivators." Such spirit was in line with the colonial state's efforts to strike a balance in affording protection to both European economic interests and day-to-day cultivation requirements of Indian peasants.

The government experts mooted a clear plan for agricultural experiments at Pusa whose direction was somewhat similar to the agricultural line of indigo trials under way at Dalsingserai. Inspector general

[55] "Confidential, Minutes of the Proceedings of a Meeting held at the office of the Revenue Secretary to the Government of Bengal, on March 6, 1901, to consider the question of the establishment of an Agricultural Research [Station] at Pusa in the District of Darbhanga," Govt. of India, Proceedings of the Dept. of Rev. and Agriculture, February 1903, "Utilization of Government Estate of Pusa in Darbhanga for Agricultural Experiments," No. 9, File No. 72 of 1902, India Office Records, P/6592.

of agriculture James Mollison's perspective note on developing the Pusa estate stressed improving varieties of indigo, sugarcane, or rice by selection, an approach that was reminiscent of Coventry's selection work on indigo. Mollison further wished for all of the agricultural side of the work including teaching to be under the charge of an agriculturist and the "chemical-technical" aspects of the work to be under a chemist. In addition, he wanted the station to have a botanist who would teach at the agricultural college as well as serve as a "gardener," engaging in propagation of fruits, vegetables, and trees. The note clearly put the practice of breeding and agricultural botany at the center of future plans for crop improvement in the colony.[56]

The grand plan for Pusa was hardly feasible solely with the funds of the provincial government, but its backers nonetheless insisted that opening the center was important. Bengal wrote to the central government asking for support through imperial funds to start the prestigious center, which would address many issues of local and supralocal import. The planned science would improve the "present country crops" in Bengal and beyond, whereas the center's cattle-breeding arm and agricultural college would be useful to the urgent need for cattle improvement and agricultural training in the province. The indispensability of the science at the center was highlighted. In addition, the Bengal officials suggested that the center would offer the newly appointed inspector general of agriculture a "means of keeping in close and practical touch with agricultural research and progress." Selling it to the central officials the backers of the project further noted that the idea of the scientific center in Bengal had been approved by the finance member of the Viceroy's Council, Sir Edward Law, who had earlier paid a visit to the proposed site for the center in Pusa.[57]

The plan for the regional agricultural center in Bengal coincided with the effort to establish an imperial agricultural institute at the center. There

---

[56] From James Mollison, Esquire, Inspector General of Agriculture in India to Government of India, dated, December 3, 1901, including the April 1, 1901, note on Pusa, "Note on the Utilization of the Pusa Estate as an Agricultural Research Station for Bengal and as a Center for Teaching the Principles and Practice of Agriculture to Students," Government of India, Department of Revenue and Agriculture, File No. 115 of 1901, Serial No. 2, Proceedings for December 1901, Part B, National Archives of India.

[57] Letter from W. C. Macpherson, Secretary, Govt. of Bengal, to Secretary, Rev and Agriculture, Govt. of India, dated June 30, 1902, *op. cit.*; another letter from Macpherson to the center, dated January 30, 1903, Government of India, Department of Revenue and Agriculture, File No. 55 of 1903, July Proceedings, Nos. 6–9, National Archives of India.

had been an independent, long-drawn-out effort at the center to employ a group of scientists to work on colony's agriculture. The suggestion for scientific research on agriculture had been made by the members of the committee established to look into the devastating famines in 1878 and by the 1893 report on the improvement of Indian agriculture by J. A. Voelcker. These processes had picked up momentum recently during the tenure of Viceroy Lord Curzon, who appointed India's first inspector general of agriculture in 1901 and followed it up by adding a cryptogamic botanist, specializing in fungal diseases of plants, and an entomologist to the latter's staff. A government agricultural chemist had existed for more than a decade. The viceroy wished to take this momentum in the establishment of a local scientific infrastructure to its logical conclusion by creating colonial India's first agricultural science center. The latter would house all the agricultural experts appointed so far, serve as the Agricultural Department of the central government, as well as have an auxiliary agricultural college and cattle research center.

The Bengal proposal for Pusa was appropriated by Curzon's administration for its own plans for establishing an imperial institute at the same site. While Bengal continued to fine-tune its plans for its own center at Pusa, sending a letter to that effect as late as January 1903,[58] on the sidelines imperial officials began to discuss the suitability of Pusa for their own center. There was no question of the imperial government's involvement in establishing two centers at a time, one of their own and another for Bengal. Still, if there was going to be only one agricultural center, it would not be appropriate to leave it under the control of one of the provincial governments. The scope of the institution planned by Curzon and his team was simply beyond the financial and supervisory resources of a provincial government. The central officials more or less unilaterally decided to shelve the Bengal plan in favor of their own. James Mollison, now allying with imperial bureaucrats, was one of the influential voices favoring the establishment of Pusa as an imperial center. Officials in Bengal willingly handed over the estate and the funds earmarked for the project to the center. There was, after all, local pride involved in the establishment of an institution of such a stature in Bengal. The imperial plan also accommodated the interests of Bengal, appreciating the needs of the province for cattle research as well as agricultural education. It

[58] Letter from W. C. Macpherson to the Secretary, Government of India, dated January 30, 1903, Revenue Department, Agriculture – No. 508, Proceedings, Government of India, Department of Revenue and Agriculture, file no. 55 of 1903, July A files, Nos. 6–9, National Archives of India.

was hoped that the agricultural college at Pusa would serve "the needs of Bengal," whose own college at Sibpur had been a failure. The proposed cattle farm would devote itself to "the improvement of the local breeds of cattle."[59]

Lord Curzon's team approached the metropolitan government for permission to move ahead on the project for Pusa, justifying the need for an imperial agricultural center in terms of the new policy in the colony for "scientific and practical enquiry and experiment" and citing examples attesting similar efforts by progressive nations elsewhere. A center of such a scale was legitimized by wider reference not only to metropolitan examples, but to models in continental Europe and the United States. Thus the viceroy emphasized that the "necessity for such an institution has been recognized not only in England – as in the institution founded by the late Sir J. Lawes at 'Rothamsted,' in Hertfordshire – but also in America and most Continental countries."[60] The Rothamsted of England was nationally accepted as Britain's oldest center for agronomical investigations. The allusion to continental Europe and the United States of America in the official communiqué was deliberate and in a way even presented as a contrast to the way Rothamsted had been originally set up. It was intended to single out and move into open view the idea of agricultural laboratories and centers as national state works. If anything, the Rothamsted of 1843 had been set up in the Victorian era of the British state's overall laissez-faire attitude toward agricultural research. Such policies had seen the establishment of an institution like Rothamsted through the private initiative of John Bennet Lawes, a wealthy squire and amateur scientist, rather than through state patronage. On the other hand, governments in Prussia and the United States lately were obvious examples for their creation of agricultural institutions and research bodies through direct state patronage. The viceroy's rhetoric was aimed at establishing connections between the embracing of key agricultural institutions by nations like Germany and the United States and their recent progress in particular while also gesturing toward the importance of England's Rothamsted.[61]

---

[59] Letter from the Viceroy's Council to George Hamilton, Secretary of State for India, dated June 4, 1903, Proceedings, Government of India, Department of Revenue and Agriculture, July 1903, No. 7, India Office Records, British Library, P/6592.

[60] Letter from Viceroy's Council, dated June 4, 1903, *op. cit.* p. 3.

[61] As some have argued, after half a century of seminal work Rothamsted experiments were in a poor state due to lack of funds. The institution was awaiting its next round of reorganization under its next famous patron, Sir Daniel Hall. Joining Rothamsted in 1905, Hall would turn it into a "Continent-style research station." Also, this next round of modernization at Rothamsted would benefit subsequently from a fresh policy of state

The bypassing of an imperial British connection and the emphasis on "the American connection" were further occasioned by the news shared by Curzon that Henry Phipps, the Pittsburgh-based industrialist, conservationist, and science philanthropist, had donated a sum of £20,000 for the benefit of Indian people that Curzon wished to use for the establishment of Pusa. This American connection with Pusa was probably facilitated by Curzon's personal and family links. The paucity of sources forces us to speculate here. But it is quite likely that Curzon had come to know of Phipps through his wife's family. In 1895 Curzon had married Mary Leiter, the daughter of a Chicago-based millionaire. She went to India with Curzon when he was appointed viceroy four years later, and this personal alliance would have likely put Curzon in contact with others prosperous industrialists in the United States. In 1902–3 Phipps visited India on the personal invitation of Lord Curzon. It was during this visit that he offered the sum for the purpose of establishing a scientific laboratory to study the economic and medicinal value of the country's flora.[62]

The expeditious approval of the Pusa proposal by secretary of state in London showed not only the concurrence of the British metropolis with the initiatives on agricultural science in the colony, but also the latter's association with extraimperial forces.[63] The expansive force of

aid for agricultural research in Britain that was inaugurated by the Development Act of 1909. *Cf.* Keith Vernon, "Science for the Farmer? Agricultural Research in England, 1909–23," *Twentieth Century British History* 8 No. 3 (1997): 310–33, quote on p. 317; see, also, Alison Kraft, "Pragmatism, Patronage and Politics in English Biology: The Rise and Fall of Economic Biology, 1904–1920," *Journal of the History of Biology* 37 No. 2 (2004): 237–40.

[62] The biography of Henry Phipps and his family's history appear in a book written by his granddaughter. No specific reference to his philanthropy on Pusa appears in the book, but it is still helpful in generally describing Phipps's interests and other personal attributes, the family's social circle, and their travels: Peggie Phipps Boegner and Richard Gachot, *Halcyon Days: An American Family through Three Generations*, New York: Old Westbury Gardens and Harry N. Abrams, 1986; a direct reference to the charity appears in correspondence between Henry Phipps and Lord Curzon and his establishment: letter from J. O. Miller to Henry Phipps, dated January 28, 1904, No. 225; letter from J. O. Miller to Henry Phipps, dated January 30 1904, No. 227; letter from Henry Phipps to Curzon, dated March 12, 1904, No. 301, The Lord Curzon: Correspondence with Persons in England and Abroad Commencing from July 1901 (confidential), European Manuscript, F111/182, India Office Records (the British Library, London), henceforth, India Office Records, Curzon Papers; letter from Henry Phipps to Curzon's office recommending Carruthers's appointment in India, dated February 28, 1903, National Archives of India, GOI, R&A, Proc., June 1903, Part B.

[63] Soon after the submission of the plan by the viceroy, the secretary of state's office in London sanctioned the setting up of Pusa. Secretary of State's letter to the Viceroy in India, dated August 14, 1903, Selections from the Despatches addressed to the Several

American philanthropic ideas around science and conservation found a liminal presence in the project of defining an Indian agricultural science around Pusa. Phipps' wide travels gave him an opportunity to propagate his ideas including those of conservancy. As he described in a letter addressed to an official in Viceroy Curzon's establishment, he had an avid interest in plant conservancy. He was the founder of the botanical garden and conservancy center in Pittsburgh. From India Phipps traveled to Ceylon, where he visited the famous Peradeniya Botanical Garden and met with the assistant director of the botanical garden, the mycologist John Bennett Carruthers. He was so impressed with Carruthers's credentials in plant sciences and management that he dispatched a letter from Ceylon to Curzon's office, recommending Carruthers for appointment "in some important place in the future" in India. The reference was duly considered by colonial state officials and an offer made to Carruthers even though he preferred instead to accept another job in the Federated Malay States, a British imperial protectorate. Indeed the core laboratory within the Pusa complex would later be named the Phipps Research Laboratory, after the benefactor, symbolically representing the imprint of American philanthropy on a colonial agricultural station in South Asia. From the perspective of Indian science these developments illuminate the impact made by lines of influence originating outside the familiar imperial-colonial framework of the Raj.

The material and symbolic importance of the "Phipps connection" is immense for understanding the institutional growth of agricultural science in late colonial India. It needs highlighting here that Henry Phipps had only contributed to meeting the initial expenses of the laboratory while, as Curzon disclosed in a follow-up letter to Henry Phipps, the colonial government was investing £110,000 in constructing the rest of the buildings in the upcoming Pusa complex of which the laboratory was just one part.[64] Indeed Phipps had made much of his fortune in steel with the industrialist Andrew Carnegie and in real estate and on the face of it could barely serve as an idol inspiring the practice of science-intensive agriculture in the colony. Still, the bursary for a "scientific laboratory for plants" by an American industrialist contributed to the making of important progressivist discourses around Pusa. The disproportionate stress on Phipps in the official campaign for Pusa in India often elided the idolizing of Phipps as representing the

Governments in India by the Secretary of State in Council, 46th Series, part II, July 1–December 31, 1903, letter no. 122, V/6/350, Indian Office Records.

[64] Letter from Curzon to Henry Phipps, dated April 27, 1905, No. 28, India Office Records, Curzon Papers.

renowned American success at the turn of the century with his support on this occasion for laboratory-centered agricultural development.

As the Viceroy's Council of Lord Nathaniel Curzon picked up the "Pusa scheme" in earnest, they anticipated transferring most of the staff and infrastructure at Dalsingserai to the new institution at Pusa. Such a measure was planned in the wake of the fact that BIPA was tottering and the joint Research Committee of indigo planters, traders, and government officials formed to oversee the joint indigo research program was hardly in a position to provide funds for or to supervise the stations at Peeprah and Dalsingserai. As planters' funds to the two stations dried up, government officials at the center stepped up efforts to assimilate the Dalsingserai work within their emergent center at Pusa. John O. Miller, the revenue secretary, contacted Bernard Coventry at Dalsingserai and asked him to submit a budget for funds to keep his operations running for the current year, showing a willingness to arrange for alternative funding to salvage his experiments in the absence of revenue from other sources. Denzil Ibbetson, member of the council, followed up by getting in touch with the Bengal governor quite early on. Even before the submission of a request to London for Pusa, he tried to enlist the governor's support for his intent to preserve the Dalsingserai Center. The two important officials had apparently previously discussed the advisability of saving the "good" work at Dalsingserai in case the planters faltered in their commitment. He reminded him how important it was that the provincial funds earmarked for indigo research should be expeditiously awarded to Coventry, who was on the last legs trying to keep his staff and work. There is little doubt that the future of research at Pusa was uppermost in his mind. Showing his preference for the type of work at Dalsingserai, he disclosed that they "do not propose to take over or in any way to subsidise [sic] the work at Pipra, which is concerned solely with manufacture of indigo." On the other hand, saving the Dalsingserai work was of utmost importance, "as we know them to be good men, and, above all, they have the experience of Indian conditions, the want of which so hampers a man new from England." If he did not transfer funds to the station immediately, Ibbetson warned, "the work must be stopped, the staff dispersed, and a great opportunity lost." The Bengal governor obliged and ahead of the formulation of any alternative plan to continue indigo laboratories in Bengal transferred funds to Bernard Coventry.[65]

---

[65] See the reference to John O. Miller in the letter from Bernard Coventry to Commissioner, Patna, dated May 1, 1903, Government of Bengal, Agriculture, Bihar State Archives,

Important central officials had set their hearts very early on hiring Bernard Coventry to lead the imperial station at Pusa, enamored as they were with his experience, his completed work, and the acclaim that he enjoyed among planters, officials, and scientists alike. Ibbetson in his letter had hoped "to transfer the Dalsingserai work (and, we hope, Coventry with it) to Pusa." Although Coventry lacked a formal degree from a college or university, he had taught himself well on scientific matters. David Prain, now at the Botanical Survey of India, commented that he had met very few people who knew the practice of agriculture and of scientific theories underlying them better than Coventry. Coventry was a person who had been trained in "the school of practical experience," he offered. Prain's note was given weight in the official circle and was considered "a remarkable testimony," coming as it did from a botanist of his stature. Coventry's experience of Indian agriculture was considered appropriate to the task of the future administrator of Pusa, with Mollison, too, insisting that he had "the required attainments" and Ibbetson echoing a similar sentiment in saying that "it is of the very *greatest* importance to us to secure Mr. Coventry as the head of the new Pusa institution."[66]

The extraordinary interest shown in the candidacy of Coventry reflected officials' wish to embrace and emphasize aspects of the work at Dalsingserai. The emergent blueprint for the work contemplated at Pusa involved a broader interest in plant experiments relating to Indian crops of all sorts. Coventry was the nerve center of Dalsingserai trials and had supervised all the work there. His long experience in the agricultural aspects of indigo and rhea seemed appropriate for purposes of developing an Indian crop science. Similarly Coventry, more than anyone else in the colony, had proven experience and capacity in selection and breeding trials with indigo, which the officials wanted to include in the research program at the upcoming Pusa facility.

The officials' interest in the Dalsingserai work was selective and consequently they screened the entry of experts from there to Pusa. They forestalled the entry of William P. Bloxam of Dalsingserai, for instance, the scientist who had been making analytical experiments on indigo. This

December 1903, File, 2I/8 5–17, Nos. 67–8; Sir Denzil Ibbetson's letter to Lt. Governor, J. A. Bourdillon, dated March 10, 1903, Government of Bengal, Agriculture, Bihar State Archives, December 1903, File, 2I/8 3, Notes and Orders; letter of A. Earle, Revenue Secretary, to Commissioner, Patna, dated March 21, 1903, Government of Bengal, Agriculture, Bihar State Archives, December 1903, File, 2I/8 3.

[66] Notes and Orders, Proceedings, A, Government of India, Revenue and Agriculture, Branch Agriculture, File No. 55 of 1903, Nos. 6–9, pp. 2, 7, 25, National Archives of India.

despite the fact that Coventry wrote to Mollison emphasizing the qualification of Bloxam as a chemist and the fact that Indian chemical sciences were in a poor state and would gain from the latter's experiments. But Mollison rejected the suggestion, saying what India needed was "a more general enquiry into the agricultural chemistry and requirements of our field and garden crops." Bloxam did not fit into this plan as he was "a pure chemist." Depending upon such criteria the officials went on to hire Cyril Bergtheil, the bacteriologist, and H. M. Leake, the biologist, and some assistants from Dalsingserai.[67]

The prioritizing of selection and breeding by officials in Pusa plans reflected broader global trends in the West and their extension into the colony. This emphasis derived from new advances in the fields of agricultural botany and horticulture in Britain, as well as continental Europe and the United States. Such emphasis was not unusual in the context of excitement with the promise of early genetics after the "rediscovery" of Mendelism in the West in 1900, as well as the awareness of the accomplishments of breeders in Britain and the United States in the last quarter of the nineteenth century. The possibility of transforming commercially useful traits in plants as proven by the concrete successes of breeders and horticulturists was probably weighing on the minds of colonial actors. There was clear convergence among the planners as far as emphasis on selection and breeding was concerned. Mollison's note emphasized that the work of plant breeding and seed growing was "of great importance for India." These practices would not only enable improvement of Indian crop varieties, but also help with the tasks of the cryptogamic botanist and entomologist because selection offered a chance to breed varieties that could resist fungal disease and insect pests. Coventry joined the discussion to point the omission in current plans that made no provision for the appointment of a horticulturist. He explicitly emphasized the need for "an expert European (or American) Horticulturist." Such a person, Coventry opined, "can undertake improved cultivation of valuable tropical plants and carry out plant selection and cross-breeding." The botanist whom Mollison wished to employ was also expected to "watch variations and investigate improvements induced by cross breeding."

[67] Semiofficial letter from Bernard Coventry to J. Mollison, dated February 13, 1903; semiofficial letter from J. Mollison to J. O. Miller, dated February 15, 1903, including "Notes on Utilization of Pusa Estate"; Notes and Orders, Proceedings, A, Government of India, Revenue and Agriculture, Branch Agriculture, File No. 55 of 1903, Nos. 6–9, National Archives of India.

Tappping into European Science So That the Nation May Benefit

Viceroy Curzon's certitude about a direct and inevitable connection between science and agricultural improvement when he raised the demand for Pusa with the secretary of state discursively hid many aspects of uncertainties and conflicts inherent in the colonial program of agricultural and scientific modernity. As an important palimpsest in the chain of the state's initiatives on agriculture, the drive to establish Pusa bore the signs of the underlying, festering tensions in the colonial order. Thus even as colonial administrators scurried to finalize plans for establishing the agricultural resource center, the highest imperial authority in India, Viceroy Lord Curzon, expressed misgivings over the choice of the site for the new institution amid the indigo tracts in Bihar, which were owned by European planters, and the selection of Bernard Coventry, former manager of a private indigo research station, as its future director. He reportedly cautioned another member of the viceroy's council that in doing so, "we shall expose ourselves to the accusation, which the Native Press is sure to make, that the whole thing is a huge job for the benefit of the European indigo industry and [European] indigo planters, rather than of native staples and cultivators." Over apprehension of such an allegation he asked his subordinates to convince him of why Pusa and Coventry were the most appropriate choices. Obviously, Curzon did not wish the measure to be delegitimized in public view, for he was just as concerned to introduce the material benefits of agricultural science to the colony, or for that matter to accommodate European economic interests, as he was to enlist the maximum common aspiratory support of those Indians to whom science mattered. The finance member, Sir Edward Law, expressed similar misgivings as he asked to be satisfied why Coventry, after all, only "an experienced agriculturist," should be appointed to lead a scientific center, a post that would ordinarily be reserved for civil servants or scientific officials.[68] The viceroy's qualms and Law's circumspection provide us a window on the conflicted path of science in the colony's agricultural landscape. The evident caution of imperial figures reflects the contested nature of the terrain that the colonial project of scientific modernity had to traverse in South Asia.

Indigo science itself had escaped scrutiny by the nationalists up to now, although such an omission does not mean that indigo science was

---

[68] Notes and Orders, pp. 6, 9, Government of India, Department of Revenue and Agriculture, File No. 55 of 1903, July Proceedings, Nos. 6–9, National Archives of India.

an "enclave science" of inconsequential import.[69] Indigo laboratories in fact had a direct connection with the effort to save a major colonial industry that was built on the labor of the Indian peasantry. But connections between indigo science and colonial plantations were not self-evident, at least not immediately. The nationalist critique first focused on the scientific dimension of the indigo enterprise after colonial officials openly moved to appropriate that science for a program for agricultural development. As plans were made to employ erstwhile indigo experts in the name of creating a science for the development of India, the nationalist critique challenged the authenticity of such efforts.

Such sensibilities that wished to appropriate European science for progress had a history of their own that went back to the late eighteenth century and were about as old as the Raj itself. These perspectives stressed the "neutrality" of science, distinguished the latter from the colonial state, and did not see it as quite a direct instrument of rule. Intellectuals like Raja Rammohun Roy, who defended European science and thought of even the indigo industry in the 1820s as a marvel of modernity, as well as the loyalist *taluqdars* (landlords) in Oudh in the 1860s – all creation of the empire's own urbanization and land policies, no doubt – were willing partners in the project of agricultural modernity. In the postmutiny era many sections of the Indian elite expressed the desire to embrace the European science that the state was trying to popularize by holding grand agricultural exhibitions. Among them were the loyalist landlords and others with strong ties to agriculture. Thus, ahead of the famous 1864 Agricultural Exhibition of Calcutta, the members of the British Indian Association (hereafter, BIA) called a meeting to emphasize the participation of Indians in the upcoming exhibition. Baboo Rommaunauth Tagore explained the significance of the exhibition to Indian agriculture, maintaining that "there could be no question more worthy of ... [our] attention than this." The demonstration of the science of crops and cattle could accomplish two functions considered important by the landlords: "engendering a spirit of emulation, and ...

[69] David Arnold has deployed the notion of "enclavism" to spotlight both the targeted deployment of Western medicine by the colonial state for maintaining the well-being of Westerners in enclosed spaces such as army barracks as well as the simultaneous realization by the colonialists that such efforts were futile given the mobility of pathogens between Europeans and Indians. More broadly, he has also disputed the use of the concept by other historians for making arguments about the limited reach of Western medicine and its purported irrelevance to colonial historiography. David Arnold, *Colonizing the Body: State Medicine and Epidemic Disease in Nineteenth-Century India*, Berkeley: University of California Press, 1993.

awakening a spirit of improvement," as the meeting emphasized.[70] In the loyalist perspective of Bengal landlords the political and economic contingencies of a fundamentally "foreign" rule in India were written out as absolute comparisons were drawn at the meeting between the courses of agricultural improvement in England and in India: "The history of agricultural improvement in England places, beyond dispute, the utility of Agricultural Exhibitions, and what has been achieved there, the committee [of BIA] see no doubt, may not be achieved in India under like circumstances and like conditions."[71] The rhetoric neglected to consider the fact that agricultural improvement in the West had materialized under a sovereign government, while India was under foreign domination.

Similarly, after attending another Agricultural Exhibition held in Lucknow in neighboring Oudh later in 1864, important *taluqdars* of Oudh belonging to another landlord's organization, the British Indian Association of Oudh, appreciated "European Science and Ingenuity" on display there. The general extolling of colonial exhibitions by Indians signified their imagining of colonialists as cohorts with them in achieving development through the medium of science. It involved acknowledging that science introduced by foreign rule had the potential for improving native lands and generating "Indian" progress. Even the landlords' description of the physicality of the exhibition was laden with idioms that represented science's ability to circumvent political and social barriers in its instrumentality: "Little did we think that this plot of ground would one day be adorned with these graceful courts; for so noble a purpose; would gather in friendly communion all ranks of the European and Native community, and unfold to the wondering gaze of the admiring crowd those arts of Peace, with which European Science and Ingenuity have contrived to increase the happiness of mankind."[72]

Indians continued to engage and contest the nature of agricultural farms and stations in the late nineteenth century as a new political economy of nationalist insurgency began to emerge. The nationalist program

---

[70] The motion of Baboo Rommaunauth Tagore, "A Special General Meeting of the British Indian Association, November 25, 1863," *Publications of the British Indian Association,* 1863, p. 2.

[71] Letter of Joteendro Mohun Tagore, Hon. Secretary of British Indian Association, dated February 5, 1864, to F. R. Crockerell, Secretary, Government of Bengal, p. 4, *Publications of the British Indian Association,* 1863.

[72] "Address on the Occasion of the Lucknow Exhibition," Memorandum to Charles John Wingfield, Chief Commissioner of Oudh, dated January 2, 1865, *Proceedings of the Meetings of the British Indian Association of Oudh, 1861–1865,* Calcutta: Thacker, Spink, & Co., 1865, pp. 192–3.

of seeking autonomy and the search for an ideal science appropriate to the needs of an emergent nation shaped these agrarian visions. The "vernacular publics" writing in the regional language newspapers were best placed to address these institutions. Such visions certainly could not claim to represent all Indians. Rather the vernacular literature was fraught with tensions representing contrasting viewpoints among different elements within the middle class, the class of landlords, and subalterns. Thus one report asked the Agricultural Departments to follow up efforts in demonstrating the utility of science at the experimental farms with firm measures to make the "traditional" landlords adopt them. The government must take steps "to stir them out of their attitude of indifference."[73] A report in *Samvad Prabhakar* offered a scathing critique of the entire class of landlords, holding them squarely responsible for the degeneration of productivity in Indian soil: "Like the British Government in India, nay with greater zeal even than that Government, the zamindars are busy collecting rents, and find no leisure to the state or the improvement [of] the soil." The foreign government had no long-term interest in improving productivity, and therefore their lack of interest was explainable. But the indifference of *zamindars* was indefensible. As the "children of the soil" they had the ultimate responsibility to invest in land improvement.[74] Another report expressed indignation about a peculiar episode regarding the improper use of agricultural shows for amusements and theatrical performances at Khulna. It conveyed the shocking news that prostitutes had been invited to perform at one show. The newspaper thus stood for restoring the show to its pristine purity of pursuit of scientific goals. But the newspaper also saw such events as socially threatening. The invitation of prostitutes was immoral to begin with. But it was additionally dangerous because the participants were "an illiterate people whose morality is not of a superior order." It represented an attempt on behalf of the Indian elites to recover the site – one created by the colonial state – from the clutches of the subalterns who had taken it over for their immoral revelry.[75]

At the time of the unfolding of the "Pusa scheme" the level of Indian intelegentsia's diverse evaluation of modernist agricultural institutions

---

[73] *Sri Sri Vishnu Priya-O-Ananda Bazar Patrika*, January 18, 1905, Native Newspaper Reports, Bengal, No. 4 of 1905, p. 82.
[74] *Samvad Prabhakar*, August 15, 1893, Native Newspaper Reports, Bengal, No. 33 of 1893, p. 703.
[75] *Sanjivani*, March 1, 1890, Native Newspaper Reports, Bengal, No. 10 of 1890, pp. 224–5; February 11, 1893, Native Newspaper Reports, Bengal, No. 7 of 1893, p. 133.

had continued. Some argued that the improved machinery shown at the agricultural exhibitions would only have a real impact if the state provided high-quality seeds and appropriate implements to the farmers free of cost or at a nominal price. Indeed a few rejected outright the relevance of model farms in the context of the peasantry's resource-poor condition. But many others voiced clear support for the model farms. *Bangavasi*, the Calcutta-based newspaper, made a case for opening of additional experimental farms: "If the experimental farms are intended to be of any use to the Bengal raiyats [peasants], there should be at least 24 such farms established in the 48 districts of the Province. In the meantime, the existing ones should be well-worked."[76] The *Jyoti* of Chittagong reprimanded the government officials for their neglect of the local experimental farm, writing that its correspondent who had visited the Government Experimental Farm two miles from Chittagong was "sorely disappointed" with what he saw there. "The farm was overgrown with grass and shrubs, and so much neglected that the crops grown on it proved unsuccessful," the report complained.[77] Such discerning responses of Indians to government measures show a cautious optimism for scientific improvement of agriculture at a local level, even if they also inhered with a suspicion of foreigners as agents of that progress.

It is the mix of such nationalist visions that cautioned Curzon as he expressed anxiousness over the choice of indigo tracts as the site for the Pusa station and an ex-planter as the director of colonial India's first imperial agricultural institute. While he could not have met all of the nationalist aspirations, he wished to reach out to those appreciative of scientific initiatives on agriculture by colonial state. The Indian opinion on the advisability of Pusa and its science was mixed. Thus, while many Indians welcomed the establishment of scientific institutions for agriculture, others were critical of them. *Hitavadi* supported the measure and the invitation of scientific experts to Pusa, saying, "We are in favour of having a large number of agricultural chemists and bacteriologists in India for ... [agricultural improvement]."[78] But there were also other voices that doubted the judgment of the colonial state in introducing such high-end, sophisticated science to India in a situation where the basic needs of agriculture were lacking: "Those [i.e., the Indian peasants] who cannot

---

[76] *Bangavasi*, April 9, 1904, Native Newspaper Reports, Bengal, No. 16 of 1904, p. 377; May 7, 1904, Native Newspaper Reports, Bengal, No. 19 of 1904, p. 433.

[77] *Jyoti*, July 28, 1904, Native Newspaper Reports, Bengal, No. 32 of 1904, p. 730.

[78] *Hitavadi*, dated March 10, 1905, Native Newspaper Reports, Bengal., No. 11 of 1905, p. 256.

purchase seeds for their fields and properly cultivate them for want of strong bullocks, this institution is for them! Bravo, Lord Curzon, bravo! What more shall we say?"[79] Such contrasting assessments together reflect not only the differences in estimating the appropriateness of science launched by the colonial state but, more important to our social analysis, the diversity of the category of such "Indians" or "natives."

The diverse sensibilities of imperial and nationalist forces shaped the fraught nature of colonial science. As against the current practice of treating any science in the colony as "colonial science," identified with the power of colonial ruling classes as a whole, science in the colony in reality reflected the difference in priorities within the imperial and "nationalist" classes.[80] The study of both indigo science and crop science as part of the broader patterns of science in the colony reveals the variegated nature of science, which bore the imprint of diverse priorities of colonial state officials, economic entrepreneurs, imperial figures, international philanthropists, experts of diverse training, and Indian landlords and nationalists.

[79] *Bangavasi*, dated April 8, 1905, Native Newspaper Reports, Bengal., No. 15 of 1905, p. 376.

[80] Ann Laura Stoler and Frederick Cooper have called for a consideration of internal dissensions within the ruling class in a colonial context. Stoler's emphasis on attention to the "fault lines" of interest between different categories of colonial actors has immense value for this study. Ann L. Stoler, "Rethinking Colonial Categories: European Communities and the Boundaries of Rule," *Comparative Studies in Society and History* 31 (1989): 134–61; Ann L. Stoler and Frederick Cooper, "Between Metropole and Colony: Rethinking a Research Agenda," in Frederick Cooper and Ann L. Stoler (eds.), *Tensions of Empire: Colonial Cultures in a Bourgeois World*, Berkeley: University of California Press, 1997, pp. 1–56, especially see synopsis on p. 4.

5

The Last Stand in Science and Rationalization

The hold of colonial locality on the developing indigo science only grew stronger in the years leading up to the First World War as the locus of decision making with regard to indigo experiments devolved to the numerous strata in the colony. About the middle of the first decade of the twentieth century a good number of indigo planters had divested from indigo and either combined indigo planting with the cultivation of sugarcane and other crops or deserted planting altogether. A note in the planters' key journal while beckoning planters to desert the "Busted" Bihar and relocate to America put a spotlight on the current trend and growing anticipation of exodus of planters from the indigo tracts. It informed the British diasporic entrepreneur, "the farmer and home-seeker," about "opportunities offered by several of Southern States [in the United States]."[1] But even at a vastly attenuated size the indigo plantations were still an important regional industry of Bengal. Thus as metropolitan and imperial forces put indigo on the back burner, the will of local planters ensured the continuation of indigo experiments. They also sought and received substantial support from the provincial officials in the colony. Much of their combined efforts focused on improving the plant in the field. They were inspired by the promise of Mendelian biological selection that had received renewed emphasis worldwide in the early twentieth century.[2] It was Mendelian science that was to instill a sense of self-belief among planters.

[1] *Indian Planters' Gazette*, Beltsville, January 20, 1906, p. 83.
[2] Gregor Mendel's seminal work on sudden changes in plant characteristics through mutation and the inheritance and stabilization of such characteristics through succeeding plant generations was first published in 1866. However, the realization of its potential for crop improvement by scientists had to await its "rediscovery" simultaneously by

But in the end even the best efforts of scientists and rationalizers failed to stop the slide in the fortunes of natural indigo. In the decade leading up to the First World War a sense enveloped a number of proponents of agricultural indigo that perhaps the dye was fast reaching the limits of improvement. But even in these years when failure seemed imminent, some hoped for a turnaround in the fortunes of agricultural indigo. These advocates of natural indigo had a separate vision of what the natural product was and where its future lay. They contested the claims that agricultural indigo had reached a "limit" to its improvement based on parameters that were extraneous to the natural product. The study of these discrepant views alerts us to a separate vision that a section of planters and their supporters optimistically espoused over more than a decade. Those alternative visions that have been hidden by the forward march of the specific modernization process to which natural indigo finally fell victim constitute an important segement of indigo's history in the pre–World War I era.

The Promise of Mendelism in the Colony

The years 1903–4 constituted a period of transition for indigo experiments in the colony. The period was marked by the closure of old laboratories and centers and opening of new ones, a major relocation of experts, and ultimately the unfolding of a new organization of indigo science. A new plan for the science began to emerge within the government establishment as Bengal officials made a renewed bid to start a second round of indigo experiments. The lieutenant governor of Bengal led from the front. He was personally sympathetic to the needs of the planting community. There was rethinking among state officials about the desirability of a province-specific initiative on indigo in addition to the planned indigo experiments at the upcoming Imperial Agricultural Institute at Pusa. As the planters' laboratories faltered and central officials retained some of the previous indigo experts for agricultural experiments at Pusa, it began to seem to many that the Imperial Institute might not, after all, meet the needs of indigo in as focused a manner as was required. The

three European biologists between 1900 and 1901. For a contemporary horticulturist's response to this rediscovery and its impact on current breeding practices in the West, see, Liberty Hyde Bailey, *Plant Breeding: Being Five Lectures upon the Amelioration of Domestic Plants*, New York: Macmillan Company, 1904, 3rd ed., first published 1895, pp. 143–226.

officials feverishly discussed the subject of starting indigo experiments on a fresh footing as soon as possible in order to prevent a hiatus in colonial indigo science after the impending closure of Dalsingserai. The way was opened for indigo experiments after a delegation of officials and planters led by the lieutenant governor of Bengal met the inspector general for agriculture, J. Mollison, at Pusa on February 18, 1904, and secured his approval.[3]

A generalized euphoria with regard to the promise of Mendelism provided the template for the emerging science. Reflecting the spirit of the times, a planter in Bihar in January 1905 categorized as misdirected the past and current efforts of chemists working within laboratories in the colony to improve the manufacturing process of indigo dye. He was commenting specifically on experiments directed at streamlining the vat processes that were used for extracting color from the leaves. These scientific efforts appeared futile in hindsight partly because of lack of substantial results from them. But such criticisms were also connected with the new worldwide optimism over the potential of biological selection work in horticulture and agriculture, which made manufacturing experiments appear unworthy in comparison. Thus, in a spirit of expectancy over plant selection, this planter wrote: "It is now quite evident that we began at the wrong end of the stick. We ought to have begun by trying to put more produce into the plant itself. We see that almost every cultivated plant has been improved and developed by selection in the exact direction striven for; and it is almost certain that any good variety of *indigofera* could be selected and cultivated to contain five times as much colouring matter as it now does." Assured that the planter's salvation lay with biological selection, he emphasized that "it is in this direction that our Government can give us real assistance."[4] Plant selection as a commercial craft and scientific principle had reached a new level of influence in the early twentieth century, and the colony was not untouched by this wave of optimism. And this spirit provided direction to the efforts of improvers too.

Focus on the Plant in the Field

The inducement of Mendelism led toward an overall emphasis on improving the plant in the field in all research programs for indigo, whether at

---

[3] "Note on the Prospects of Indigo, Sugarcane and Cotton," No. 19, File 2-I/2 13, Proceedings of the Government of Bengal, Revenue Department for the month of July 1904, India Office Records, GOB (Rev), P/6793.

[4] *Indian Planters' Gazette*, January 21, 1905, p. 77.

Pusa or at Bengal's own indigo center. This was to be done no doubt with the help of both chemists and agricultural botanists. The chemists were given a role in enhancing the color-bearing ability of the plant by adding their expertise on plant physiology and soil chemistry. At the very first Pusa meeting everyone agreed that the previous indigo chemist, Christopher Rawson, was a fine, "practical" chemist who could potentially play a positive role in future experiments. Aside from perfecting vat processes, Rawson had previously worked on the agricultural aspects of indigo cultivation, and it was believed that his prior experience would be useful. An invitation was sent out to Rawson to return to India. But he rejected the offer, stating that there was not much he could do to help natural indigo's fight with the synthetic as he had done his work. The officials then decided to employ Cyril Bergtheil, the other chemist with prior experience on indigo. Bergtheil had already committed to join Pusa. But officials contemplated that perhaps he could work on indigo till the general bacteriological work for which he had been appointed at Pusa got off the ground. Or, he could join Pusa and then be sent on deputation to the proposed indigo center in Bengal for the indigo work.[5] In addition, at these meetings, the planters emphasized with government officials the need to establish a seed farm in the northwest, where the climate and soil were ideal. They sought the government's help in finding an expert seed grower. They also envisioned a role for the government chemist at this farm for his assistance with leaf analysis and the identification of the best plants for raising seed.

The new indigo research program was also sensitized to the possibility of biological "selection" to improve indigo varieties by the important imperial official James Mollison, the inspector general of agriculture. A career agricultural official who had worked for a decade in the Agricultural Department in Bombay, Mollison was involved with the substantive biological selection project on wheat already under way in the colony.[6] Just before taking up the office of the inspector general of

---

[5] "Note on the Prospects of Indigo, Sugarcane and Cotton," No. 19, File 2-I/2 13, Proceedings of the Government of Bengal, Revenue Department for the month of July 1904, India Office Records, GOB (Rev), P/6793; "Proposal to Continue Mr. Rawson's Work and the Establishment of an Indigo Seed Farm," Note by the Director of Land Records and Agriculture, Bengal, No. 59, File, 2-I/2 15, Proceedings of the Government of Bengal, Revenue Department for the month of July 1904, India Office Records, GOB (Rev), P/6793.

[6] The colony had a history of a wheat improvement program that began with the introduction of exotics and pedigree wheat at specific farms in different provinces in the 1870s and later expanded to include selection. In the 1890s J. Mollison in Bombay,

India he had visited the United States with the specific object of studying selection methods in practice there. A firm believer in the potential of these processes, he stressed the importance of selection to the planters for generating a stock of indigo plants that would be optimal to their requirements in the colony. He thus became the most important link between the long-running wheat selection work in the colony and the emerging focus on selection in indigo work.

In meetings with S. L. Maddox, the director of the Land Records and Agriculture Department in Bengal, and the BIPA secretary, Mollison tried to persuade everyone to embrace a prospective plan for indigo selection. He even suggested the criteria on which the native and foreign varieties should be selected. The stock of native variety should be built with races giving higher leaf growth since the local plant was otherwise known to give a relatively poor yield of leaf. The foreign Natal-Java/Java variety was to be selected for its ability to germinate successfully. The Bihar planters had previously reported problems with the germination of the Java seed. Some planters had tried thinning the seed coat of Java by treating it with acid with some success. Mollison instead suggested selecting specific races of Java indigo that did not need such a treatment to germinate. This was because he believed germination to be an inheritable quality. In addition, for the long term, Mollison also advised the planters to undertake the more ambitious task of hybridization between races to develop a variety of indigo that would be most suitable for their purposes in the colony.

Colonial officials took a number of steps to set up institutions in pursuit of the long-term and short-term goals implicit in the preceding vision. A new farm was set up at Dasna near Delhi in the North Western Provinces for selection work on the Sumatrana variety of indigo. Trials on the Java plant were proposed to be conducted in the short-term at Pusa, where agricultural fields were already available, even as new stations were slowly coming up. The Bengal government acquired an existing indigo factory at Sirsiah and turned it into the new indigo research station for additional experiments. On the basis of an understanding between the central government and the Bengal government, Bergtheil

---

W. H. Moreland in North-Western Provinces/United Provinces, and in 1905–6 Albert Howard at Pusa were the three main pivots of this evolving program on wheat selection. *Cf.* Albert Howard and Gabrielle L. C. Howard, *Wheat in India: Its Production, Varieties and Improvement*, Calcutta: Thacker, Spink & Co., 1909, pp. 113–60.

was called from England to join Pusa and then immediately allowed to proceed on deputation to Sirsiah.[7]

Analytical Experiments Moved to the Metropolis

Meanwhile other experiments, especially the analytical experiments by chemists, were relocated to the metropolis because there was no hope of "practical" results from those efforts in the near term. There was also a belief among local officials that such experiments could be best carried out in the metropolis, where resources were available aplenty. The analytical work in progress on indigo chemistry by the Dalsingserai-based chemist William P. Bloxam was clearly running out of favor. Bloxam had been trying to determine the actual content of color in the leaves and the efficiency of processes used to extract it. Like other indigo scientists, he too believed that the ultimate goal was to maximize the yield of color. But he was convinced that knowing the theoretical value of recoverable color in the leaf and the percentage of color found in samples at different stages of manufacturing would help pinpoint with accuracy where recoverable color was being left behind or wastefully being converted into other substances and ultimately help plug those losses. His analytical explorations did not stop there. Bloxam's quest for precision set him on course to checking the accuracy of tests being used to measure color percentages. Indeed he argued that the task of improving the manufacture of indigo must await a fuller understanding of the "pure chemistry of indigo."

Bloxam's time-consuming work involving collection of hundreds of samples and validation of existing tests did not inspire confidence among planters or colonial officials who wanted results that could be put to immediate use. Colonial officials refused to sponsor Bloxam's work any further in the colony. While Bloxam was confident that he was on the right track and on the verge of a novel discovery that would bring unmatched "scientific reputation to all ... concerned" and save the imperiled indigo

---

[7] Notification – by the Government of India (No. 953, dated July 26, 1904), No. 22, Serial No. 27; Notification – by the Government of India (No. 954, dated July 26, 1904), No. 23, Serial No. 28, Government of India, Proceedings of the Department of Revenue and Agriculture for 1904, May, 1904, India Office Records, GOI, Proc. Rev & Agr, P/6826; No. 3337 Agri., Letter from Revenue Secretary, Government of Bengal, to Director of the Department of Land Records and Agriculture, dated August 10, 1904, No. 68, File 2-I/20; No. 2154A, Letter from S. L. Maddox to Revenue Secretary, Government of Bengal, Proceedings of the Revenue Department for the month of August 1904, India Office Records, GOB, Rev (Agr.), P/6794.

industry, others in the colony were unimpressed. The officials were criticizing the logic of Bloxam's analytical experiments, saying, "It would seem more promising if Professor Bloxam would hold out hopes not so much of determining the quantity of the blue in the plant nor of proving that the present methods of extraction of the blue were faulty, as of inventing new methods for the extraction of a far higher percentage of blue."[8] A planter at Dalsingserai at whose precincts the agricultural station was located also wrote to express his disappointment with Bloxam's work, which "could probably not be put to immediate practical use."[9] Bloxam's experiments on examining questions of a theoretical nature evidently had no place in the colony. State officials conveyed to him unequivocally that they had no intention to continue this line of query.

Bloxam himself was committed to taking his analytical work – including his most significant work on tests – to a successful end. This quest followed from his expert judgment about the importance of tests and his professional ambition to stay the course so that he would be the originator of the new findings. He firmly believed that accurate tests would give an appropriate edge to indigo science and would advance it closer to meeting the competition from synthetic indigo. In line with his professional ambition he was offended by the suggestion from officials to submit an interim report on his experiments. He insisted on first presenting the major findings of his work to the Chemical Society in England. Bloxam believed that he had obtained the "most valuable and far-reaching experimental results" in his current work. Since he was a member of the Chemical Society, the columns of the society's journal were the "proper place" where such chemical discoveries should be first published. He feared that he might lose "priority of claim to be considered the discoverer of the novelties" if his findings were put in an interim report to the Bengal government. The officials were forced to agree with that. Bloxam left for England full of hopes of completing his work in the metropolis.[10]

[8] Letter from S. L. Maddox, Director, Land Records and Agriculture, Bengal, to Revenue Secretary, Government of Bengal, dated January 25, 1904, Proceedings of the Government of Bengal, Revenue Department for the month of July 1904, No. 144A, Nos. 4–5, File, 2-I/8 1, India Office Records, GOB (Rev), P/6793.

[9] Letter from F. M. Coventry to the Director, Department of Land Records and Agriculture, Bengal, dated January 20 1904, Proceedings of the Government of Bengal, Revenue Department for the month of July 1904, No. 144A, Nos. 4–5, File, 2-I/8 1, India Office Records, GOB (Rev), P/6793.

[10] Extracts from Bloxam's letter in Letter of S. L. Maddox to Revenue Secretary, Government of Bengal, dated March 6, 1904, No. 1T-A, No. 9, File 2-I/2 4, Proceedings

William P. Bloxam had close ties with the network of prominent academic chemists in the metropolis that ensured his smooth reemployment in London.[11] He spent the first six months after returning to London writing a comprehensive report, which he sent to various specialists who could be counted among the leading lights in the field of chemistry at the time, such as William Ramsay at University College and Edward Thorpe at the Government Laboratory in London. Both Ramsay and Thorpe praised his report, authenticating its scientific merit. On Bloxam's request Ramsay also made a case with the India House in London to employ Bloxam for a year so that he could complete the work he had carried over from India. Ramsay played the role of an effective interpreter for Bloxam's work in soliciting official patronage for it. First of all he mentioned that Bloxam had obtained pure indigo in the laboratory, which in itself indicated his mastery of a very complex aspect of dye chemistry. Bloxam had also isolated red bodies present in the dye. It was "not unlikely" that this red body was an intermediate compound from which the coloring matter itself was formed in the leaves. Thus Ramsay raised expectations that Bloxam might be close to uncovering the very process of formation of color in the plant, an insight that might prove critical in the effort to salvage the colonial indigo industry. Ramsay also highlighted Bloxam's current efforts to devise a correct method of estimating color percentages that could help optimize indigo manufacturing processes. Work of such a nature, Ramsay argued, could better be completed in

---

of the Government of Bengal, Revenue Department for the month of July 1904, India Office Record, GOB (Rev), P/6793; Bloxam's letter to the Director, Department of Land Records and Agriculture, Bengal, dated March 27, 1904, Proceedings of the Government of Bengal, Revenue Department for the month of July 1904, India Office Record, GOB (Rev), P/6793; letter from S. L. Maddox to Revenue Secretary, Bengal, dated March 29, 1904, Nos. 14–15, File, 2-I/2 9; letter from A. Earle to S. L. Maddox, dated March 29, 1904, No. 16, File 2-I/2 10, Proceedings of the Government of Bengal, Revenue Department, for the month of July 1904, India Office Records, GOB (Rev), P/6793.

[11] Indeed Bloxam was a product of the academic establishment in England. He was born into a family of chemists who held professorial positions at the King's College and at the Goldsmiths' Institute. He first worked for two decades as a demonstrator at the Royal College in Greenwich. In the early phase of his career he published several papers in the *Journal of the Chemical Society* and in *Chemical News*. He then took up work in succession at prestigious laboratories, first at the Davy Faraday Laboratory and then at a laboratory of Royal College of Physicians and Surgeons. In search of a more permanent position, Bloxam later moved to India to assume the position of the chair of chemistry at Presidency College in Calcutta. But when the latter position came to an end abruptly, he decided to join the indigo station at Dalsingserai. All along his relationship with the academic chemists within the establishment at home stayed on a firm basis. On his return to England in 1905, he again reached out to them in order to have his work validated.

one of the fine laboratories in England, a country "where advice from skilled chemists is easily obtainable, and where the literature on the subject is easily accessible, than in India."[12] Ramsay's letter spurred interest in Bloxam's work at the India House and ultimately led to his engagement at the Clothworkers Laboratory of the University of Leeds. The Leeds laboratory was home to A. G. Perkin, the leading British expert on natural dyes, and his presence meant that Bloxam at Leeds would be in the most appropriate surroundings.[13] After the initial hesitation about the merit of Bloxam's work by colonial officials in India, a belief had gradually emerged there as well that a program for indigo science could be rationally laid out in the metropolis and the colony with its two arms complementing each other.[14]

Planters Fight Back: Reevaluating the Scientific Tests

The era of the second round of indigo experiments was also marked by an astonishingly earnest debate on the nature of tests by scientists and market players in the colony and England. These tests were used to check the percentage of color in scientific processes as well as to settle the price of indigo in the market. The sudden resurgence of interest in the test question reflected a new optimism about the natural indigo industry and consequent heightened stakes in the industry, however temporary it might

[12] Letter from William Ramsay to the Under-Secretary of State for India, India House, dated November 6, 1904, "Employment by the India Office of Mr. W. P. Bloxam for the purpose of carrying on further researches regarding the methods of production of natural indigo," Government of India, Proceedings of the Department of Revenue and Agriculture for May 1905, No. 25, Serial No. 1, India Office Records, GOI, Proc. Rev & Agr, P/7069.

[13] Letter from A. Goldby, Under-Secretary of State for India, to Sir William Ramsay, No. R&S 2662, dated November 11, 1904, No. 25, Serial No. 1, Government of India, Proceedings of the Department of Revenue and Agriculture for May 1905, India Office Records, GOI, Proc. Rev & Agr, P/7069.

[14] Letter from R. W. Carlyle, Revenue Secretary, Government of Bengal, to Secretary, Revenue and Agricultural Department, Government of India, dated March 27, 1904, letter no. 1717, No. 28, File 2-I/2 6, Government of Bengal, Proceedings of the Revenue Department for the month of November 1905, India Office Records, GOB, Rev (Agr.), P/7034; letter from the Governor General in Council in India to the Secretary of State in England, dated May 4 1905, letter no. 15, Nos. 29–30, File 2-I/2 7, Government of Bengal, Proceedings of the Revenue Department for the month of November 1905, India Office Record, GOB, Rev (Agr.), P/7034; letter from A Godley, Under Secretary of State for India to W. P. Bloxam, dated July 27, 1905, letter no. R&S 1911, No. 1, Serial No. 8, File No. 18 of 1905, Government of India, Proceedings of the Department of Revenue and Agriculture for September 1905, India Office Record, GOI, Proc. Rev & Agr, P/7069.

prove to be ultimately. The new higher-yielding Java indigo had been in cultivation for a second year running, and by all appearance the new variety seemed promising. An early 1906 entry in the *Indian Planters' Gazette* accordingly noted that "the trade has been interested in some very optimistic reports concerning the ultimate prospects of indigo production and if all that is claimed for the Natal-Java seeds proves true, there will be plenty of use for indigo mills and vats for many years to come."[15] The total production figure for indigo also got a bump in 1906–7. The Season and Crop Report for 1906–7 published by the office of Director of Agriculture C. A. Oldham confirmed the increase in output, adding that it "was partly due to the better indigo made from the Java-Natal plant."[16] There was general optimism now, after a long interval when cultivation and production of indigo had been on a downward spiral. In the current wave of optimism an attempt to renew and even reinvent the older arguments of efficiency of manufacturing and those over the quality of the natural dye did not seem out of place. One market watcher thus welcomed "the revival of interest in the relative merits of natural and synthetic indigo" and was pleased at this outcome because it seemed to indicate to him that "the fight has not ended."[17]

Bloxam and Bergtheil: Professional Rivalry and Different Logic of Experiments

The hope that indigo work in the metropolis and the colony would proceed efficiently in conjunction was belied from the very beginning. The relationship between the metropolitan and colonial arms of the indigo research apparatus was fraught on account of several factors. The personal aspect of this fractious relationship became apparent with the two scientists disagreeing over division of work on the indigo program. Some of this was related to their effort to stake priority in publishing on what had become a critical area of science as a result of the ongoing market competition between natural and synthetic indigo. As personal letters exchanged between Bloxam and Bergtheil revealed after the fact, Bloxam had first proposed "mutual agreement" between the two on the "time and place of publication" for their respective works. Bergtheil was open to the suggestion. He stated that while he was bound to communicate

---

[15] *Indian Planters' Gazette*, January 6, 1906, p. 17.
[16] *Indian Planters' Gazette*, August 24, 1907, p. 259.
[17] Letter to the Editor of *Manchester Guardian*, reproduced in *Indian Planters' Gazette*, April 6, 1907, p. 398.

results of "practical value" obtained at the Sirsiah laboratory to BIPA, he was free to publish work of a "more scientific nature" in professional journals under his own name. On the latter work he was willing to reach an informal agreement with Bloxam so that they both focused in separate areas "with no disadvantage to one another." However, in the end, an understanding eluded them. Fundamental disagreement stemmed from Bloxam's insistence on reserving analytical chemical work relating to color estimation and the nature of the "red bodies" (indirubin) for himself. Bergtheil was not ready to concede to Bloxam the "sole authority" to work in those areas as the latter desired. Indeed Bergtheil disclosed that he was also working on color estimation processes and was about to present the results at a professional meeting. The two made different interpretations of the government's directive on division of work. Bloxam claimed that it was he who had first started working on the problem with color testing processes. And the government's order had entitled him alone to continue working on analytical aspects relating to manufactured indigo. Bergtheil disagreed with that interpretation.[18]

Both Bloxam at the University of Leeds and Bergtheil in colonial India continued to work in parallel on color estimation processes. Bloxam's work was unique in exclusively focusing on them. He was away from India and in any case could not engage in real time experiments on cultivation and manufacturing processes. His scientific materials comprised specimens of indigo leaves and of the actual dye in different stages of manufacturing that were amenable to analytical investigation of this sort. The superior scientific resources at Leeds aided his efforts. At the Clothworkers Laboratory he could draw on the expertise of some of the most esteemed chemists of vegetable dyes, most notably A. G. Perkin. Additionally, he had access to the latest research on plant indigo elsewhere in the world, especially that relating to indican and indigo enzyme by the Dutch chemists. Bloxam also obtained the rare samples of indican, the color-bearing glucoside in indigo, and indigo enzyme for his ongoing analytical investigation, all of which aided his quest for developing very precise tests for measuring color percentages. In contrast, at Sirsiah, Cyril Bergtheil designed his experiments more broadly on all aspects of

---

[18] Letter from Cyril Bergtheil to W. P. Bloxam, dated September 2, 1905; letter from Bloxam to Bergtheil, dated October 12, 1905, enclosure in letter from C. A. Oldham, Director of Agriculture, Bengal, to Secretary, Board of Revenue, Lower Provinces, dated July 11, 1906, letter no. 2755A, No. 19, Serial No. 4, File No. 123 of 1906, Government of India, Proceedings of the Department of Revenue and Agriculture for April 1907, India Office Records, GOI, Proc. Rev & Agr, P/7613.

cultivation and manufacturing. But he too gave more than trivial attention to developing experiments to check the validity of color estimation processes.[19]

Bloxam and Bergtheil ultimately disagreed over the accuracy of color estimation processes. The dispute specifically revolved around the accuracy of the persulfate and permanganate tests for measuring color percentage in the leaf and the finished dye, respectively. In his published report on prior work at Dalsingserai Bloxam highlighted the inaccuracy of the permanganate test that was used in the trade for checking color percentages.[20] This test had sufficed as a rough and ready test for those interested in measuring the dye's purity and in fixing its value in the market. Manufacturers, drysalters, and buying and selling brokers had unreservedly employed the test so far. Easy to administer and relatively quick in giving results, the test lent itself well to the requirements of commercial transactions. But now in the new context of market competition and elaborate government programs to increase yield, the accuracy of the permanganate test as well as the persulfate test assumed new significance. They were potential tools in settling the question of accuracy and validity of the manufacturing process. And there was a deadlock between the two scientists on the question of efficiency of the two tests.

In the summer of 1906 Bloxam and Bergtheil began arguing against each other's interpretation of color tests at the meetings of the Society of Chemical Industry in Manchester and London. Going first before the society, Bergtheil argued that the two existing tests with modifications gave reasonably precise results. If these tests were indeed accurate, as Bergtheil claimed, they clearly confirmed that the manufacturing process extracted a very high percentage of color present in the leaf.[21] Bloxam

[19] Cyril Bergtheil, "An account of the scientific investigations which have been and are being conducted in India," *Indian Planters' Gazette*, December 23, 1905, pp. 771–2; Cyril Bergtheil's account of scientific work at Sirsiah during 1905–6, *Indian Planters' Gazette*, December 22, 1906, pp. 750–1. See also, Report of the Indigo Research Station, Sirsiah, 1905–6, Calcutta: Baptist Mission Press, 1906, pp. 4–5, British Library, Oriental Collections ST 1882; Report of the Indigo Research Station, Sirsiah, 1906–7, Calcutta: Baptist Mission Press, 1907, pp. 4–12, British Library, Oriental Collections ST 1882; Report of the Indigo Research Station, Sirsiah, 1907–8, Calcutta: Baptist Mission Press, 1908, pp. 3–8, British Library, Oriental Collections ST 1882. Hereafter, these reports are cited as "Sirsiah Report," suffixed by the year.

[20] W. Popplewell Bloxam and H. M. Leake, with the assistance of R. S. Finlow, *An Account of the Research Work in Indigo, Carried Out at the Dalsingh Serai Research Station from 1903 to March 1904*, Calcutta: Bengal Secretariat Book Depot, 1905, pp. 26–8.

[21] C. Bergtheil and R. V. Briggs, "The Determination of Indigotin in Commercial Indigo and in Indigo-Yielding Plants," *Journal of the Society of Chemical Industry* 25 (1906): 729–30.

argued to the contrary. Presenting results at the Yorkshire section of the Society of Chemical Industry in the same month, he pointed to major faults in the two tests. According to him the current tests gave a much lower figure for the color in the leaf and a higher figure for the color present in the manufactured dye, thus implying that the manufacturing process was inefficient. Bloxam also proposed two new methods that he claimed were more accurate, the Isatin method for the leaf and the tetrasulphonate method for the cake. These tests pointed to a higher value for color in the indigo leaf and a lower value for color in the cake. In Bloxam's argument thus a good amount of color was being left behind in the leaf or dissipated during the manufacturing process.[22]

Planters Reject the "Science" of Tests

The involvement of analytical chemists with the indigo trade was a new development that started in the last quarter of the nineteenth century. The importance of analytical tests for measuring color in the dye also rose correspondingly. The chemists replaced the earlier methods of valuation used by buyers who checked the dye for contamination with earthly materials, its texture, its amenability to grinding, its appearance (bright blue color meant a good dye, whereas a slaking texture and darkish black color indicated faulty manufacturing and poor-quality dye), and its weight (the best quality was supposed to float in water). The new class of professional analysts additionally began to subject the different brands and marks of indigo imported from India to reductive chemical tests in order to determine the percentage of color in the dye. Indian indigo was known to vary greatly in color-giving power, depending upon the region in which it had been produced, and thus the practice of checking color percentages helped with the task of general categorizing. Increasingly, the market value of separate marks was fixed on the determination of color percentage, although other subjective criteria continued to be used by analysts and purchasers alike.

In the postsynthetic indigo era, however, the system of valuation on the basis of percentages assumed a new meaning for market competition. The BASF claimed that their dye was solely indigo, the substance indigo that new chemistry could represent with a formula, which they sold in 20 percent concentration. The Bengal indigo, an agricultural product derived

---

[22] W. P. Bloxam, "The Analysis of Indigo," *Journal of the Society of Chemical Industry* 25 (1906): 735–44; I. Q. Richardson, S. H. Wood, and W. P. Bloxam, "Analysis of Indigo – Part II," *Journal of the Society of Chemical Industry* 26 (January 15, 1907): 4–7.

from leaves of a plant, varied wildly in constitution. It was known to contain on an average about 60 percent indigo, as determined by analytical tests. Gradually a reductionist system of assessment established itself in the market that compared natural and synthetic on the basis of percentage of indigo. In this system unit weight of 20 percent concentration synthetic would be equal in value to one-third the weight of natural containing 60 percent indigo. The synthetic and natural producers disagreed about the value of the balance of 40 percent of constituents in the natural dye such as indigo brown, water, indigo glutten, minerals, and resins. The synthetic's manufacturers claimed that these latter elements were of no use to the dyeing process and thus counted them as "impurity." The producers of the natural dye argued to the contrary that the remaining constituents other than 60 percent indigo also had tinctorial value. Some of the latter additionally played an important role in fixing the color onto fabric. The most passionate argument by the planters was made about indigo-red in their dye, which, they argued, imparted a unique red sheen to garments dyed with natural indigo. The different interpretations aside, the market gradually settled in favor of the system of using color percentages for comparing the values of natural and synthetic dyes. The advocates for natural indigo lost the argument on this front in the decade after the launch of synthetic indigo. As one of them noted remorsefully, "The scientists jumped to the conclusion that the 40 per cent of other constituents [in the natural dye] were valueless; and as 'science' is everything nowadays, the dyers followed suit, and started the 'centage testing."[23]

In the years of building optimism around the expansion of Java indigo, planters reopened the debate on the nature of analytical tests. By highlighting how current tests failed to gauge the "real value" of the natural dye, they wished to resituate their product in the market. Thus going against current wisdom, in 1907, T. R. Filgate, the secretary of the Bihar Planters' Association, insisted that one pound of natural indigo should be equivalent to five pounds of synthetic because the latter was admittedly sold in 20 percent concentration. In Filgate's understanding all of natural dye was indigo, not conceding any room for making a distinction between natural dye and the latter's so-called impurity. Some claimed that natural and synthetic were two absolutely different things, and therefore there was no question of comparing them for settling price. "There is no such thing as artificial indigo," one of them said. "What's meant [by that expression] is synthetic indigotin produced from the derivatives of coal tar, a very

[23] *Indian Planters' Gazette*, July 7, 1906, p. 27.

different thing. Natural indigo is an organic compound body which contains several other ingredients and colours besides indigotin, and it is *the combination of these* that gives natural indigo preeminence over all other blue colours. One might as well call albumen an egg or starch a potato as to call synthetic indigotin, indigo!"[24] The planters saw the new system of reductionist analytical chemistry and its purported support to dominant market practices as flawed and unfair to natural indigo. In a way there was nothing new about these fundamentally quality-centric arguments in favor of natural indigo. But they were getting a leg in the context of the attempt by the planters to navigate current market rationalities and resituate their product in it.

Such assertions clearly went against the system of valuation and fixation of price for the two dyes that was now prevalent in the market. Thus Jules Karpeles, a broker of natural indigo in the London markets, tried to correct Filgate's drawing of equivalence between one unit weight of natural as the same as five unit weights of synthetic of 20 percent concentration. He wrote to remind Filgate that the latter's communication had allowed an "inaccurate statement to slip in." On the contrary, he maintained that in reality in the market synthetic indigo of 20 percent concentration sold at nine pence a pound and 63 percent natural sold at thirty-six pence a pound. Market price was being fixed on the basis of percentage of color. And the tests were critical in determining percentage and comparative price of unit weight of natural and synthetic indigo. He pointed out that he was sympathetic to the argument of planters that such a system of comparisons was not fair, and that there indeed were consumers in the market who did not believe in such tests and who had never tried synthetic indigo, but that they "they are unluckily very few!"[25] The truth was that despite spirited campaigns by planters to use a different system of appraisal, the market continued to purchase on the basis of percentage determination tests.

Public Demonstration at the Cawnpore Textile Mills: Winning Back Consumers

The consumers apparently held the key to any effort to revive the fortunes of the natural indigo industry. But approaching the consumers and persuading them to go back to natural was going to be a difficult task.

---

[24] Letter to the Editor, *Indian Planters' Gazette*, October 26, 1907, p. 520.
[25] Letter of Jules Karpele to T. R. Filgate, dated January 2, 1907, *Indian Planters' Gazette*, February 23, 1907, pp. 236–7.

The planters' advocates perceived the complex nature of the market and ably made distinctions between end users, the textile using public and the commercial users of dyes, the dyers and printers, as they attempted to win back consumers. Speaking about the textile buyers at large, an indigo advocate, Keith MacDonald, noted that "the general public are [*sic*] quite indifferent to the virtues of natural … indigo."[26] MacDonald also argued that buyers were unable to get correct information on the exact dye used in manufacturing. If they ever wanted to buy garments dyed with natural indigo, and MacDonald believed there were many who did, sellers sold them garments dyed with a concoction of different dyes. To be sure, aside from woolen dyers, who preferred natural indigo for technical reasons, there was a niche market for natural indigo comprising buyers to whom price was not a determinant. They would presumably prefer natural indigo even if it were more expensive. MacDonald argued that the number of the latter was significant and that they were being duped by dyers and printers who used the synthetic. Thus there was muted demand by the planters to mandate that textile manufacturers label their products and disclose the specific dye used in them. But the planters could not build enough momentum to persuade the government or the trade to launch any such marketwide reforms. Advertisements were possibly another way to reach out to the general public. But in the end the demand for using advertisements was also not acted upon. Individual textile customers were a disparate group and somewhat incapable or inert in executing their choice in the market, and the planters read very well their limitation in this regard.

In contrast, to some planters it was apparent that the middle-level consumers or commercial consumers were powerful in the market. As large-scale purchasers of the dye these were the groups that were earlier sought out by the synthetic manufacturers and their marketing men in the early phases of expansion of synthetic indigo. Their bargaining power in the textile market in Britain was further strengthening in an era of rise of combinations in the British dyeing and printing industry. The Calico Printers' Association (CPA) of Manchester was formed by merging into a federation a total of forty-six printing companies and thirteen merchant firms in 1899. In subsequent years many more joined the CPA. The British Cotton and Wool Dyers' Association was another conglomerate launched in 1899. These combines emerged as significant bloc purchasers of dyes,

---

[26] Keith MacDonald's letter to the Editor, *Indian Planters' Gazette*, December 7, 1907, p. 704.

who purchased their supplies jointly and distributed them to individual members internally. They exercised influence in the market as major purchasers of dyes. The planters realized that they would have to sway the class of commercial users to re-capture the market for blue dye.[27]

These planters saw the changes to the old order and wanted to navigate the rationalizing forces in the market on their terms. MacDonald partly represented this spirit among the planters. He pointed to the attributes of the new system of competition with masculine allegories: "In these days of fierce competition in all trades, the weaklings who fail to adopt business habits ... are certain sooner or later to be ousted from the field of commerce, to make room for more energetic men." Planters found the system of valuation around percentage tests to be disadvantageous to natural indigo. And they saw a set of forces in the current market that supported those tests. They would have to circumvent these forces in order to be able to relaunch their product on new terms. One possibility was to make their case with the market players on tests. This is the message that Keith MacDonald tried to give out to the planters in asking them to "get around public bodies, and scientific journals." The other possibility was to make their case with commercial consumers while still working with the percentage tests, the dominant method of valuation. It is the second option that the planters embraced in arranging for a grand public demonstration of the superior dyeing properties of agricultural indigo in the colony. The Cawnpore Textile Mills in the neighboring United Provinces was the venue for this trial. The Sirsiah scientist Cyril Bergtheil supervised the trial supported by J. Scott, the dyeing manager of the company, who had sufficient experience in using synthetic indigo. Cyril Bergtheil was a believer in the percentage theory and yet asserted that natural gave more color than the synthetic when equivalent weights of the two dyes were compared in practical trials. Bergtheil was also a scientist of international credibility. His experiments would be considered reliable and valid in design, and their results would be acceptable to the chemists in the West and the dyers and printers. Thus the planters

---

[27] A short reference to the process by which the German dye companies effectively used a network of salesmen to approach such major buyers is available in John Joseph Beer, *The Emergence of the German Dye Industry*, Urbana: University of Illinois Press, 1959, pp. 95–6. The records of the Calico Printers' Association at Manchester Archives reveal the correspondence of German firms with CPA, who offered special "contract" prices, made the supply of other dyes conditional on the purchase of synthetic indigo, and guaranteed lower prices on commitment to purchase future supplies. These records are available at the Manchester Archives and Local Studies, Manchester: M464.

asked him to display to everyone what he had deduced in the laboratory. Bergtheil's extensive laboratory experiments had confirmed the superior dyeing power of natural. But, as he pointed out, because these assertions were opposed to the understanding of many practical dyers, he was repeating them as public trials "under absolutely practical conditions." Thus these trials were scientific experiments, carried out publicly. The planters were drawing on the legitimacy of the same empirical realism that science was known to draw on for establishing truth.[28]

The Cawnpore demonstration was aimed at unshackling consumers from their current product by proving the superior dyeing properties of natural indigo. The planters' primary argument was that natural indigo in the hands of a seasoned dyer and when used under actual conditions of dyeing gave more color. The argument was going beyond the previous claims of uniqueness of color and appeal of natural indigo to a demonstration of the economic "value" of natural indigo in order to enchant commercial consumers. Organized at a textile mill and under the supervision of an expert proficient in the use of synthetic indigo, the planters were taking the battle to the home grounds of synthetic and to the "market." In the trial a sample of 20 percent BASF indigo was tested for percentage of indigotin found to contain in reality 19 percent indigotin. Another sample of natural indigo was then tested for indigotin percentage and an equivalent weight set aside for the comparative experiment. Three pieces of serges of identical weight, length, and texture were dyed in two similar vats simultaneously using natural and synthetic separately. These trials consistently showed that the vats containing natural indigo allowed for more dyeing. Bergtheil clarified that "in each case the colour produced was considerably deeper and of a richer, 'bloomier' shade from the vat containing natural indigo."

---

[28] There is extensive literature on the historical emergence of laboratory science and experimental design as valid tools for generating scientific truth. Some important palimpsests in this field include Steven Shapin and Simon Schaffer,. *Leviathan and the Air Pump: Hobbes, Boyle, and the Experimental Life*, Princeton, N.J.: Princeton University Press, 1985, and Steven Shapin, *A Social History of Truth: Civility and Science in Seventeenth Century England*, Chicago: University of Chicago Press, 1994; for contributions by sociologists, see Bruno Latour, "Give Me a Laboratory and I Will Raise the World," in Karin Knorr-Cetina and Michael Mulkay (eds.), *Science Observed: Perspectives on the Social Study of Science*, Thousand Oaks, Calif.: Sage, 1983; Bruno Latour and Steve Woolgar, *Laboratory Life: The Construction of Scientific Facts*, Princeton, N.J.: Princeton University Press, 1986; for a focused analysis of scientific "practice" in a more contemporary context, see, Harry Collins, *Changing Order: Replication and Induction in Scientific Practice*, Beverly Hills, Calif.: Sage, 1991.

The Cawnpore trials revealed planters' efforts to counter the prescriptions of existing percentage tests and the particular scientific premise that underlay them. According to Bergtheil the Cawnpore results confirmed that natural indigo had a "marked advantage" over synthetic for dyeing woolen goods specifically. He also emphasized, very importantly, that the other constituents present in the natural over and above indigotin had a definite role in imparting deeper color. He admitted that all these conclusions remained to be validated further through additional experiments. For instance, he could not yet disclose in quantitative terms the "advantage" of natural indigo over synthetic indigo. Additional trials would reveal "the extent of this advantage, its consistency, and its causes." But his inferences were in line with and validated the planters' quest to reinsert an alternative model of valuation in the market. Bergtheil disclosed that he was trying to develop equipment for easy administration of alternative tests that could measure the actual coloring potential of the dye going beyond mere ascertaining of color percentages.[29]

Indigo, the Scientific Paradigm, and Progress

Cyril Bergtheil's public demonstration at Cawnpore accomplished the task of putting planters' perspective on tests on a "scientific pedestal." The trial caught wide attention all around. Many in the scientific world were impressed that "science is lending her invaluable, if belated aid" to the task of reviving natural indigo. They were referring to the latest effort to align natural indigo science with the logic of analytical tests. In the din of voices that responded to Bergtheil's trials, one was that of J. Grossman, a dye chemist, a Yorkshire resident, a fellow of the Institute of Chemistry, and a longtime associate of the Society of Dyers and Colourists. Dye testing was his professional specialty. He saw himself as standing on the side of "accurate" science and welcomed the latest turn of events, in which science seemed to be prevailing over art in the natural indigo industry. Writing in the key newspaper in the Yorkshire region, *Manchester Guardian,* he expressed admiration for the contributions of science in explaining the manufacture and use of natural indigo in recent times. He was happy to emphasize that "the recent controversy on indigo must have convinced the most skeptical that indigo dyeing is a *science.*

---

[29] Cyril Bergtheil, "The Dyeing Principle of Natural Indigo," *Indian Planters' Gazette,* January 12, 1907, p. 53; "Natural versus Synthetic Indigo: Practical Dye Test," *Indian Textile Journal* (September 1907): 385; Sirsiah Report, 1906–7, pp. 4–12.

And yet it is less than thirty years since it was considered an *art* [italics added]." Grossman offered the path forward for natural indigo. "If we follow the development of any chemical industry, we shall find that its ultimate aim has always been to produce the purest article," he said. Thus if indigo planters made sufficient effort to remove "the browns and gluttens," in the agricultural product, they would be left with indigo "greatly resembling the artificial product," which would readily sell in the market. He asserted that the success of every industry today was dependent on "practical application" of scientific principles and that was how the real improvement on indigo might occur, laying down what he thought needed to be done for improving natural indigo.[30] He was prescribing a future path of improvement for natural indigo within the parameters of rationalizing forces in the market.

Bergtheil's trials drew the attention of BASF itself. A scientific study of this scope could not be allowed to pass. Its chemists and Bombay agents, Messrs. Ostereyr and Co., responded to the publication of Bergtheil's results even as they admittedly awaited full disclosure of the process followed in the trial. They did not contest the methods of his tests, but only its results. In their early response, based on the information they had so far, the BASF trade representatives pointed out that the version of BASF indigo used by Cyril Bergtheil was a very specific one containing additives such as lime. Presence of such chemicals was meant to enhance synthetic's applicability on specific yarns. But the same would have compromised its general dyeing prowess in Bergtheil's comparative test on serges. Thus they argued that the experiments at Cawnpore had given an undue advantage to the natural and should not be taken as a valid proof of its superiority.[31]

Bergtheil presented planters' arguments within the norms that were acceptable to analytical chemists and traders. He had assumed the role of a "planter's scientist" who was willing to act as the interlocutor for the planters' world with the world of synthetic indigo in the market. But, in a way, Cyril Bergtheil, J. Grossman, and the BASF representatives shared a platform in placing agricultural indigo in a specific vision of science and progress. It is true that Bergtheil and Grossman disagreed on the utility of elements other than indigotin in the natural dye, the former arguing

---

[30] Several letters including J. Grossman's and one of T. T. Reynolds's, which appeared in the *Manchester Guardian*, were reprinted in the *Indian Planters' Gazette*. *Indian Planters' Gazette*, May 4, 1907, pp. 519–22.

[31] Reference in Jules Karpeles's letter to the editor, dated October 17, 1907, *Indian Planters' Gazette*, November 12, 1907, p. 611.

that they also aided dyeing and the latter that they did not. Bergtheil and the BASF chemist might have squabbled about the design of experiments in the former's trials and the nature of his specimens. But they agreed in pointing a way forward for natural indigo into a certain paradigm that was in step with the march of reductive science and a corresponding market rationalization.[32] This was a path that seemed to spell doom for the natural. Most planters very well realized that competing with synthetic on the latter's terms was taking on a formidable foe in a less than level playing field. There was no way natural could compete with synthetic on the criteria of purity and price determined on the basis of percentage of constituents identified by analytical chemistry.

Aside from the "rationalizers" there were clearly others who wished to redefine indigo's place in the world on completely separate parameters. Many planters thus distanced themselves from this preceding paradigm suggested for natural indigo. Some of this opposition almost verged into expressions of aversion for "science" with the latter becoming a sign for the rationalizing system that was seen to be ultimately subverting natural indigo. At other times, such opinions tried to attack the consensus that was seen to be existing between the beliefs of scientists, including planters' own scientists like Cyril Bergtheil, and the norms followed in the trade. Thus some planters charged scientists with tacitly giving approval to the "centage test," which was outright discriminating against natural indigo. As one planter wrote in the *Gazette,* the planters had wrongly conceded ground to this science. As evidence he pointed to the fact that the planters had right under their wings a scientist, the lead scientist Cyril Bergtheil, who believed in these tests and had sold out to the other side. What this planter asked for was a complete rejection of the system of tests for the task of restoring indigo's past glory, saying the planters "were infatuated enough to select a 'centage tester as their chemist, and with still more infatuation are selling their indigo in Calcutta on a 'centage test, and so playing directly into the hands of the synthetic

---

[32] These contestations might be characterized as what the sociologists of science Trevor Pinch and Wiebe Bijker have called "orderly" disagreements among scientists. They argue that orthodoxy and heterodoxy conceal a more radical philosophical censorship in scientific practice. Their characterization of the norms of science can be extended to our purposes of distinguishing what counted as science/modern and nonscientific/nonmodern. Trevor J. Pinch and Wiebe E. Bijker, "The Social Construction of Facts and Artifacts: Or How the Sociology of Science and the Sociology of Technology Might Benefit Each Other," in Wiebe E. Bijker, Thomas P. Hughes, and Trevor J. Pinch (eds.), *The Social Construction of Technological Systems: New Directions in the Sociology and History of Technology,* Cambridge, Mass.: MIT Press, 1995, pp. 17–50.

people."[33] These were the planters who wanted to renounce completely the system that favored synthetic indigo in the marketplace and wished to continue the improvement of indigo on separate criteria.

The dyer Alexander Playne of Dunkirk Mills in Gloucestershire similarly challenged the basic premise of the percentage test and its "fallacy" that more color could be obtained from synthetic indigo. Playne announced that he had "repeatedly appealed to the chemists to show ...[him] how to get the 40 per cent more colour they say is in indigotin than in indigo." And he had failed in eliciting any response. Playne put the practical skills of a dyer against the assertions of formalistic science. Attacking Cyril Bergtheil's report from Sirsiah later, he would highlight the fact that Bergtheil admitted to getting a mere 10 percent of indigotin from the dye bath on fabric in his trials. As against that, he claimed that he could normally get 62.5 percent of the color from a first dip of ten minutes in the dye bath at his factory.[34]

Playne's voice represented the old order. He was representing the strongest traits of the old order – skill, practical experience, and artistry. Despite his individual tenacity – and his longevity and survival in the market was partly also based on the fact that he was a woolen dyer and thus worked in a sector where natural indigo had technical advantages over the synthetic – he was a member of a quickly depleting class of natural indigo loyalists. For a long time he had offered to hold public demonstrations revealing the better returns dyers could get from natural indigo. He was supported with funds by some planters, including W. Hudson and L. MacDonald. Playne issued challenges that appeared in several journals (Fig. 12) asking anyone to prove the cheapness of dyeing with synthetic indigo in a public contest. There were signs that actors like Playne were becoming irrelevant. The consumers were refusing to buy on any basis other than that of "percentage tests" and showed a clear preference for the cheaper and purer synthetic. Playne's numerous challenges were paid little heed to by the competition. The dominance of synthetic indigo was so well established that its supporters saw no need to respond to a dissenting voice among dyers.

W. B. Bridgett, the natural indigo dealer in London, clearly saw people like Playne as part of the order that was being overshadowed. Synthetic indigo was part of the new industrial order. Bridgett also highlighted the

[33] *Indian Planters' Gazette*, July 7, 1906, p. 27.
[34] *Indian Planters' Gazette*, June 30, 1906, pp. 768–9; *Indian Planters' Gazette*, March 13, 1909, p. 345.

A CHALLENGE.

German Chemists have made an organised attack upon the Planters of natural Indigo and for 15 years have taught in the Text Books and Universities that 60lbs. of their " Synthetic Indigotin " made out of Coaltar is equal to 100lbs. of good Natural Indigo chemically testing 60% " Indigotin."

For 15 years I have repeatedly asked in vain for any proof of the truth of this deliberate mis-statement, I now challenge anyone to dye as much colour honestly with 1lb. of 100% "Synthetic Indigotin" as I can dye with 1lb. of Natural Indigo chemically testing 48% Indigotin.

ALEX. PLAYNE,
Stroud.

FIGURE 12. Playne's advertisement challenging the chemists.
Source: *Indian Planters' Gazette*, April 16, 1910, p. 678.

fact that synthetic's advance in the market was undergirded by a network of intersecting factors, highlighting the complementarities of technical criteria like purity with other nontechnical factors. He pointed to the fact that industrial schools were producing a new class of dyers who only knew how to operate the chemical vat and not the traditional fermentation vat used for dyeing with natural indigo.[35]

The Colony Rejects Metropolitan Diktat on Tests

The appearance by Bergtheil at the meeting of the Society of Chemical Industry and the publication of the meeting's proceedings in the society's journal left Bloxam very indignant. Intending to block Bergtheil from working in the area of color estimations he approached the officials at

[35] *Indian Planters' Gazette*, December 25, 1909, p. 926.

India House, complaining that Bergtheil had ignored legitimate work on the bacteriological aspects of indigo improvement that he had been assigned. Such dereliction of duty by Bergtheil was going to prove detrimental to the future of the natural indigo industry. He also contended that Bergtheil's unjustified pursuit of chemical aspects of estimation of indigotin had led to an "unseemly" situation when an important chemistry journal in England carried publications on the same subject by two experts, both of whom were employed by the colonial government. Such a duplication of work on the government's money was wasteful. Last, he stressed that under current rules the results of research completed by government chemists in the colony could not be published in Europe except by special permission. Bloxam thus raised a number of procedural issues in order to monopolize color estimation research.[36]

Bloxam's attitude reflected a fundamental metropolis-centered mindset about a normative division of work between the scientists based in the metropolis and the colony. This specific vision wished to reserve the practice of high-end science for experts in the metropolis, where better infrastructure and access to information existed, while leaving the task of data collection to the staff based in the colony. It thus seemed very much in the nature of things that only metropolis-based experts should contribute to frontier areas in the process of knowledge creation. A. G. Perkin also gave expression to this perspective when he wrote to India House in support of Bloxam's idea of division of work. As he explained, Bergtheil was present "on the spot" in India, where the cultivation and manufacture of indigo were taking place. Thus he could observe the processes and obtain relevant data about them. On the other hand, Bloxam in England had access to "many more facilities and advantages for working out the scientific side of the question."[37]

These visions emanating from the metropolis were often resisted by the scientists and officials based in the colony. In this particular case,

[36] Letter from Bloxam to T. W. Holderness, Secretary, Revenue and Statistics, India House, dated August 23, 1906, No. 20, Serial No. 5, Government of India, Proceedings of the Department of Revenue and Agriculture for April 1907, India Office Records, GOI, Proc. Rev & Agr, P/7613.

[37] Letter from A. G. Perkin to T. W. Holderness, Secretary, Revenue and Statistics, India House, dated December 3, 1906, No. 21, Serial No. 6; letter from W. P. Bloxam to Cyril Bergtheil, dated October 12, 1908, enclosure in letter from C. A. Oldham, Director of Agriculture, Bengal, to Secretary, Board of Revenue, Lower Provinces, dated July 11, 1906, letter no. 2755A, No. 19, Serial No. 4, File No. 123 of 1906, Government of India, Proceedings of the Department of Revenue and Agriculture for April 1907, India Office Records, GOI, Proc. Rev & Agr, P/7613.

the metropolitan actors failed to enforce collaboration between Bloxam and Bergtheil on such asymmetrical terms. The London-based officials at India House complained to officials in India that Bergtheil had failed to collaborate with Bloxam judiciously. However, officials in India defended their colonial scientist, asserting that "Mr. Bergtheil does not appear to be to blame in his action."[38] There were countervailing forces at work in the colony that ensured an autonomous line of decision making on scientific matters. The physical distance from the metropolis ensured a certain degree of autonomy a priori for colonial science. Bureaucrats responded to local needs, while the colonial scientists ran the risk of losing their patronage if they did not respond adequately to the demands placed on them by their local patrons. The planters and local officials in this controversy stood solidly behind Bergtheil to justify his work in the colony on color estimation.

In 1907 as Bloxam and Bergtheil continued to argue against each other on the nature of indigo tests metropolitan and colonial forces took sides, evaluating their claims differently and interpreting their significance in a contrasting way. Bloxam and Bergtheil presented their papers back to back at the society's meetings, both making counterarguments without conceding any ground to the adversarial viewpoint.[39] The larger significance of these technical debates was not lost on anyone. On November 4, 1907, the chair of the London section's meeting while introducing the two presenters asserted that "the matter dealt with in the two papers was mainly one of analysis and assay, but it had a very far reaching and indeed an imperial implication, for it affected the attempts to maintain one of the great manufactures of the Empire." The presentations at the Society of Chemical Industry were followed by a spirited discussion of finer details by the members in attendance. These debates also caught the attention of leading metropolitan chemists more generally. But beyond

---

[38] Letter from J. H. Seabrooks, Assistant Secretary, Revenue and Statistics Department, India House, to Revenue Secretary, Government of India, dated August 31, 1906, R&S No. 2423, No. 20, Serial No. 5, Government of India, Proceedings of the Department of Revenue and Agriculture for April 1907, India Office Records, GOI, Proc. Rev. & Agr, P/7613; letter from E. D. Maclagan, Revenue Secretary, Government of India, to Secretary, Revenue and Statistics, India House, dated January 17, 1907, letter no. 73, No. 23, Serial No. 8, Government of India, Proceedings of the Department of Revenue and Agriculture for April 1907, India Office Records, GOI, Proc. Rev & Agr, P/7613.

[39] C. Bergtheil and R. V. Briggs, "The Determination of Indigotin in Commercial Indigo," *Journal of the Society of Chemical Industry* 26 (March 15, 1907): 182–4; (November 30, 1907): 1172–4; W. P. Bloxam, "Analysis of Indigo (Part III) and of the Dried Leaves of *Indigofera arrecta* and *Indigofera sumatrana*," *Journal of the Society of Chemical Industry* 26 (November 30, 1907): 1178–9, 1182.

the community of scientists, the controversy also caught the attention of politicians and policy makers, who discussed the two different scientific viewpoints in their own ways and reached separate conclusions.

The bureaucrats at the India House were intrigued by the controversy and turned to A. G. Perkin to seek clarification. A. G. Perkin was a much-respected expert working on the natural dyes. Besides, not only was he supervising Bloxam's current work at the University of Leeds, he was also in attendance at the presentations of Bloxam and Bergtheil and follow-up discussion at the meeting of the Society of Chemical Industry. Thus he had firsthand information on what had transpired at the meeting. Metropolitan officials wanted him to interpret for them the difference in the accounts of the two experts, to provide an evaluation of the arguments presented by them, and to enlighten them as to which test the commercial classes preferred. The India House apparently trusted Perkin to provide an objective assessment on a scientific question of imperial importance.[40]

Perkin was evidently convinced about the superiority of Bloxam's new tests over those favored by Bergtheil. He assured officials at India House that leading chemists in the country including those present at the meeting had overwhelmingly expressed themselves in favor of the higher accuracy of tests presented by Bloxam and his associates. The objections raised by Bergtheil to the validity of tetrasulphonate and isatin methods had been published in the *Journal of the Society of Chemical Industry* and in turn been repudiated by Bloxam in the same journal very recently. Besides, he mentioned that the Bergtheil-Briggs method for color estimation in the cake was demonstrated to him on two or three occasions and he did not think it was any more accurate than the existing permanganate test. He positioned himself on the side of Bloxam in the latter's claim that Bergtheil's tests were faulty and gave lower than actual estimate in the leaf and an inflated estimate of color in the cake. And he backed Bloxam's claims that his tests were more accurate.[41] India House forwarded this information to the imperial government in the colony to consider.

[40] Letter from T. W. Holderness, Secretary, Revenue and Statistics, India House, to A. G. Perkin, dated November 12, 1907, R&S No. 3295, No. 58, File No. 27 of 1907, Government of India, Proceedings of the Department of Revenue and Agriculture for the month of January 1908, India Office Records, GOI, Proc. Rev & Agr, P/7896.

[41] Letter from A. G. Perkin to T. W. Holderness, dated November 15, 1907, Government of India, Proceedings of the Department of Revenue and Agriculture for the month of January 1908, India Office Record, GOI, Proc. Rev & Agr, P/7896.

The agricultural officials and experts in Bengal were notably unimpressed by Bloxam's results, remaining indifferent and showing no immediate concern to examine the questions raised by them. J. Mollison represented the sentiments at the local level in the colony accurately in stating that work like Bloxam's did not have any relevance for the agricultural needs of Bengal. He was dismissive of the relevance of Bloxam's efforts, saying, "From a practical point of view no great importance needs to be attached to Mr. Bloxam's indigo research work in England. He may have worked out problems in the laboratory which are of interest to pure scientists, but I am strongly of the opinion that for the time being, and probably for all time, his results (so far as I know them) will be of no value to the ordinary growers of indigo in India."[42]

The Cosmopolitan Grounding of Colonial Indigo Science

The indifference of planters and scientists in the colony to the results obtained by Bloxam at Leeds have earlier prompted the historian of science, Peter Reed, to speak of Bihar planters as an agricultural class innately "hostile" to science.[43] Reed's criticism of planters for apparently ignoring Bloxam's counsels appears to be incorrect as it fails to take into account planters' priorities and the local conditions of indigo cultivation and manufacturing in Bihar. Participating in a follow-up discussion on Bloxam's paper at the Society of Chemical Industry the prominent metropolitan chemist Professor Raphael Meldola rightly argued that the actual impact of Bloxam's results on colonial indigo experiments was still indeterminate. Meldola pointed out that from all appearance Bloxam's isatin test seemed to be accurate. But he added that "the end of the matter was not reached when the most perfect of analytical processes had been devised." It was now the bounden duty of the planters and the government to get to the bottom of the matter while investigating at which point, and by how much, potential color was being lost. They also had to determine whether it was possible to plug the holes in the manufacturing processes and recover the extra color whose existence the new test had predicted. It was one thing to prove analytically in the laboratory that

---

[42] Letter from J. Mollison to Revenue Secretary, Government of India, dated October 2, 1907, letter no. C 672, No. 6, Serial No. 15, Proceedings of the Department of Revenue and Agriculture for the month of October 1907, India Office Records, GOI, Proc. Rev & Agr, P/7614.

[43] Peter Reed, "The British Chemical Industry and the Indigo Trade," *British Journal for the History of Science* 25 (1992): 113–25, quote on p. 116.

the indigo plant could yield more color, and quite another actually to recover that extra color on a manufacturing scale viably using available techniques. In short, the indigo advocates would have to conduct a new set of trials if they wished to absorb and benefit from Bloxam's results.

But the planters were simply not persuaded by Bloxam's conclusions. They did not sponsor any follow-up experiments to take their science in a direction counseled by Bloxam. Evaluating such a reception of Bloxam's results requires an additional consideration of the momentum of colonial indigo experiments leading up to 1907–8. Planters were very well aware of Bloxam's experimental results and the implications of his experiments, which encouraged a return to the chemical line. However, planters, Bengal officials, and their scientists had by now moved toward experiments of an agricultural kind. They strongly believed that agricultural experiments focused on the plant in the field offered the best way forward. This trust reflected legitimate priorities shaped by local conditions, and the optimism of Mendelism in the colony, all of which Peter Reed's study fails to consider.

The inability of the Leeds-based scientists to change the direction of indigo experiments in Bengal also clarifies the cosmopolitan moorings of "colonial science," which was apparently not just open to influence from a single dominant "center." In the end the momentum of the local science was determined by a complex set of inspirations. The general promise of biological selection cut across the usual categories of "metropolitan" and "local" in the colony. Agricultural science in the British metropolis in this period was itself being recast with forces aligned not with chemistry but with botany.[44] And Mendelian influence was important to national science in Britain. It had also gained currency in programs for the selection of wheat and beet sugar in the larger British imperial space including India. At the same time the local science of indigo planters in Bengal looked up with hope to the success of horticulturists in several places in the West including Europe and the United States. This general wave of

---

[44] The rise of agricultural botany and economic botany in the British metropolis was prominent in the early twentieth century. See relevant chapters in E. J. Russell, *A History of Agricultural Science in Great Britain, 1620–1954*, London: George Allen & Unwin, 1946; Alison Kraft has documented the rise of economic botany in Britain during this period. Kraft in fact considers the role of colonies and botanists stationed in the colonial outposts to explain the rise of economic botany in Britain. Kraft's study, if anything, reflects the recent trend of highlighting the reciprocity of influence between the metropolis and the colony in studies of colonial science. Alison Kraft, "Pragmatism, Patronage and Politics in English Biology: The Rise and Fall of Economic Biology 1904–1920," *Journal of the History of Biology* 37 No. 2 (2004): 213–58.

expectancy created the groundwork for initiating attempts to engage in selection at the Sirsiah Indigo Station and at Pusa. It would also eventually lead to the involvement of botanists who had knowledge of selection and hybridization with the indigo program.

The Reign of Expert Biologists (1905–1908): Selecting Robust Plant Types

The scientific work on indigo at the Sirsiah Station began with an equal focus on the manufacturing process and the plant but soon began to take on an overwhelmingly agricultural orientation. Year after year the emphasis on manufacturing experiments began to fade until they were given up completely.[45] The emphasis on experiments focused on the plant, in contrast, kept deepening. At the time of the launch of experiments at Sirsiah Cyril Bergtheil had said that most promising results on augmenting the yield of the dye were likely to result from improving the plant in the field. Pointing to selection and hybridization in particular, he emphasized, "This is the most important aspect of the work [of improvement], and perhaps the direction in which the chief hope of permanent assistance to the industry lies."[46] Such a belief among experts only became stronger with time.

The Sirsiah experts launched an ambitious seed program to purify and isolate superior "types" of indigo plants in order to change the character of plants under cultivation in Bihar. The different "types" of indigo were of "very different values" corresponding with intraspecies distinctions. Some possessed more leaf than others, some could be distinguished on the basis of whether they were flowering and developing early or late, and so on. The native variety of Bihar, Sumatrana, for instance, was found to be "a mixture of several subvarieties," and therefore the experts planned that "the best of these subvarieties should be selected and grown and the others excluded."[47] As many as nine distinguishable types of Sumatrana were distinguished over time, out of which the Delhi type was considered the best for vigor of growth and the Multan type for higher leaf percentage.[48] The foreign species of indigo in Bihar, collectively called Java indigo

---

[45] Bergtheil noted on Sirsiah Station's report for 1908–9: "we have practically disposed of all questions connected with manufacture." Sirsiah Report, 1908–9, p. 1.

[46] Cyril Bergtheil, "An account of the scientific investigations which have been and are being conducted in India," *Indian Planters' Gazette*, December 23, 1905, pp. 771–2, quote on 772.

[47] Sirsiah Report, 1905–6, pp. 19–20, 28.

[48] Sirsiah Report, 1906–7, pp. 20–1.

in order to distinguish from the indigenous variety, was also found to be composed of different races. For instance, the Madagascar indigo existing among Java indigo was "a comparative poor colour-yielder." There were probably more of such poor varieties in the stock waiting to be identified, and a plan was drawn out "to eliminate all such impurities." As better understanding of botanical characteristics accrued, experimenters understood and appreciated that Natal and Java, both belonging to the same taxonomical species *Indigofera arrecta*, had subsequently digressed from each other. This happened after Natal indigo brought in to Java in Southeast Asia was selected and acclimatized locally by the Java-based planters. The imported variety from Java probably represented a cross between the original mother variety and local varieties in Southeast Asia. Experts identified as many as six races among the Java indigo grown in Bihar.[49]

The isolation of ideal types for both sumtrana and Java was achieved through a supervised program at seed farms managed by the Sirsiah staff. Bergtheil thought seed growing to be a specialist's work, which he insisted should be implemented under the close supervision of experts. He counseled against the practice current among some planters of growing seed on their own farm as part of a regular cultivation cycle wherein some of the plants after being stripped of leaves were allowed to mature and flower. He favored separating the process of planting and seed growing. In line with these beliefs three farms for Java seed were established in Bihar. Bergtheil also preferred to establish farms for Sumatrana seeds in Bihar in order to monitor and regulate the process of recurring selection year after year. But he asked for the existing Dasna farm that was growing Sumatrana seeds to be retained before the Bihar farms were set up. He and his assistants in the meantime constantly monitored the seed program at Dasna.

The momentum toward seed farms in Bihar also developed in the context of a demand from planters to secure the supply lines of seeds for Java indigo, whose cultivation was expanding very rapidly in Bihar. The planters were somewhat upbeat about the new species as it gave a much higher yield than the native variety. Indeed to many planters Java indigo appeared to be the only real hope for saving the natural indigo industry from absolute doom. In that context of hope many planters wanted Sirsiah scientists to turn focus to the new variety, in particular to perfecting methods of growing Java seeds that would germinate without failure,

[49] Sirsiah Report, 1907–8, pp. 17–18.

as some in the past had been known to fail, and to make sure that the new species did not become "mixed" over time and lose its vigor. Thus Jules Karpeles, a trader with substantial interests of his own in plantations, wrote to Bengal officials suggesting, as H. H. Baily had done four years ago, that the mother variety of Java plant, known to grow wild in the Natal Province in South Africa, should be imported directly to Bihar.[50] Karpeles wanted this to be done in order to guard against possible "deterioration" in the stock of Java in Bihar. His views on importing Natal seeds to enrich the stock of Java in the colony were echoed by another planter, who also argued that "we ... know from experience that most foreign plants deteriorate in Behar, and that although they sometimes grow even better the first year, every succeeding year sees them gradually become worse."[51]

Despite demand from a section of planters that seeds should be imported from South Africa to enrich the stock of foreign species, both Bergtheil and Coventry steadfastly maintained that the seeds should be grown locally. H. H. Baily again wrote on this occasion that Coventry's assumption that the Java seed "will retain its characteristics [in the colony] surprises me." While there was a possibility that Coventry might be proved right in the end, he wondered why Coventry was not willing to incur the rather small cost of keeping the supply line of fresh seeds from Natal open to guard against any eventuality. After all, that is precisely what the Java planters who also grew the same strain had done before. "The Behar planters have, generally speaking, been pennywise and pound foolish," he concluded.[52] The government scientists, however, remained opposed to the idea of importing seeds from South Africa. Coventry wrote back to Karpeles explaining why he thought importing Java seeds or Natal seeds from South Africa would not be rewarding. He mentioned several instances where seeds imported from Java were giving poor germination of barely 10 to 12 percent, and in that light acclimatization in

---

[50] For correspondence of Jules Karpeles, the important indigo broker and the relentless campaigner for Java indigo in Bihar, see letter of Karpeles to L. Hare, Member, Board of Revenue, Bengal, dated January 3, 1906; to secretary of BIPA, dated February 19, 1906, *Indian Planters' Gazette*, February 24, 1906, pp. 229–31; also letter of H. H. Baily, the Natal-based trader, to Jules Karpeles, dated April 9, 1906, *Indian Planters' Gazette*, February 24, 1906, pp. 229–31; These letters together refer to Karpeles's initiatives to start import of Natal indigo from South Africa.

[51] "Natal Indigo Seed: Letter to the Editor, Indian Planters' Gazette," *Indian Planters' Gazette*, March 3, 1906, p. 255.

[52] Letter from H. H. Baily to Jules Karpeles, dated April 9, 1906, *Indian Planters' Gazette*, May 26, 1906, p. 614.

the local environment of Bihar seemed the right step. He also preferred Java seed to Natal seed. The conviction of the two experts carried the day as the seed program on Sumatrana and Java developed at seed farms on the subcontinent.

The experts were now also ready to take "selection" to the next level. In line with these goals, they hired a breeder, J. G. Turnbull, and asked the government to place a permanent botanist on the staff at Sirsiah.[53] The planters' scientists in the colony had earlier practiced simple mass selection that involved collecting seeds of better types and planting them together in the hope that they would breed true and their progenies would bear the same overall superior characteristics. This limitation was admitted in the station's report (1906–7), which stated that "at present no scientific selection is being attempted."[54] More advanced selection involving possible identification of individual plants with clearly demarcated qualities, their controlled reproduction, and directed inheritance of specific characteristics belonged in the domain of works by specialist botanists and breeders. The next year (1907–8) with the assistance of J. G. Turnbull individual plants were isolated on the basis of color content, leaf content, and "botanical characteristics." The experts also reported the important discovery that large difference occurred in the dye-yielding ability of plants of apparently the same type. Whether these differences were of a "varietal" nature that could be "selected" and whether plants could be managed specifically to inherit the higher color-bearing characteristic remained to be seen. It is in this direction that the Sirsiah experts saw the potential for the next level of selection. As Turnbull departed from Sirsiah, Bergtheil continued to emphasize the need to appoint a trained botanist in order for the selection program to be moved forward.[55]

The next year, Bergtheil reiterated the demand for a botanist. He reflected that in the absence of a botanist they were still "groping in the dark." What was needed was an astute application of knowledge of

[53] BIPA requested the secretary of state in London through the provincial government that a botanist be sent from England. Apparently some delay occurred in the process. Becoming impatient, the BIPA appointed the botanist J. A. Turnbull, without obtaining prior official approval from the government. Government officials were not pleased with this development and refused to consider Turnbull as a government appointee. But they agreed that the expert's salary could be paid out of the grant given by the government to BIPA for indigo research. Turnbull had already joined Sirsiah Station in the summer of 1906 but left after a few months. Letter of R. W. Carlyle, Revenue Secretary, Bengal, dated December 3, 1906, to Revenue Secretary, Government of India. Bihar State Archives, GOB, Rev (Agr.), May 1907, File 2I/2 5–6, No. 1.

[54] Sirsiah Report, 1906–7, p. 35.

[55] Sirsiah Report, 1907–8, p. 16.

heredity by a trained botanist who alone could conduct "scientifically-controlled breeding." He had noticed that seeds picked up from plants were not breeding true, likely on account of cross-fertilization. But even if they bred true in terms of morphological characteristics, the need of the hour was to make sure they bred true in terms of indican content. He emphasized that such botanical selection carried out patiently for a few years at the hands of a botanist and a chemist seemed the only way "that any further material advance is to be made by scientific aid."[56]

Given the current preoccupation of indigo experts and their patrons in Bihar with "the seed question" it is no surprise that Bloxam's advocacy of return to manufacturing experiments found few takers. Bloxam's conclusions were published in the form of a report in 1908.[57] But by then the planters, scientists, and the government officials most closely connected with the indigo effort had moved on to the agricultural line, which seemed more promising to them. They were busy resolving the riddle of acclimatizing the Java variety, producing seeds of the native and foreign varieties, and using the higher principles of selection to improve the indigo plant. In short, they were completely absorbed with biological experiments and trials with the plant. Under such circumstances Bloxam's call for a return to vat processes – especially as he was still falling short in suggesting a concrete way to increase yield – did not carry enough conviction to make colonial indigo science change tracks. Thus, in contrast with the alacrity with which the officials at the India House pursued the controversy sparked by the opposite viewpoints of Bloxam and Bergtheil at the Society of Chemical Industry, the colonial state officials in Bengal were quite lukewarm. That a claim of such import did not meet with adequate interest of stakeholders – the planters, indigo scientists, and relevant policy makers in the colony – had much to do with the trends in colonial indigo science and its momentum. The practical concerns and challenges of the indigo industry in the colony had moved in a direction in which Bloxam's report did not hold out a lot of promise or mean much.

1908–1909: A Year of Mixed Messages for Natural Indigo

The political economy context for natural indigo manufacturing began to change somewhat both in the metropolis and the colony. On the

---

[56] Sirsiah Report, 1908–9, pp. 16–19.

[57] William P. Bloxam, *Report to the Government of India Containing an Account of the Research Work on Indigo Performed in the University of Leeds, 1905–1907*, London: His Majesty's Secretary of State, 1908.

metropolitan end, the national government oversaw the passage of critical patent reforms in response to the long-held concern of British chemical manufacturers who blamed the English patent laws as being too flaccid to stop flooding of English markets with foreign manufactured products. They held this leniency at least partly responsible for the decline of the national capacity in chemical manufacture in general and in synthetic dyes in particular. The new legislation amended the patent laws and mandated that foreign firms should source their raw materials locally and engage local personnel and capital while executing their patents. This measure at once created British national stakeholders in the manufacture of synthetic dyes and qualitatively changed the dynamic of support for colonial agricultural indigo. On the colonial end, the Bengal government, while still resolute in its support to the industry, was beginning to express itself against any instance of "coercion" of Indian indigo growers by the planters. Jacques Pouchepadass has studied the agrarian context of planters' response to price competition in the colony. The changes in the circumstances of indigo plantations were generating friction between the planters and Indian peasants of all stripes. The planters were resorting to the collection of illegal *abwabs* from ordinary cultivators (a payment claimed by the planters in their capacity as landlords), as well as demanding adjustments in prior contracts in order to grow more remunerative food crops on the contracted land instead of indigo. The move by some of the planters to sugar manufacturing put them in conflict with the Indian rural oligarchy who had so far dominated the local sugar industry. The district of Champaran in particular witnessed widespread peasant discontent on the plantations. A wave of court cases involving disputes between the planters and indigo growers put the latter in contact with lawyers in the divisional city of Patna, indicating the potential for connections that could be formed between local conflict on indigo tracts and the growing nationalist movement. The colonial officials were increasingly unwilling to tolerate any open infractions by the planters at a time when the latter were inclined to exploit loopholes to subvert rights of the cultivators to drive costs down. In a confidential report the lieutenant governor was quoted as expressing the new sensitivity of adminiatration to open violations of laws by the planters: "India has none too many industries, and no one wishes to destroy or discourage any that she has; indigo as little as any. But the indigo industry must stand on a commercial footing and be dependent in no degree whatsoever on compulsion or even pressure. I am ready to help the planters in every legitimate way. But if indigo cannot subsist without coercion, indigo must perish." The tone of the administration reflected wariness with regard to bearing any political costs emerging out of action by the planters.[58]

The officials and scientists involved in determining national and imperial policies with regard to indigo experiments in the metropolis were not untouched by these winds of change in the political economy context of the natural indigo industry and were likely influenced by these currents as they turned somewhat lukewarm to scientific experiments on indigo. The eminent chemist Raphael Meldola, an influential voice in policy making, stopped short of recommending continuation of Bloxam's experiments at Leeds. On February 20, 1909, Meldola was speaking in front of the Indian Government Advisory Committee of the Royal Society about Bloxam's research at Leeds. While he proposed that experimental work on indigo should be continued and asserted that it seemed clear to him from Bloxam's conclusions that the natural indigo industry had a glimmer of hope, he did not favor the continuation of his work in Britain. "What is now wanted is field work carried out in India both from the chemical and biological point of view," he recommended.[59] The bureaucrats at India House agreed with Meldola's arguments and refused to extend Bloxam's tenure at the University of Leeds even though Bloxam had expressed his interest in pursuing those experiments. That decision lowered the curtains on indigo experiments in the metropolis. Bloxam was personally distressed by the apparent lack of interest in his work. He suffered a paralytic attack two years later and died in 1913. The wider

[58] W. R. Gourlay, the director of the Department of Agriculture, conducted the inquiry. Gourlay's report submitted in April 1909 stated in no uncertain terms that the purchase price given to indigo growers by the planters was not remunerative. It recommended that the prices be enhanced by 12.5 percent. Gourlay also made a set of recommendations to streamline the working of the indigo contract system. The departmental report suggested that the government initiate necessary measures to limit the term of indigo contracts to a maximum of nine years, reduce the obligation to grow indigo from 3/20 to 2/20 (that is, from three to two *katthas* per *bigha*) of the peasant's holding, and forbid the efforts to force peasants to switch from indigo to food crops, sugarcane, and so forth. Jacques Pouchepadass, *Planters and Gandhi*, pp. 150–66, quote on p. 154–5.

[59] "Memorandum by Professor R. Meldola, FRS, upon the Present Position of the Indigo Question – to the Indian Government Advisory Committee of the Royal Society," Proceedings of the Indian Government Advisory Committee of the Royal Society for 1909, CMB/59. These records are located at the archives of the Royal Society, London. The same report also appears as an Appendix in Selections from dispatches addressed to the several governments of India by the Secretary of State in Council, 52nd Series, Part 1, January 1–June 30, 1909, V/6/361. These records are available at the British Library in London. Henceforth, IOR, Dispatches. For the role of institutions like the Royal Society in England and the Board of Scientific Advice in India, see Roy MacLeod, "Scientific Advice for British India: Imperial Perceptions and Administrative Goals, 1898–1923," *Modern Asian Studies* IX No. 3 (1975): 343–84.

community of chemists working on dyes in Britain was amazed too at the indifference to Bloxam's evidently accurate experiments whereby no proactive attempt had been made to pursue the openings suggested by Bloxam's experimental results at Leeds.[60]

The imperial officials in India responded with their own biases while assessing the implications of Bloxam's report from scientific and practical angles. The central arm of the colonial government set up a committee comprising three agricultural experts – J. Mollison, inspector general of agriculture in India; B. Coventry, director of the Agricultural Research Institute at Pusa; and J. Hector Barnes, agricultural chemist in the province of Punjab. The committee was asked to report specifically on the implications of Bloxam's report for the future conduct of indigo experiments at Sirsiah and expeditiously submit a report on three questions: what portions of Bergtheil's work at Sirsiah were affected by the results of Bloxam's investigations; how far Bloxam's characterization of Indian manufacturing processes as inefficient was justified, so far as it was possible to form an opinion on the basis of materials available and without undertaking additional research; what was the prospect for work currently being done at Sirsiah? They did not intend to analyze in detail the scientific import of Bloxam's chemical experiments or to settle the controversy raised by Bloxam's statements about the efficiency of manufacturing. Their agenda was limited to assessing the impact of Bloxam's findings on the colonial government's current research program at Sirsiah, that is, examining what parts of the current research remained unaffected by Bloxam's assertions and should be judiciously carried forward and which parts of the experiments had become of doubtful efficacy in the wake of Bloxam's assertions.[61]

The committee endorsed Bergtehil's experiments being conducted at Sirsiah and recommended their continuation. It explained that only a detailed investigation by a separate body could resolve the controversy regarding the efficiency of manufacturing operations raised by Bloxam's report. The members of the committee did not quite have the qualifications to carry out an investigation of that nature. Thus it only addressed the utility of the Sirsiah experiments in its current incarnation rather than addressing the larger question of relative utility of agricultural and

---

[60] Bloxam's obituary notice by A. G. Perkin, "William Popplewell Bloxam," *Journal of the Chemical Society* 105 (1914): 1195–1200.
[61] Letter of Revenue Secretary, Government of India, R. W. Carlyle, dated February 3, 1909, to Revenue Secretary, Bihar State Archives, GOB, Rev (Agr.), File 2–1/2 1–73/4, Board's File, 114 of 1909: 3.

chemical experiments. And they let themselves be guided by the opinion of the planters in forming an opinion on this subject. Many planters had testified that the ongoing agricultural investigations at Sirsiah were beneficial. Bloxam himself had not questioned the utility of such experiments but only suggested that the chemical route would be more fruitful. His empirical work at Leeds had scarcely addressed agricultural experiments of the type current at Sirsiah. The report in Bengal spoke favorably of the contributions made by the scientists at Sirsiah and saw merit in their continuation. The central officials agreed with that recommendation.[62]

Sirsiah Indigo Station: Promoting Science in Bengal

In 1907 the Bengal officials were a little surprised when their request for the extension of Cyril Bergtheil's deputation to Sirsiah was rejected by the central government. For the last few years the Bengal government had supported the indigo experiments at Sirsiah with Bergtheil as lead scientist. On two earlier occasions they had requested and successfully received an extension to Bergtheil's deputation. If Bergtheil left at this point, the officials feared disruption to the existing infrastructure. They made a strong case for retaining Bergtheil at Sirsiah as revealed in their hard bargaining with the center that followed. These correspondences illustrate the new emerging perspectives around the future of indigo in the colony.

The center's denial of extension to Bergtheil on procedural grounds reflected a shift from their earlier approach as they were now clearly stressing the importance of research at Pusa over that of indigo experiments at Sirsiah.[63] In a follow-up rejoinder they pointed out that the appointment of Bergtheil at Sirsiah had been of "a purely temporary nature"originally. They reminded them that back in 1904 Bergtheil had

[62] J. Mollison, B. Coventry, J. Hector Barnes, "Report of the committee held at Pusa in October 1908 to consider the research work carried out at Leeds University by Mr. Bloxam, as set forth in his report to the Government of India, and how far it affects the investigations being carried out at Sirsiah in India"; letter of Revenue Secretary, Government of India, R. W. Carlyle, dated February 3, 1909, to Revenue Secretary, Government of Bengal, Bihar State Archives, GOB, Rev (Agr.), December 1908, File, 11 – A/12 24–25 of 1908, No. 2; letter from F. W. Duke, Revenue Secretary, Government of Bengal, to Secretary, Revenue and Agriculture, Government of India, dated April 17, 1909, letter no. 61 TR, No. 37, Serial No. 7, File No. 18 of 1909, Bihar State Archives, GOB, Rev (Agr.).

[63] Letter from J. Wilson, Revenue Secretary, Government of India, to Revenue Secretary, Government of Bengal, letter no. 1114, dated July 1, 1907, No. 27, Serial No. 2, Government of India, Proceedings of the Department of Revenue and Agriculture for June 1907, India Office Records, GOI, Proc. Rev & Agr, P/7613.

accepted the appointment of imperial bacteriologist offered to him by the secretary of state. But on the Bengal governor Andrew Fraser's request he had been allowed to work out definite experimental issues related to indigo at Sirsiah. The understanding reached at that time was that Bergtheil would revert to his position at Pusa as soon as the laboratory and other infrastructure there were ready. Thus they were justified in demanding the recall of Bergtheil. They also contended that a bacteriologist was more appropriately required at Pusa. Research and education work had been going on at Pusa in every branch of agricultural science except bacteriology, the field for which Bergtheil had expertise. The imperial mycologist, for instance, had to put aside some potentially collaborative experiments in the absence of the bacteriologist. The officials at Pusa had also received representation from a provincial government to start instructions in bacteriology for native apprentices as soon as possible. The central government had so far held back its original plans at Pusa in order to protect the interests of the planting community. But they had done enough. They argued that Bergtheil must return now in the "general interest" of agricultural research at Pusa.[64]

The Bengal officials were, however, steadfast in their defense of the current arrangement at Sirsiah. The indigo industry was still a major industry of the province, and the Bengal government its primary patrons. These officials were much more cognizant of the needs of the community of planters. They also had an intimate knowledge of the direction of current indigo experiments. They heard the planters expressing their appreciation of the Sirsiah work and of Cyril Bergtheil. Thus they fought to preserve the experiments at Sirsiah under Cyril Bergtheil.

The Bengal officials defended the science they had cultivated at Sirsiah for the last few years, both the merit of the work in progress and the appropriateness of Bergtheil for undertaking that work. Talking past the center's argument that Bergtheil as a bacteriologist was hardly suitable for Sirsiah, they maintained that there was "great hope" from the indigo experiments that he was supervising. They minimized the relevance of Bergtheil's formal training in chemistry and bacteriology. Instead they based their support for him on a consideration of his practical experience of six years with indigo planting and manufacturing in Bihar. The current focus on the plant in the field required many skills including estimating

[64] Letter from R. W. Carlyle, Revenue Secretary, Government of India, to Revenue Secretary, Government of Bengal, dated February 3, 1909, letter no. 152, No. 20, Serial No. 2, Government of India, Proceedings of the Department of Revenue and Agriculture for February 1909, India Office Records, GOI, Proc. Rev & Agr, P/8174.

the percentage of leaf on different plants in the field and the color content of leaf in the laboratory. They argued that Bergtheil was ideally suited to split work between the laboratory and the field. All that he needed was the assistance of an additional botanist. With the latter's assistance, the officials argued, Bergtheil was well placed for the futuristic program of biological selection of indigo plants planned at Sirsiah.

The provincial officials also fought for saving a local scientific institution by defining local needs. Bergtheil was an expert who was in touch with the ground realities of Bihar and was trusted by planters. He was the "man on the spot," a local man, who had made "hundreds, even thousands" of determinations on leaf content and the indigo plant in the local environs. Sirsiah and Bergtheil had proven their relevance to the evolving needs of the planting community. Sirsiah had lately turned out to be a very useful "advisory center" for the planters. The experts at Sirsiah routinely conducted scores of analyses with samples of dye, water, leaf, and so on, that the planters took to them and provided expert advice on cultivation practices. The need for such a center was especially critical when the planters were adopting a new variety of indigo plant – the Java variety. To them it only seemed natural that the infrastructure at Sirsiah should be preserved. They were even willing to turn Sirsiah into a government center with Cyril Bergtheil as in-charge. In consultation with J. Mollison they prepared a blueprint of such a plan and submitted it to the center. This second center, they proposed, would address the agricultural needs of north Bihar specifically while also meeting the needs of the planters.[65] But none of this moved the center, which still demanded that Bergtheil must join his services at Pusa.

Just when it seemed that the Bengal officials were on the verge of losing their argument with central officials, the planters stepped in, adding their voice to the demand to retain Cyril Bergtheil at Sirsiah. After the decision of the central government was received in Bengal, the planters met in an extraordinary session of their association to discuss the implications of the central government's decision. They unanimously resolved that the work at Sirsiah under Bergtheil must continue and requested that the director of agriculture in Bengal intercede on their behalf with the central officials. They also offered to pay a major part of Bergtheil's salary and laboratory expenses if their request was granted. Bergtheil on his part

[65] Letter from F. W. Duke, Revenue Secretary, Government of Bengal, to Revenue and Agriculture Secretary, Government of India, dated June 17, 1908, letter no. 1182T.-R, Government of India, Proceedings of the Department of Revenue and Agriculture for October 1908, India Office Records, GOI, Proc. Rev & Agr, P/7896.

offered to resign his position of imperial bacteriologist and instead work as an indigo specialist at Sirsiah if the planters could assure his employment for a reasonable time. He apparently believed in the indigo work that he was pursuing and was confident that he could deliver results.

The revenue secretary of Bengal again wrote to his superiors at the center highlighting the strong support at the local level for continuing the current work at Sirsiah. He pointed to the exceptional commitment of planters and of Bergtheil himself. On his part, he assured that the Bengal government was willing to commit funds for this work for up to five years given the fact that experiments on selection and breeding currently being planned at Sirsiah could only start to bear result after an extended trial over several years.[66]

Finally the central government relented in deference to the wishes of Bengal officials after assessing that the removal of Bergtheil from Pusa would not disrupt their research at Pusa. They consulted the inspector general, James Mollison on the possible effect of Bergtheil's resignation. As the agricultural lead official at Pusa, Mollison was also asked to state the measures he contemplated taking in order to fill the void created by Bergtheil's resignation. Mollison replied that he did not think that the absence of Bergtheil would pose insurmountable problems. The station already had a mycologist, Dr. Butler. Mollison expressed confidence that the current mycologist would be able to perform the functions expected of the bacteriologist in the short-term as the boundary between the fields of agricultural mycology and agricultural bacteriology was a blurred one.[67] Mollison's accommodative stance and tacit support for Bengal experiments on selection was critical for the continuation of Bergtheil's work at Sirsiah.

The path was thus opened for the inauguration of a new round of indigo experiments at Sirsiah. The central government communicated to

[66] Letter from F. W. Duke, Revenue Secretary, Government of Bengal, to Revenue and Agriculture Secretary, Government of India, dated April 17, 1909, letter no. 61TR, No. 37, Serial No. 7, File No. 18 of 1909, Government of India, Proceedings of the Department of Revenue and Agriculture for June 1909, India Office Records, GOI, Proc. Rev & Agr, P/8174.

[67] Letter from Secretary, Revenue and Agriculture, Government of India, to the Inspector General of Agriculture, dated May 15, 1909, letter no. 543, No. 40, Serial No. 10, Government of India, Proceedings of the Department of Revenue and Agriculture for June 1909, India Office Records, GOI, Proc. Rev & Agr, P/8174; letter from the Inspector General of Agriculture to Secretary, Revenue and Agriculture, Government of India, dated May 18, 1909, letter no. 1939, No. 41, Serial No. 11, Government of India, Proceedings of the Department of Revenue and Agriculture for June 1909, India Office Records, GOI, Proc. Rev & Agr, P/8174.

Bengal that it had no objection to releasing Bergtheil from his contract, and thus he could be employed in Bengal with immediate effect. The Bengal government sanctioned a recurring grant of Rs. 32,500 to the Sirsah Station for the next five years that included a provision to retain Bergtheil and to hire a new botanist. The government also received a commitment from BIPA that they would contribute Rs. 10,000 per annum for the experiments at Sirsiah.[68]

Nature Disrupts the Implementation of Mendelian Selection at Sirsiah

In 1909 the scientists at Sirsiah were hopeful of results from a long-term program of improving the Java variety of indigo through the method of biological selection. The exclusive focus on Java variety reflected the fact that the latter had more or less completely replaced Sumatrana in Bihar. The Dasna farm specializing on Sumatrana indigo was closed down in 1910, making Sirsiah the only dedicated center on indigo in the colony. A new botanist with knowledge of plant breeding, F. R. Parnell, was appointed to Sirsiah in October. The government had given them the charge to develop a long-term research program. In the first report after the resumption of indigo trials at Sirsiah, therefore, Bergtheil announced that he would submit only abbreviated reports for the station's work for the next two or three years. That is because biological selection work was not expected to produce outcomes as rapidly as the erstwhile chemical experiments. But he was also hopeful that at the end of this period of latency the experts at Sirsiah might "have something of a definite practical nature to report."

But the task of biological improvement of indigo in Bihar proved to be even more arduous on account of factors that were both internal and external to the developing science. The scientists were equipped to deal with challenges of a scientific nature even if the path forward was not

---

[68] Telegram from Secretary, Revenue and Agriculture, Government of India, to Revenue Secretary, Government of Bengal, dated May 26, 1909, No. 42, Serial No. 12, Government of India, Proceedings of the Department of Revenue and Agriculture for June 1909, India Office Records, GOI, Proc. Rev & Agr, P/8174; for sanction of Rs. 32,500 to the BIPA, see "Order – by the Government of Bengal, Revenue Department," Agriculture, October 1909, File 2-I/3 1–7 ¾, Board's file 114 of 1909, No. 10, Bihar State Archives, GOB, Rev (Agr.); for reference to BIPA's acceptance regarding payment of 10,000 per annum, see the letter of Director of Agriculture, Bengal, W. R. Gourlay to Revenue Secretary, Bengal, dated March 31, 1909, Agriculture, October 1909, File 2-I/3 1–7 ¾, Board's file, 114 of 1909, No. 5, Bihar State Archives, GOB, Rev (Agr.).

going to be easy. But it was the challenge posed by flood and diseases, the "externalities," that further complicated their task. From the very beginning floods and diseases severely compromised the efforts of Sirsiah experts. In a way, these "external" impediments were a reflection of local climatic conditions and disease environment thwarting the acclimatization of an imported strain. Not being indigenous, Java indigo turned out to be more vulnerable to these cycles of flood and disease when the latter finally caught up with it. As against the short life cycle of Sumatrana, the Java indigo stood in the field for two years and had to withstand the whole cycle of seasonal changes over the entire year as well as the entire range of seasonal pests. Bergtheil was perceptive enough to realize this aspect of Java's vulnerability as soon as a "disease" of as yet an unidentified type made its very first appearance on a small scale in 1907. His report for that year mentioned "some disquietude" on account of this malady and connected the latter to the unique aspects of Java's requirements in a new agroclimatic context. His report for the year said, "By the introduction of Java plant a condition of affairs has been set up under which indigo is standing all the year round, and harbouring places for parasitic insects and fungi are afforded in the winter where formerly they did not exist [in the case of Sumatrana]."[69] Also Java, unlike Sumatrana, which was harvested by June, had to endure the annual flood caused during the ensuing monsoons.

But despite challenges the in-house botanist, Parnell made a number of determinations that gave some direction to the trials and boded well for results in the future from selection and hybridization experiments. He first found out that the indigo plant was naturally cross-fertilized, a fact that explained why Bergtheil had so far failed in generating plants that were morphologically or analytically uniform from one generation to the next. The plant's morphology prevented self-fertilization. Parnell also identified the Indian bee as the key vector that facilitated cross-fertilization among indigo plants. Thus the following work on selection of specific attributes of the indigo plant was performed on plant specimens that were artificially prevented from cross-fertilizing and instead "selfed," or self-fertilized. Parnell achieved this by covering the plants with muslin and by manually placing pollens in contact with the ovary on the same flower. There was a caveat to be borne in mind that repeated "selfing" could result in sterile plants that produced flower but did not lead to the formation of seedpods. In addition, Parnell reported that morphological

[69] Sirsiah Report, 1907–8, p. 26.

characteristics did not have any correlation with color-bearing ability among indigo plants, an attribute that called for additional laboratory analysis for each specimen and prolonged the work.[70]

Parnell's early trials led him to believe that the Java leaf's color content was varietal and an inheritable quality. For instance, in the first year of trials, twenty plants were analyzed and found to contain between .43 and .84 percent color. Seven of them containing the highest percentage of color were picked out and their seeds collected and sown together. While most seedlings were destroyed in a devastating flood at Sirsiah, seven survived, all progenies of the highest-content plant in the original sample. Selfed seed was obtained from each of them and planted. Only one plant in the F1 (or first hybrid) generation gave analyzable results, but it was found to contain .50 percent color, which was higher than the usual. This trial – though very limited in scale and thus of questionable validity – suggested that color content was an inheritable quality possibly amenable to manipulation through controlled reproduction.[71]

The next set of results in 1911, more extensive by comparison, seemed to confirm the belief about the inheritance of color potential. This time a total of five hundred and fifty-five plants with a range of color content and other characteristics were singled out for further observation. The plants suffered from attack by an unknown disease because of which most specimens died prematurely. Even though the size of specimens was now drastically reduced, the Sirsiah experts continued to work with what was left of them. All of the latter were selfed and seeds obtained from them were planted. Two batches from the highest-color-containing plants gave progenies in the F1 generation that consistently showed high color content. All other high-content parents gave progeny averaging more than the average, while the only surviving low-content parent gave an F1 hybrid averaging less than normal. Parnell admitted that the number of specimens was still limited. An even bigger problem was that the "controls" in this round had perished. So the yields of the specimens were compared against a random figure of .40 percent or what was known to be the average yield of plants in Bihar. Yet in Bergtheil's opinion these results had returned "clear indication that indigo content is a hereditary character."[72]

The total devastation by the disease and the growing problem of sterility among "selfed" plants the next year broke the backbone of the selection

---

[70] Sirsiah Report, 1909–10, 1910–11, 1911–12, 1912–13, *passim.*
[71] "Appendix," by F. R. Parnell, Sirsiah Report, 1912–13, pp. 20–2.
[72] "Appendix," by F. R. Parnell, Sirsiah Report, 1912–13, p. 22.

program at Sirsiah. What followed in the year was a near-total failure of all specimens belonging to current and past years. Further experimentation with a new batch of Java plants came to a complete halt as all plants selected for observation perished or did not bear seeds. Parnell's report for the year gloomily summed up the serious consequence of the yet unidentified disease, saying, "In 1912 analyses were made of 572 plants … for further original selections but it was impossible to obtain seed from any selected plants owing to disease." The previous year's specimens fared equally poorly "owing to the failure of a large proportion of the plants to set seed." A mere total of six selfed plants could produce seeds of the F2 hybrid generation.[73] Overall the experts were left with too little to carry forward the selection trials.

Amid these circumstances a decision was made to close down the Sirsiah center in 1913. It could be perhaps said that indigo – "the nature's product," as one of the planters had called it – had foiled planters' efforts to alter it along a predestined direction of higher yield. The attack by the disease at Sirsiah proved to be demoralizing, especially so because the Sirsiah experts failed to understand the cause of the disease and thus had no clue as to how to prevent its recurrence. As Bergtheil reported, it seemed impossible to carry on the program at Sirsiah. One possibility was to relocate the station and its selection program to a district where the plants might be expected to grow healthily. But apparently the disease had broken the will of the Bengal government as the primary patrons of the trials. Apparently, the planters too now wished to wash their hands of the work at Sirsiah. The Sirsiah program did not inspire their confidence any more.

The imperial government offered to absorb indigo experiments within its ongoing colonial research program on Indian crops at the Pusa Institute and the planters' association gladly accepted that offer. The planters' acceptance of the offer brought down the curtains on the decade and a half old program of scientific improvement by the indigo industry. Indigo experiments were henceforth to be conducted as part of the government's broad program for agricultural science in the colony.[74] Indigo science would get a fresh lease of life one last time with the outbreak of the First World War. This last upcoming initiative would be born out of the contingencies of the war.

---

[73] "Appendix," by F. R. Parnell, Sirsiah Report, 1912–13, p. 22.
[74] Sirsiah Report, 1912–13, p. 2.

6

A Lasting Definition of Improvement in the
Era of World War

Keith MacDonald had argued in 1907 that natural indigo would prevail over the competition of synthetic indigo:

Nothing made in a chemical laboratory will even come near what is created in the laboratory of nature. The incessant work going on in the cells of the plant – God's laboratory – can't be equalled by any human agency, which, at best, can only be an approximate imitation, without substance; the ghost and shadow of what has stood the test of ... thousands of years. The dye which coloured the ribbons that bound the mummies deposited in the tombs of Egypt more than 2,000 years ago, that coloured the beautiful carpets of Achilles, 1,100 B.C. and has given luster to the magnificent colours of Oriental textile fabrics for countless generations, is not going to succumb.[1]

As an advocate for the colonial industry Keith MacDonald represented the views of those who believed that natural indigo had a future. He was right in arguing that indigo had a history comparable to human civilization itself in its worldly span. In the modern times, it thrived in the wild in Zululand, was a major crop of choice in the emerging Atlantic complex in the sixteenth and seventeenth centuries, and then became a favored commodity of colonial plantations as they were established in Spanish Central America, French Saint Domingue, Dutch Southeast Asia, and the Indian subcontinent in the eighteenth and nineteenth centuries. As the old order began to change again, the indigo planters based in Bengal led the battle against the synthetic substitute manufactured in German factories. In the first decade of the twentieth century the indigo plantations

---

[1] Keith MacDonald's letter to the editor, *Indian Planters' Gazette*, December 7, 1907, p. 704.

in colonial India remained the last major holdout against the marauding artificial dye, and Keith MacDonald's assertions represented that spirit of defense and resilience. At the time when MacDonald wrote the note, exactly a decade after the launch of synthetic indigo, the natural indigo industry in colonial India was a pale reflection of its previous self. But the planters were continuing the fight. However, it would have been far more difficult for him to stay equally positive about natural indigo's promise at the beginning of the world war. The planters had walked away from sponsoring indigo experiments at designated stations, which were now being conducted at a slow pace at the Pusa Institute. The assertion that natural indigo would not "succumb" would have appeared overly optimistic to anyone. Only the outbreak of the First World War and the needs of dyers in the allied market led to the revival of indigo experiments on an earnest basis.

During these years questions about the improvement of the natural dye and the precise parameters of improvement rose again as efforts were made to revive natural indigo for meeting demands in the West. Those debates were sealed in the context of wartime needs and an aggressive demand by the English dyers to supply natural indigo in a form similar to its synthetic competitor. These contingencies were critical arbiters in the history of agricultural indigo before its demise as no serious, large-scale effort would ever be made to revive natural indigo for industrial use.[2]

Search Continues for the Best Science

The Plant or the Vat?

The momentum of previous years determined the direction of indigo experiments in the colony before the rumblings of the war would begin to galvanize colonial officials and metropolitan consumers to take charge. The carryover experiments from Sirsiah were put under the charge of the Botanical Section at Pusa headed by the imperial economic botanist, Albert Howard. Howard was well qualified to continue those experiments. He had earlier provided key input to Sirsiah scientists as they struggled with the problem of "wilt," the disease that ravaged specimens picked up for selection at Sirsiah, and thus had a level of familiarity with

---

[2] Balfour-Paul has spoken, though, of the recent revival in the use of "natural" products by aficionados and craft practitioners and depicted a romantic hopefulness about their future. See, Jenny Balfour-Paul, *Indigo*, London: British Museum Press, 1998, Chapter 10, "Into the Future," pp. 229–33.

experiments there. The Bihar Planters' Association also gave its full backing to the biological line. The association that had met to discuss the future of indigo experiments averred that "if as a result of Mr. Howard's [selection] work, the present yield from indigo can be improved by (say) 25 per cent, there is a good chance of its being produced at a price which would compete with the 'Badische [indigo].'" The planters also specified that selection among plants should be implemented for higher indican content. In contrast, they categorically stated that any further research into manufacturing experiments would be "practically useless."[3] To the association members, increasing color content in the plant through selection seemed the most appropriate course for reducing its price and surviving the competition of synthetic indigo. Their backing helped solidify the emerging paradigm of selection experiments at Pusa in the short run.

Against the overwhelming support for biological experiments among the planters in the colony, at least one notable effort was made from the metropolis that sought to revive chemical experiments for streamlining the vat processes. This initiative was attributable to the ex-planter Sir John Lewis Hay. Hay marshaled the support of key natural dye experts in England in an attempt to reinsert chemical experiments into the colonial program for indigo improvement. A baronet and an indigo planter, he had retired to England after spending much of his active life in Bihar. His close relatives remained in the indigo business in India. He was recently knighted and as a person of distinction in British society, he used his privileged access to important figures including the secretary of state to push his agenda.

Hay had a unique vision of a holistic relationship existing among the indigo plant, its surroundings, and its internal processes, and of the science underlying indigo culture. In maintaining a view that others found somewhat fanciful, he argued that the indigo plant received energy from sunlight and stored it within for future use. He put special emphasis on sunshine as the source of some form of energy and the role of hairs on indigo leaves in capturing that energy. As a planter in India he had diligently recorded the changes in color that took place in the indigo liquor as

---

[3] Note by the agricultural adviser to the government of India on the discussions at a meeting arranged with the representatives of the Bihar Planters' Association in connection with indigo, enclosed in letter from F. Noyce, Under-Secretary, Government of India, to Secretary, Revenue and Statistics, India Office, dated September 4, 1913, letter no. 1104, Serial no. 4, No. 56, File No. 142 of 1913, Government of India, Proceedings of the Department of Revenue and Agriculture for September 1913, India Office Records, GOI, Proc. Rev & Agr, P/9215.

it passed through the various stages of manufacturing. On their basis he prepared "color charts" and hypothesized that these color changes represented chemical changes. He analyzed color changes in the liquor to infer wasteful dissipation of plants' amassed energy resulting in loss of color that the plant could potentially yield. He thus wished for manufacturing processes to be tuned to optimize the process of dissolution of the plant's amassed energy.

Utilizing the good offices of the secretary of state at India House, John Hay first sent his chart to J. Mackenna, the new agricultural adviser to the government of India, and to the general secretary of the Bihar Indigo Planters' Association in India, arguing that it was possible to improve manufacturing yield vastly. The communication only elicited a placid response from Mackenna in India initially. The BPA secretary also disagreed with the claims made in Hay's chart, being of the view to the contrary that only 12–15 percent of the dye was being wasted during the manufacturing cycle. Such a loss was generally "unavoidable." The secretary also pointed out that there were ways to plug the residual wastage, but that it was "not worth the cost."[4]

But John Hay found some common grounds with the metropolitan chemists only in the conclusion that substantial loss of color was taking place during manufacturing. He reached out to the natural dye expert at Leeds A. G. Perkin in particular. Some of Hay's hypothesis about color changes ran against the inference of Perkin's formulaic analytical chemistry. Thus, for instance, in communications with Hay Perkin expressed complete inability "to account scientifically or to advance a theory" for Hay's observations on color changes or to attest to the presence of some of the compounds Hay believed existed in the liquor. Perkin also expressed disagreement with Hay's hypothesis in giving a place to "sunshine" in the decomposition of indoxyl in the vat. But both agreed that maximum loss of color was the outcome of less than optimal conversion of indoxyl. Hay asked for Perkin's assistance in analyzing the waste from indigo manufacturing in order to trace products of such wasteful conversion. Hay also persistently wrote to the colonial government, claiming that Perkin supported the theory of loss of color during manufacturing. He also argued that Perkin's estimate of loss was lower because the chemist's

---

[4] Letter from J. MacKenna, Officiating Agricultural Adviser to the Government of India to Secretary, Revenue and Agriculture, dated May 19, 1913, letter no. C-169, Serial no. 2, No. 56, File No. 142 of 1913, Government of India, Proceedings of the Department of Revenue and Agriculture for September 1913, India Office Records, GOI, Proc. Rev & Agr, P/9215.

laboratory tests failed to detect many additional, secondary processes that took place during actual manufacturing.

In the end, John Hay received a positive response from central government officials in India. Hay was, after all, asking for a more thorough consideration of all possibilities in determining the course of improvement. He made a case that all "alternative methods" including the way suggested by his own intuitive "science" should be at least given consideration by the experts in India. He tactically used Perkin's authority to evoke interest in manufacturing experiments and attacked the colonial officials for limiting themselves to the advice of "2 chemists and 1 bacteriologist." The government should form a new committee of advisers comprising "real experts," he urged, so that the finest dye in the world did not have to surrender to the synthetic substitute. If the Germans owned the natural indigo industry, they would have worked on every path possible, he argued rhetorically. Finally the officials responded by expressing interest, not in Hay's color charts, but in Perkin's experiments on the nature of substances in *seet* and requested that they be acquainted with the outcome of the researcher's investigations in England.[5] It is important to point out that in 1913 Lewis Hay, backed by supporting assertions from Perkin, was demanding something similar to what Bloxam had demanded in 1908, that is, revival of manufacturing experiments. But, five years later, in the changed circumstances when war clouds were hovering around, there seemed more interest among state officials in investigating all and any possible path that might have a chance of success.

Albert Howard's Experiments
Albert Howard's measures succeeded in solving the immediate problem of paucity of seeds. Java indigo crops across Bihar were increasingly succumbing before the onset of the seeding season. The problem became so widespread that there was a scarcity of seeds to sow in the next season in

---

[5] Letter from A. G. Perkin to Lewis J. E. Hay, dated May 14, 1913, enclosed in letter from Francis C. Drake, Secretary, Revenue and Statistics, India Office, to Secretary, Revenue and Agriculture, Government of India, dated June 6, 1913, letter no. R&S 1891, Serial No. 3, No. 57; letter from F. Noyce, Under-Secretary, Revenue and Agriculture, Government of India, to Secretary, Revenue and Statistics, India House, dated September 4, 1913, letter no. 1104, Serial No. 4, No. 58; letter from Under Secretary, Revenue and Agriculture, Government of India, to Secretary, Revenue and Statistics, India House, dated December 31, 1913, letter no. 1647, Serial No. 6, No. 44, Government of India, dated October 30, 1913, letter no. R&S 3943, Serial No. 5, No. 43, Government of India, Proceedings of the Department of Revenue and Agriculture for September 1913, India Office Records, GOI, Proc. Rev & Agr, P/9215.

Bihar. Records from one of the largest plantations in Belsund confirmed the menacing scale of wilt in Bihar. On the important plantation belonging to D. J. Reid, admittedly, "the season of 1911–12 was the worst season for wilt yet recorded."[6] Howard tried a rather simple solution to the problem of raising seeds by sowing the Java plant in August, two months ahead of the regular sowing month of October, and successfully obtained seeds. This approach was based on the premise that the Java plant could not tolerate rainy conditions for more than two months. The planting of seedling in August at Pusa, Howard assumed, allowed the young plant to avoid major rainfall during the period of its most vigorous growth. The plants were also positioned on highlands to ensure that adjoining subterranean water did not accumulate underneath the crop. Howard's operation at two government farms at Pusa and Dholi resulted in bountiful production of seeds.

But Howard failed to pinpoint with accuracy the root cause of wilt. In two reports that reflected his experience in successfully raising Java seeds he counselled planters to adopt better cultivation practices and hypothesized that waterlogging might be the cause of the disease. A well groomed plant grown under ideal conditions was the key to its ability to bear seeds. Maintaining that "the [current] practice of growing seed from an old worn out plant cannot be too strongly condemned,"[7] he recommended changes to the current cultivating practices. He asked for planting seedlings two feet apart to allow for enough space between branches for insect visitation and pollination and "pruning" of leaves instead of heavy cutting at the first harvest so that transpiration currents from the root and transportation of nutrients for further growth were not disrupted. He also suggested maintaining proper drainage of the soil. In his second report on indigo at Pusa Howard elaborated that the indigo root nodules containing the rhizobium bacteria needed an appropriate supply of food, water, and air in the soil for proper functioning. The root nodules were primarily responsible for fixing atmospheric nitrogen that was used to produce indican by the plant. In his view a persistently waterlogged condition of the soil caused the onset of wilt as it cut off air and thus inhibited the functioning of nodules. When the air spaces of soil were filled with water for a prolonged time the

---

[6] D. J. Reid, "Ten Years' Practical Experience of Java Indigo in Bihar," *Agricultural Journal of India* xii No. 1 (1917): 15.

[7] Albert Howard and Gabrielle L. C. Howard, *First Report on the Improvement of Indigo in Bihar*, Bulletin No. 51, Calcutta: Agricultural Research Institute, Pusa, 1915, p. 5, India Office Records, V/25/500/121.

nodules stopped functioning and the formation of indican too stopped. The plant then fell back on its reserve of indican, thus exhausting the amount of color in the leaves. Prolonged waterlogging ultimately caused the death of the plant. Howard devised the "Pusa method of drainage" to overcome the problem of waterlogging and wilt in the indigo tracts.[8]

The problem of wilt in Java indigo continued to defy solution. It did not seem that waterlogging was the primary cause of wilt even though everyone was grateful that Howard had solved the immediate problem of seeds.[9] Wilt made planters despondent by reducing the output of indigo leaf. Howard had fundamentally succeeded in raising seed by using early maturing varieties of the indigo plant. The onset of wilt was known to occur in plants after the first or the second cutting. As wilt viciously continued to attack Bihar indigo in the later stages, planters' total output across Bihar fell in the absence of the possibility of second and third cuttings. Thus Java indigo forfeited its primary advantage of extra output of leaves through additional cuttings. Some planters resumed the cultivation of the native Sumatrana as they were despairing over the effort to contain wilt in Java indigo. The plantation of D. J. Reid also saw a resumption of Sumatrana cultivation after the particularly vicious attack of wilt on Java in 1912. Reid was one of the planters who had most unreservedly taken to cultivation of Java since 1904–5.

The promise of concrete applicable results from biological selection also remained unfulfilled in the immediate term. The efforts of experts since 1913 were mostly directed toward the biological selection of early maturing varieties of both Java and Sumatrana plants. The longer the plant stayed in the field the greater were the chances that floods, pests, or wilt could harm it. The considerations in favor of having an early harvest seemed to outweigh all others. Howard's trials confirmed what F. R. Parnell had maintained earlier, that indigo was a "natural" cross-pollinator. Howard also found that Java plants raised from self-fertilized seeds showed a marked falling off both in the size of the plant and in general growth even in a single generation. If individual plants with high

---

[8] Albert Howard and Gabrielle L. C. Howard, *Second Report on the Improvement of Indigo in Bihar*, Bulletin No. 54, Calcutta: Agricultural Research Institute, Pusa, 1915, pp. 2–5, India Office Records, V/25/500/121.

[9] Writing much later, D. J. Reid maintained, "My experience is that indigo is not killed outright by standing water, and that wilt is not affected to any great extent by climatic influences." *Cf.* D. J. Reid, "Ten Years' Practical Experience of Java Indigo in Bihar," *Agricultural Journal of India* (1917): xii 1, p. 26.

indican content were artificially self-fertilized, they were likely to turn sterile. Thus there was no point in avoiding cross-fertilization. Under the circumstances it was going to be preferable "to control crossing rather than to attempt to prevent it." But this was admittedly the "least satisfactory" of all methods of selection and would require "continuous" effort over time and over several cropping seasons to obtain positive results. Howard isolated the stock of successful plants and made them grow together. After the next crop matured the odd ones were weeded out and another round of healthier varieties was planted. Two specific methods were tried out – the "selection of mixed early types" and the "selection of single early type." In the first method seeds of good plants from a season's crop were selected for propagation. These breeds were then grown together in a separate location. In the second method all seeds of a single "successful" breed were collected and planted together. The work on the selection of the Sumatrana variety proved to be even more tortuous as characteristically the Sumatrana variety bred fewer seeds than the Java variety. The methods of selection were virtually the same as those in the case of Java indigo. There were three regional varieties of Sumatrana – the Madras type, the Northwest type, and the Cawnpore type – out of which the experts focused on the Cawnpore type. But in the end the selection process in the colony was a work in progress, offering no immediate results that the planters could apply. It continued at Pusa as part of the broader colonial agricultural research program to use selection for the improvement of Indian crops.[10]

Wartime Advocacy for Expanding Indigo Science

The buildup leading to the First World War and eventual outbreak completely altered the context of support for the fledgling natural indigo industry in the colony. In an indication of the new importance of Indian indigo in the period, key trade officials were working urgently to secure the existing indigo in the market for Britain and other Allied textile manufacturing nations. The British Board of Trade in London secretly engaged the brokers Lewis and Peat to round up all available Bengal indigo and any residual amount of Dutch indigo and prepare a list of all dyers and printers to whom the consignment could be supplied in a regimented

[10] Albert Howard and Gabrielle L. C. Howard, *Third Report on the Improvement of Indigo in Bihar*, Bulletin No. 67, Calcutta: Agricultural Research Institute, Pusa, 1916, pp. 1–32, India Office Records, V/25/500/122.

manner. The latter were also allowed to sell parts of the indigo to allies, France and Russia, after meeting the demand from national consumers. Equally urgently, the Board of Trade and the Treasury collaborated to make available all raw materials in England to the Swiss dye companies for the purpose of speeding up the production of anilines, even categorizing railway shipments of such raw materials as "No. 1 Label" for priority transit. An official consignor was appointed to manage the traffic of ships carrying dye intermediates to the Continent in the international waters effectively.[11] A directive was also sent through India House to the colony to prevent export of Indian indigo to all other countries. All these measures were indicative of the heightened importance of indigo dye in the context of the war.

The war generated a fresh wave of metropolitan interest in reviving the colonial natural indigo industry and indigo science. Disruption in the supply of synthetic indigo from Germany created a new demand in the markets in Britain and elsewhere among Allied nations. Wartime sentiment in England also generated political support for a "product of the empire." In these conditions of a changed political economy the supporters of natural indigo sensed an opening to push for renewed measures. In a letter to his brother-in-law, also a planter in India, John Lewis Hay said, "Our 'Hour' has arrived.... *Now,* I think is the appointed hour and we should strike quickly and *hard.*" The advocates for indigo science were also emboldened. The scientist A. G. Perkin was sanguine about the opportunities for indigo in the new context, in which his proposal for reviving manufacturing experiments on indigo might be put into effect. Referring to the war he said, "It is an ill wind which blows nobody any good, ... [but] in this case the war should be of great benefit to the indigo planter, and give him opportunity and breathing space to set his house in order." Perkin was of the opinion that it would be difficult to produce synthetic indigo at home immediately. Even if efforts were initiated it would be another two to three years before such attempts would succeed. After the war the Germans would also require time before they would

---

[11] Letter from Lewis and Peat to Walter Runciman, President, Board of Trade, dated November 18, 1914; letter from Board of Trade to Lewis and Peat, "Secret," dated November 20, 1914; letter from Lewis and Peat to Permanent Secretary, Board of Trade, dated January 11, 1915; letter from Board of Trade to the Railway Executive Committee, dated February 26, 1915; reply of Gilbert Szulmer, Secretary, Railway Executive Committee, to S. Whetmore, Commercial Department, Board of Trade, dated March 8, 1915, including "Note on the arrangements relating to the Swiss dye traffic." Public Records Office, Board of Trade papers, BT13/60, 100168.

be able to supply synthetic indigo again at prewar prices. Therefore now was the time for the planters to set their house in order and launch efforts to bolster natural indigo so that it might be able to compete with synthetic at par at the end of the war. He wrote to John Hay in support of manufacturing experiments, "Apart from the agricultural side, I am certain that if your industry is to be saved it must be by purely chemical means."[12] John Hay continued to push for chemical experiments while also suggesting procurement of the absolutely original strain of Java indigo, not from Natal but from Zululand, where it was known to grow wild, as a possible way out of the problem of disease among the strain currently grown in India. India House consulted the director of Kew Garden, David Prain; secured his approval for the Hay proposal; and prompted steps to import Zululand seeds in an attempt to overcome the devastation of wilt.[13]

All these efforts materialized in the context of the trade reality of dwindling supplies of natural indigo in England. As another backer of indigo, William B. Bridgett, pointed out to the Indian viceroy in the colony: "There is practically no stock [of natural indigo] available." Bridgett was an important market player, a major importer and distributor of natural indigo in Britain. He owned the East India Indigo Company of London, which had been in the business of importing indigo for more than thirty years. He sent to the viceroy a copy of the monthly statement of his firm reflecting how the demand for natural indigo had soared. His company had sold every bit of stock of indigo within weeks of the declaration of the war. Whatever remained with him was already committed to the buyers while cables were being received in London from overseas locations requesting supplies that could not be met. He added that the stocks of natural indigo in all London warehouses had similarly dried

---

12 Letter from John Lewis Hay to planter L. W. MacDonald at Hathwa, Saran, in Bihar, dated August 19, 1914; letter from A. G. Perkin to John Lewis Hay, dated September 20, 1914, enclosed in letter from E. J. Turner, Assistant Secretary, Revenue and Statistics, India House, to Secretary, Revenue and Agriculture, Government of India, dated September 4, 1914, letter no. R&S 3465, Serial No. 6, No. 33, Government of India, Proceedings of the Department of Revenue and Agriculture for March 1915, India Office Records, GOI, Proc. Rev & Agr, P/9726.

13 Letter from E. J. Turner at India House to David Prain, Director, Botanical Garden, Kew, dated September 1, 1914; reply of David Prain to E. J. Turner, dated September 2, 1914, enclosed in letter from E. J. Turner, Assistant Secretary, Revenue and Statistics, India House, to Secretary, Revenue and Agriculture, Government of India, dated September 4, 1914, letter no. R&S 3465, Serial No. 6, No. 33, Government of India, Proceedings of the Department of Revenue and Agriculture for March 1915, India Office Records, GOI, Proc. Rev & Agr, P/9726.

up. Bridgett wanted to apprise the viceroy of the new trade opportunity presented to natural indigo in the markets of the West. He clearly had a vested interest in promoting natural indigo. He was not pleased that the industry had fallen onto bad times. With the outbreak of the war and the disruption in the supply of synthetic indigo he sensed his chance and tried to impress upon the Indian viceroy the need to revive the colonial indigo industry.

Bridgett received a favorable response to his endeavors to promote the interests of the natural indigo industry from most important officials in the colony as well as the metropolis. The viceroy's office in India promptly acknowledged Bridgett's letter and forwarded it to the secretary of the Commerce and Industry Department for further action.[14] In England Bridgett campaigned in favor of the need for providing patriotic support for the natural indigo industry of India. The English consumers had earlier made a mistake by preferring cheaper synthetic indigo and thus promoting a German product at the cost of natural indigo from within the empire. Now the same Germans were at war with their country. The Indians continued to be loyal, having sent their troops to fight side by side with the English soldiers against the Germans. The nation should act together in standing by its imperial interests, and the government should do everything possible to resuscitate an imperial product. Such demands for support to an "imperial" industry were not new. But in the context of the war such rhetoric assumed added meaning and appealed to popular sensibilities and the government alike. Bridgett wrote to the secretary of state that he was about to demonstrate through public trial what he believed – that the natural dye had stronger dyeing power than the synthetic. A personal note from Secretary of State Sir Thomas W. Holderness was communicated to him suggesting that he should try to arrange for as much publicity as possible for his upcoming public trials. The secretary of state also encouraged Bridgett to submit a memorial to different government departments in England seeking their commitment to promote the use of natural indigo in Britain. Bridgett accordingly took the lead in organizing a joint appeal on behalf of the supporters of natural indigo. It asked the British consumers and government offices including the War Office, Board of Trade, Admiralty, and Post Office to support the indigo

[14] Letter from W. B. Bridgett to the Viceroy, dated September 11, 1914, letter no. 2968, Serial No. 12, No. 39; acknowledgement of the letter by the Office of the Viceroy, dated October 12, 1914, letter no. 11222–32, Government of India, Proceedings of the Department of Revenue and Agriculture for March 1915, India Office Records, GOI, Proc. Rev & Agr, P/9726.

industry in India. The India House also warmly received a memorial from Bridgett on the ways to help the natural dye industry, which was duly forwarded to officials in India.[15]

Final Word on the Parameters of Indigo Improvement

Writing at the cusp of the world war in September 1914, William B. Bridgett recapped that unfair terms of trade had led to natural indigo's current ruinous state.[16] The blame lay with the "percentage theory" and its backers, the entire trading system and those scientists who supported it, including planters' own scientists, because it assessed natural dye on the basis of percentage of "indigo" in it. Bridgett at this time was reviving the old debate about the authenticity of natural indigo, a product "created in the laboratory of nature," as Keith MacDonald had characterized it, which, Bridgette asserted should not be assessed on the basis of inappropriate scientific assays. The dominant principles of the market did not assign appropriate value to the natural dye's intrinsic and unique qualities. The natural dye had indigo and many other constituents that gave it a character as a unique dye and made it preeminent among all available blue colors. Others had argued in support of natural indigo that the real economic "value" of natural indigo could best be determined by color tests, which alone could gauge its coloring power. Any seasoned dyer could attest to the higher dyeing potential of natural dye as compared to synthetic dye. Almost a decade earlier, one of them had rhetorically compared the planters' situation to "makers of, say, Highland malt whisky [who might be called] to sell their brands by testing them for alcohol, and thereby bringing them on a level with German potato spirit, which undoubtedly contains more alcohol, and can be made at a penny a

---

[15] "Confidential," letter from Bridgett to the Viceroy and the Secretary of State, dated September 9, 1914, letter no. 3028, Serial No. 13, No. 40; letter from Francis C. Drake, Secretary, Revenue and Statistics Department, India House, to Bridgett, dated September 24, 1914, letter no. R&S 3735; letter from Bridgett to the Secretary of State, dated October 27, 1914; letter from E. J. Turner, Secretary, Revenue and Statistics, India Office, to Secretary, Revenue and Agriculture, Government of India, dated November 13, 1914, letter no. 4471, Serial No. 16, No. 43, File No. 67 of 1914, Government of India, Proceedings of the Department of Revenue and Agriculture for March 1915, India Office Records, GOI, Proc. Rev & Agr, P/9726.

[16] "Confidential," letter from Bridgett to the Viceroy and the Secretary of State, dated September 9, 1914, letter no. 3028, Serial No. 13, No. 40, Government of India, Proceedings of the Department of Revenue and Agriculture for March 1915, India Office Records, GOI, Proc. Rev & Agr, P/9726.

gallon."[17] Such voices had been sidelined in the past. The backers of synthetic were just too content simply winning the market inch by inch to bother about arguing the logic of their product's winning qualities.

A very important moment in the process of defining improvement for natural indigo occurred at the Delhi Conference in 1915, which was called to discuss the future direction of indigo experiments. This meeting was attended by planters, government officials, and trade representatives. The deliberations between these different stakeholders would set the tone for the future efforts on indigo. Views like Bridgett's attacking the fundamentals of the system of international trade, pricing, and competition were sidelined at the meeting. Such voices were categorized as nothing more than partisan and in any case out of line with the view of users. On the contrary, the perspective that emerged victorious seemed to echo the diktats of dominant forces in the metropolis, particularly metropolitan chemists and dyers who wanted natural indigo to be produced as an exact replica of the synthetic product. This perspective definitively established a certain path forward as the only possible path for improvement and laid out its parameters definitively.

The two significant interlocutors for the authoritative viewpoint were Bernard Coventry, the director of the Imperial Agricultural Institute, and Henry E. Armstrong, Professor of chemistry at Central Technical College in London. These two had emerged as key persons in the new dispensation in the colony and the metropolis respectively. On matters related to indigo science the two were particularly influential. Bernard Coventry was the head of the institution where experiments on indigo were being conducted. Armstrong was influential in the policy-making circles within the national government. He had recently visited the plantations in India at the behest of India House to advise planters on the ways and means to resurrect the natural indigo industry. They made a pitch for chemistry-based experiments for fine-tuning natural indigo into a form similar to that in which synthetic indigo was sold. While chemistry had been used by the planters earlier to make vat processes more efficient, this time around it was to be used for the entirely different purpose of turning natural indigo into a form desired by the commercial users. Under the new plan scientists would run experiments and trials to turn indigo cake into paste, the same form in which synthetic was sold to consumers. The new plan also aimed to produce natural indigo in a standardized strength

---

[17] Letter of a planter, Y. Z., dated January 9, 1906, *Indian Planters' Gazette*, January 13, 1906, p. 46.

containing a fixed percentage of indigo. It was believed that a predictable strength of the dye would improve its usability.

At the Delhi Conference, Bernard Coventry's perspective carried the day and established the lasting paradigm for indigo improvement in the last round. He challenged the dissenters, stating: "Now, it is maintained by some of our friends that natural indigo is superior to the synthetic; that the indigotin of the synthetic is not true indigotin; and that the artificial product has been taken up by the dyeing trade not on account of its superior intrinsic merits but to the bribery and intrigue of the Germans. I hope to show this to be an utter delusion."[18] The gist of his argument was that the market had spoken. The partisans on the side of natural indigo were in a denial, "blind" to the fact that natural indigo was now being used by only 10 percent of the consumers worldwide while 90 percent of dyers and printers had switched to the synthetic. Implying that the supporters of natural indigo's intrinsic qualities were on the wrong side of the forward march of progress, he characterized the current users of natural indigo as backward. The use of natural indigo had survived in "semi-civilized" countries like Russia and Persia, among the carpet makers of Smyrna, who practically knew only of natural indigo and no other dye, and among those in whose case the momentum in the use of traditional wooden fermentation vats had ensured a continuing demand. But on the basis of the claims of the latter it was incorrect to claim that natural was better. On the contrary, most dyers and printers considered natural indigo to be "the inferior dye" as compared to the synthetic. The latter's strength lay in its cheapness and usability. Thus Coventry not only contradicted the dissenters' claims and grounds for optimism, but also asserted that wartime effort would deploy science to improve indigo on existing dominant terms of the market. This dominant view lastingly settled the debate over what improvement for indigo meant.

The Delhi conference also discussed purely commercial factors and market-oriented steps that planters must undertake. Producing indigo in a paste form for the consumers was not the end of the matter. Attendees admitted that the indigo planters did not have any trade organization to promote their product with the consumers. It was important "to get in touch with different markets." The members discussed the possibilities

---

[18] Appendix A, "Note on Indigo Research in India," by Bernard Coventry: 9–10, "Memorandum of Proceedings of the Indigo Conference held at Delhi on 22nd February 1915," No. 44, Serial No. 17, Government of India, Proceedings of the Department of Revenue and Agriculture for April 1915, India Office Records, GOI, Proc. Rev & Agr, P/9726

of establishing such organizations to push natural indigo in the Western markets. Such a step to facilitate "distribution" as against "production" was again a new aspect, which indigo planters in India were addressing frontally now, at the instigation and even persuasion of elements in the metropolis.

Metropolitan Power and Rationalization

At no point in the history of indigo experiments in Bengal had the metropolitan forces had as much power as they had now. These forces stood solidly behind a plan for consumer-friendly science and a "rational" production system for the market. These initiative were rooted in wartime contingencies. But the goal of these efforts was to bolster the prospect of natural indigo and make it stand on its own legs in the hope that plant indigo would compete against synthetic indigo on its own merits during and after the war. These forces were the main determinant of a refurbished indigo science that emerged during the era of the world war.

The process to initiate a new form of rational, reductive intervention for the indigo commodity had started to unfold with the determination by Coventry, chemists, and officials at the Delhi Conference that the aid of chemistry was needed. It was decided that a new indigo research chemist should be put on the rolls of the Pusa Institute. The central government extended its wholehearted support to the proposal to hire a chemist for improving usability of agricultural indigo among consumers. A letter from the Viceroy's Council was sent to England requesting the appointment of a chemist. It also asked that the views of prominent chemists like Henry Armstrong and the natural dye expert, A. G. Perkin in the metropolis be elicited on the matter of selecting the appropriate chemist. The experts and colonial policy makers were evidently gravitating toward the metropolitan core in many ways in determining the future direction of indigo experiments.[19]

The demand for a market-friendly science and rationalization was further reinforced at a follow-up conference in London organized by the India House to which planters, traders, and chemists in the metropolis

[19] Appendix A, "Note on Indigo Research in India," by Bernard Coventry: 10, "Memorandum of Proceedings of the Indigo Conference held at Delhi on 22nd February 1915," No. 44, Serial No. 17; letter from the Indian Viceroy to the Secretary of State, dated June 17, 1915, No. 204 of 1915, Government of India, Finance Department, No. 22, Serial No. 1, Government of India, Proceedings of the Department of Revenue and Agriculture for June 1915, India Office Records, GOI, Proc. Rev & Agr, P/9726.

were invited. Most prominently at this conference Henry Armstrong, along with A. G. Perkin, who had impeccable credentials to speak authoritatively on indigo or on natural dyes generally for that matter, supported the demand for measures to make natural indigo paste in a commercially viable manner and for a reorientation of the organizational basis of its production and distribution. Armstrong argued that the production of a standardized paste required that the planters set up a centralized factory or a few factories where the indigo planters should forward their indigo in a semimanufactured state. Such a consignment would possibly be in a liquid or semiliquid state and would have to be sent in tanks. At the factories the raw product would be assessed for indigotin, bulked, standardized, and made into a paste ready for export. Armstrong thus proposed collection of indigo of different strengths at central locations. The actual number of such factories would depend on a consideration of the costs of carriage, and so on. Thereafter the indigo would be mixed and turned into standardized paste of 20 percent strength. The individual planters would then be compensated according to the quality of their consignment.

The conference participants emphasized the need for establishing a dedicated agency for improving the performance of natural indigo in the market. The solution to the problem required a tactical tapping into science, on the one hand, and reduction of transaction costs and dealing "in the most direct manner with the consumer" by the natural's producers, on the other hand. By optimizing distribution the planters could take direct control of their commodity and ensure both "economy" and "continuity of supply." The conference made concrete suggestions in terms of establishment of two marketing boards, one at Muzaffarpur (Bihar) and another in London. The primary responsibility of the boards would be "to supervise within their respective spheres all [activities] pertaining to marketing." The role envisaged for the board was to make it the overall hub of industrywide reforms in trade and marketing. But the agenda of standardization, an important aspect of marketing of the product, involved the spheres of both science and the market. Scientists were to work out the methods for turning out standardized paste to be sold to all consumers. Norms regarding color percentage, test assays, and units of measurement had to be established and popularized among the producers and the consumers. With that end in mind the members of the conference planned to staff this board with "eminent men of both business and science."[20]

[20] "Note of Proceedings of Conference on the Subject of Natural Indigo Held at the India Office on the 20th September, 1915," p. 4, enclosure, letter from India House to the

The proceedings of the conference were a curtain raiser to a new metropolis-centered drive to revamp indigo science and production in the colony along "modern lines." The planter representatives present at the meeting expressed reservations over the possibility of rationalizing an agricultural operation on the lines that mimicked the production and distribution of an industrial product. Most had qualms about how plantation manufacturing could be centralized. The planter Martin T. MacDonald was unsure how successful the efforts to induce cooperation among the planters for production and especially distribution/selling would be. John Hay was also doubtful that the planters would be able to cooperate to have their dye centrally processed. C. B. Gregson was relatively more optimistic on the issue of organization provided that the financial returns be assured in some way. He thought of government guarantees as one way to give those assurances. But any government assurance on guaranteed returns was not forthcoming. Gregson also pointed to other problems at ground level in the current times. Indigo acreage had shrunk and many factories were deserted in Bihar, and a revival of those factories would require major financial investment. Armstrong curtly responded to the planters' misgivings by saying that the latter did not have any option but to organize. He also clarified that unless the planters committed to cooperate in the production of standardized paste, there was no point in the government's sanctioning the appointment of a chemist. As a spokesman for the government, he was consciously addressing the entire planting community and not just a handful of planters in attendance there, asking them to act if they wished for the government's support.

Having inserted themselves into the decision-making process for the reorganization of science and trade, the metropolitan chemists were also inclined to exert control over content of the revamped scientific program on indigo A. G. Perkin was in favor of the formation of a committee of experts in England to supervise and direct the work of the chemist if he were appointed to work on indigo in India. "In the past young men sent out had been without expert advice or criticism," Perkin noted. Such experts had more or less worked in isolation and had made errors in judgment. Henry Armstrong agreed with the proposal to appoint a committee. In fact, the two scientists unanimously advised

Viceroy, dated November 19, 1915, Revenue, No. 123, Government of India, Proceedings of the Department of Revenue and Agriculture for May 1916, India Office Records, GOI, Proc. Rev & Agr, Z/P/1980.

"management" of research in colonial India by a team of experts in England.[21]

The views favoring production of agricultural indigo in a commoditized form received further backing as a major body representing English consumers rose to demand that planters supply indigo to the English markets in a standardized paste form. In a significant public meeting held on November 20, 1915, members of the Foremen Dyers' Guild, the largest organization of dyers in England, gathered to discuss the shortage of the blue dye that they were facing. On the occasion Rowland E. Oldroyd, the president of the guild, scathingly attacked the planters for their lack of effort in meeting the demands of users, saying, "The Indigo Planter ... has never been in touch, or taken any interest in the actual user of his product." The dyers were ready to approach the planters to persuade them, bypassing the mediation of market channels. Emphasizing that resolve, Oldroyd argued, "If these people [meaning the planters] will not come and see what the users require, the users as represented by the "Guild" must ... go to them." A resolution was passed at the end of the meeting imploring the planters in India to start making indigo paste. The natural indigo market comprised the indigo planters/producers in India, the vast number of intermediaries including the shippers and brokers, and the consumers in England. At a time like this when the war had further disrupted the operations of the market, the consumers perceived the need to bypass the normal demand and supply forces of the market and approached the secretary of state directly with their demand. In a number of communications the dyers, just like metropolitan chemists, asked the India House to persuade indigo planters in the colony to start making standardized paste of indigo and supply it to the English markets. They too called for a reorganization of their production and distribution system. They also recommended that the India House officials keep a close watch on the initiatives taken by the planters in this regard so that "the planters will not be allowed to whittle away the new opportunity." Favoring a new level of control by metropolitan officials the dyers suggested constant watchfulness over the planters to ensure that "constant pressure is brought to bear" on them.[22]

[21] "Note of Proceedings of Conference on the Subject of Natural Indigo Held at the India Office on the 20th September, 1915," p. 4, enclosure, letter from India House to the Viceroy, dated November 19, 1915, Revenue, No. 123, Government of India, Proceedings of the Department of Revenue and Agriculture for May 1916, India Office Records, GOI, Proc. Rev & Agr, Z/P/1980.

[22] Letter from Secretary of Foremen Dyers' Guild to Sir William P. Byles, dated November 25, 1915; "Report, Natural Indigo. A chat to the Dyers' Guild by the President"; letter from

In complete agreement with the new perspectives the secretary of state, the highest colonial official in London, communicated to the planters his desire to support indigo experiments along the lines suggested by key scientists and consumers in the metropolis. The support was also made conditional on the planters' reorganizing their entire industry within a year. The secretary of state expressed his dismay to the planters in noting that they were conducting their business "on old lines." If it was not evident that the planters had changed their organization, production, and distribution within one year from the start of experiments to make indigo paste, the government would withdraw support to indigo science, the secretary of state warned. The planters fell in line, accepting the offer of scientific support for making standardized paste while committing to industrywide reorganization.[23]

A close monitoring of the experiments and trials in the colony by experts and officials in the metropolis was a hallmark of indigo science in the World War era. In new institutional developments a new indigo research chemist, W. A. Davis, was appointed to be stationed at the Imperial Institute at Pusa. The planters formed an Indigo and Paste Committee within the Bihar Planters' Association to provide focus to the planned work on indigo paste by Davis. The officials in England oversaw the formation of an Indigo Paste Committee to liaison with and to monitor the work of the indigo research chemist. The secretary of state announced the formation of this committee, headed by the revenue secretary at India House, L. J. Kershaw. Two other members were the chemist Henry Armstrong and a dyer. The planters nominated their representative to this committee. The secretary of state also mandated that the committee in England correspond with W. A. Davis about his work on indigo through the agricultural adviser to the government of India, J. MacKenna, bypassing the provincial administrators.[24]

R. E. Oldroyd to the Secretary of State, dated December 17, 1915, No. 37, Serial No. 1, Government of India, Proceedings of the Department of Revenue and Agriculture for May 1916, India Office Records, GOI, Proc. Rev & Agr, Z/P/1980.

[23] Letter from the Bihar Planters' Association to Secretary, Revenue and Agriculture, Government of India, dated January 7, 1916, No. 16, Serial No. 10, Government of India, Proceedings of the Department of Revenue and Agriculture for May 1916, India Office Records, GOI, Proc. Rev & Agr, Z/P/1980; letter from J. M. Wilson, General Secretary, Bihar Planters' Association, to Secretary, Revenue and Agriculture, Government of India, dated March 13, 1916, No. 22, Serial No. 16, Proceedings of the Department of Revenue and Agriculture for May 1916, India Office Records, GOI, Proc. Rev & Agr, Z/P/1980.

[24] Two telegrams from the secretary of state in England to the viceroy in India, dated April 14, 1916, and August 6, 1916, Government of India, Proceedings of the Department of

Final Shutdown: Had Improvement Reached a Natural Limit?

Indigo had by now undergone several rounds of improvement. From diasporic efforts in the first stage in the Western Hemisphere to colonial improvement in Bengal involving agricultural institutions including refinement under the duress of competitive threat, indigo was now thrust into a third round of improvement under the conditions of war contingencies. This round was going to be the boldest of all efforts in terms of the intent to transform the shape of the commodity completely. The scientists were planning to turn the agricultural product into a replica of the synthetic substitute. That this was the right thing to do had already been settled by the dominant opinion of commercial consumers in the market and metropolitan chemists. What remained to be seen was whether this was feasible.

The chemists' work on paste-making in the colony started in true earnest but was soon stymied. First, W. A. Davis, the newly designated indigo research chemist, was sufficiently assured by the progress of his experiments to announce that making a standardized indigo paste of a reasonably high purity was possible on an industrial scale. He planned to make standardized paste at the premises of each planter separately – "factory by factory" – not in a central factory. But he encountered an unforeseen problem on account of the higher freight charges that the planters anticipated having to encumber in exporting indigo of low concentration. The planters opposed the idea of sending indigo to England in 20 percent concentration as the size of consignment increased three times compared to that of the earlier indigo of 60 percent concentration, especially when wartime freight charges were skyrocketing. Most planters thought the additional cost would be prohibitive. The BPA members met to pass a resolution highlighting the problem and imploring the London-based Indigo Paste Committee to make the paste in England itself out of cakes sent from India. The members of the committee in London agreed with the proposal.[25] They proposed that the planters establish a home syndicate in London that would serve as a hub to collect natural indigo from colonial plantations and then process reimbursement for the individual subscriber

Revenue and Agriculture for August 1916, India Office Records, GOI, Proc. Rev & Agr, Z/P/1980.

[25] Letter from J. M. Wilson to L. T. Harington, dated December 6, 1916, No. 109; Harrington's reply, dated March 26, 1917, No. 36, Government of India, Proceedings of the Department of Revenue and Agriculture for August 1917, India Office Records, GOI, Proc. Rev & Agr, Z/P/1981.

planters. The committee decided to appoint Henry Armstrong to lead the experiments on manufacturing the paste to customer specifications in England assisted by another chemist, Reginald B. Brown, a professional of German background who had earlier worked for the BASF's synthetic indigo operations in Germany and England.

Once the paste-making project had moved to the metropolis in the context of continuing global disruption in the supplies of German indigo (Figure 13), the indigo chemist W. A. Davis turned his attention to overcoming the problem of wilt. Through his study he proposed a radically different interpretation of the causes of wilt. Davis claimed that wilt was connected with the deficiency of "available" phosphates in the soil, especially those at a deeper stratum in the ground. He defined available phosphate as that component of the soil's phosphates that could be dissolved in the carbolic acid produced by the bacteria in indigo nodules and was thus "available" to the plant, as against the phosphate that could not be dissolved and therefore whose presence was irrelevant to the growth of the plant. While the alluvial plains of Bihar had some amount of "available" phosphates in the topsoil, the lower part was absolutely deficient. Indigo, especially Java indigo, possessed deep roots. The young plant showed normal growth as it kept using the available phosphates in the upper portion of the soil, but by the time the roots grew six inches or more they ran into the subsoil that was totally deficient in the mineral. The growth of the plant then suffered. In light of these assertions Davis called for a long-term project for the rejuvenation of the soil in Bihar. But planters did not quickly follow up on his recommendations. They were probably befuddled by the sharp contrast between this new interpretation and the earlier interpretation of the botanist Albert Howard. The planters noted the "sharp divergence of opinion" between experts on the causes of wilt, and were not sure what to make of it.[26]

Davis also collaborated with other experts at the Pusa Institute in an effort to optimize the process of steeping. Pusa's larger resources and the greater precision of the developing bacteriological sciences enabled the launching of these manufacturing trials. The most notable contribution was that of the imperial bacteriologist, C. M. Hutchinson, who helped organize efforts to raise the appropriate strains of bacteria to carry out

---

[26] W. A. Davis, *A Study of the Indigo Soils of Bihar: The Urgent Necessity of Immediate Phosphate Manuring If Crops Are to Be Maintained*, Calcutta: Agricultural Research Institute, 1918, pp. 1–77; letter from J. MacKenna to Secretary, Revenue and Agriculture, Government of India, dated December 8, 1916, letter no. 315-C, No. 7, Serial No. 6, Government of India, Proceedings of the Department of Revenue and Agriculture for June 1917, India Office Records, GOI, Proc. Rev & Agr, Z/P/1981.

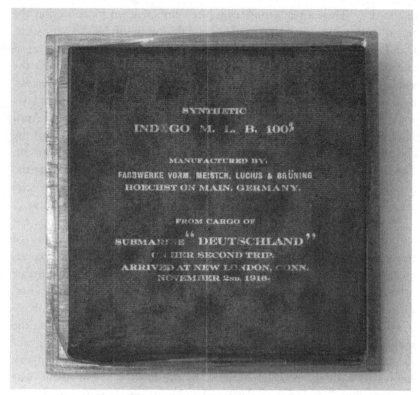

FIGURE 13. Indigo Box (From the Cargo of the German U-Boat "Deutschland" that made Two Trips to the United States in 1916 at the Behest of an Hoechst Importer) Courtesy of the Chemical Heritage Foundation Collections.

the hydrolysis of indican. In working toward that goal numerous strains of bacteria were isolated and catalogued. But Pusa experts were far from the stage where they could claim that they had the means to control bacterial action in the fermentation vats.[27]

In an initiative that showed planters' surviving faith in science to improve indigo even at such a late stage, the planters reached agreement over the imposition of a levy to fund indigo experiments. At the end of 1916, the members of the BPA passed a resolution recommending that a duty of one rupee be imposed on every *maund* of indigo shipped out of the Indian ports. They asked the government to legislate the imposition of the duty. They also wanted the government to be responsible for the collection

[27] *Scientific Reports of the ARI, 1916–17*, Calcutta: Agricultural Research Institute, 1917, pp. 107–10, India Office Records, IOR/V/24/17.

of the duty. Then after setting aside the sum spent on its collection the
rest of the money could be made available for the indigo experiments.
D. J. Reid, the chairman of BPA, calculated that at the prevailing level of
exports from India Rs. 70,000–75,000 could be collected through the
duty. If up to Rs. 10,000 were left as charges for its collection, then Rs.
60,000–65,000 could be utilized for the experiments. After relevant laws
were passed the Indigo Cess Act III of 1918 went into effect from April
1, 1918. The planters decided to use these funds to pay the salary of the
chemists working on indigo paste in London and to hire a new botanist.
The latter was expected to work on improving indican by enhancing the
plant's physiology. Some also wished to use the fund to hire a commercial
traveler to promote the sale of natural indigo among dyers and printers.[28]

It seemed that nothing that the planters attempted offered hope for
a turnaround in natural indigo's fortunes. Despite the availability of
research funds through the 1918 cess the momentum of scientific experiments slowed in the last two years before stopping completely. No major
initiatives were launched afresh in the colony by Davis and his associates.
The efforts by the metropolitan chemists led by Henry Armstrong, while
relatively successful, ran into unforeseen circumstances surrounding the
failure of crop in India. Armstrong reported that they had successfully fab-
ricated the synthetic paste whose specimens were liked by English dyers.
Impressed with the product supplied by the Paste Committee the War
Office in England placed a substantial order for ten thousand pounds of
standardized paste. But adverse climatic conditions ruined the harvest on
the plantations in 1918–19, and as a result the price of indigo soared in
the Calcutta market. The planters found it profitable to sell their indigo
in Calcutta rather than send it to London. Despite repeated reminders
from the Paste Committee, the planters did not send their indigo, stalling
the work of metropolitan chemists. Armstrong's science had apparently
encountered the logic of prices in the local market in the colony. Put
another way, the logic of agricultural production and the market was
proving to be resistant to the imposition of the new type of rationalization. The planters made a counterproposal to Armstrong to examine
whether his team could turn the low-grade indigo from India into a paste
for the English markets. Armstrong stated that making a paste out of

---

[28] Notification, Government of India, Proceedings of the Department of Commerce and
Industry for the month of June 1918, Nos. 1–6, India Office Records; No. 99, "Bihar
Planters' Association Ltd., An informal meeting of the Board of Directors of the
Bihar Planters' Association," Government of India, Proceedings of the Department of
Revenue and Agriculture for March 1919, India Office Records, GOI, Proc. Rev & Agr,
Z/P/1983.

low-grade indigo would not be viable because of the low price for such a dye. Nor did another local project to make indigo paste in Calcutta or China for the Chinese markets work out to the satisfaction of planters.[29]

As production of synthetic indigo by English manufacturers started becoming a reality, the India House lost interest in the workings of the Indigo and Paste Committee. The agricultural adviser, J. MacKenna, in India had to prompt India House officials to take a more active interest in the affairs of the indigo committees. He reminded that the officials in London were neglecting the same committees that they had helped form. He wanted authorities in London to review the work being done by the Paste Committee more carefully and to pass on more information to India in a timely fashion. But he failed to enthuse them. Devoid of any supply of natural indigo the Indigo and Paste Committee stopped their work.

The last few steps on indigo science in the colony materialized as a consequence of the momentum of decisions taken much earlier. Under the provisions of the funds supplied by the indigo cess the metropolitan government appointed an indigo botanist, Major W. R. G. Atkins, who assumed the position of indigo research botanist at Pusa from October 2, 1920.[30] Atkins's stay in the colony was uneventful on the scientific front. He did not take well to the tropical climate and asked to be relieved to return to England. He told the director of the Pusa Institute that indigo plant yield could be improved by methods of selection and propagating but that such an effort would require at least another five to seven years of dedicated work.[31] W. A. Davis had also practically stopped his work on indigo, although his departure to England was delayed by a couple of years by the terms of his contract. Henry Armstrong objected to the decision to let Davis leave.[32] Armstrong's recommendations to promote the natural indigo industry had carried weight in 1917, at a time when

[29] Letter from Moran and Company to Paste Committee, dated August 28, 1919, and November 13, 1919; letters from the Paste Committee to Moran Company in India, dated October 16, 1919, and November 20, 1919; letter from the Indigo Paste Committee to the Under-Secretary, Revenue Department, India House, dated November 25, 1919, Government of India, Proceedings of the Department of Revenue and Agriculture for May 1920, India Office Records, GOI, Proc. Rev & Agr, P/10846.

[30] "Notification – by the Department of Revenue and Agriculture," letter no. 155–337, dated October 14, 1920, No. 4, Serial No. 4, Government of India, Proceedings of the Department of Revenue and Agriculture for October 1920, India Office Records, GOI, Proc. Rev & Agr, P/10846.

[31] Letter from W. R. G. Atkins to the Director, Agricultural Research Institute, dated October 20, 1920, Proceedings of the Department of Revenue and Agriculture for July 1921, India Office Records, GOI, Proc. Rev & Agr, P/11050.

[32] Letter from Henry Armstrong, Nos. 53–6, File No. 81 of 1921, Part B, Proceedings of the Department of Revenue and Agriculture for April 1922, India Office Records.

the English dyers and printers needed their supplies of the blue dye. With the manufacture of synthetic indigo going ahead full steam in England, at factories controlled by English manufacturers, and the dyers and printers fully satisfied with the supplies, metropolitan officials in England and the colonial officials in India saw no reason to pay any attention to Armstrong's counsels. There was no commercial future for natural indigo now, and the administrators knew it.

The war-induced, metropolis-driven, and consumer-inspired indigo science failed to change natural indigo's destiny. The very best effort through science and rationalization hit a ceiling in terms of developing the agricultural dye's ability to face the competition of the factory-produced synthetic indigo. This reality became very clear to the backers of the colonial dye at the end of the war as English manufacturers enthusiastically took to the production of synthetic indigo for the home market and dyers, printers, and users unreservedly opted for the synthetic when it was available. The goal of the metropolis-inspired effort to make natural indigo stand on its own legs was roundly abandoned.

Imperial Britain's Global Role and the Death of Indigo

It is a paradox of history that victory for imperial Britain in the world war caused the final defeat of an agricultural product that had survived under the patronage of imperial Britain. This paradox can be partly explained with reference to the aspects of the global role that Britain assumed from time to time. The British imperial framework had provided a template for global interactions. Since the late eighteenth century it had provided patronage to a Bengal industry whose product was supplied to home and world markets. The national government in the metropolis that lay at the core of the imperial lattice balanced its obligations toward expatriate planters in the colony and the British dyers and printers. But the same consumers were also allowed to make a switch to German synthetic indigo under the long-held policies of free trade by the national government in Britain in the pre–world war period.[33]

---

[33] Niall Ferguson's study of the British Empire sheds some useful light on the character and implications of the global spread of the nineteenth- and early twentieth-century British Empire. While the current study does not agree with Ferguson's neo-imperial tone, it shares the author's consideration of empire as an overarching template that enabled the consolidation of structures and processes of a worldly scale. Niall Ferguson, *Empire: The Rise and Demise of the British World Order and the Lessons for Global Power*, New York: Basic Books, 2004, first published in 2002.

As the war drew to a close the global empire did not see any incentives for making exceptional efforts to keep the colonial indigo industry afloat. On the contrary, it was a key participant in the common Allied strategy of poaching on German technologies of chemicals including dyes that others have documented before. The Treaty of Versailles provided the official sanction for the inspection of German chemical plants and for the dismantling of "warfare chemical processes" in Germany. But the phrases used in the treaty were intentionally vague. Articles 168–72 of the treaty in effect provided the victors with unrestricted access to all chemical manufacturing technologies including that for the dyes. The initial inspections of the German chemical factories by the Allies started as early as late 1918, beginning in the Rhineland. The commercially motivated visit by the Association of British Chemical Manufacturers to Germany occurred in May and June 1919. Rigorous and ongoing inspection of the German chemical plants continued under the powers of the Inter-Allied Military Commission of Control, which was the principal facilitator of the process of transfer of technologies. Some specific instances of the transfer of indigo techniques are well documented and too prominent to escape notice. To give a few examples already noted by other historians, the French extracted the dyestuffs technologies through forced licensing arrangements with BASF and IG Farben. The British members of the Inter-Allied Military Commission of Control obtained the technique for producing ethylene and ethylene chlorohydrine, which were very important intermediates for the manufacture of synthetic indigo. And the mission of the Association of British Chemical Manufacturers appropriated for the British entrepreneurs the Hoechst version of the indigo process. The knowledge related to the production of indigo also changed hands through espionage and bribery by individual companies. The military/political control over the German territories continued up to 1927, but there is a general consensus among historians that by the early 1920s most relevant chemical technologies had already passed into the hands of England and France.[34] The empire stood by the English chemical manufacturers and consumers who had a common interest in synthetic indigo.

---

[34] The Treaty of Versailles' regime of political control and transfer of chemical technologies has been discussed in Jeffrey Allan Johnson and Roy MacLeod, "The War the Victors Lost: The Dilemmas of Chemical Disarmament, 1919–1926," in Roy MacLeod and Jeffrey Johnson (eds.), *Frontline and Factory: Comparative Perspectives on the Chemical Industry at War, 1914–1924*, Dordrecht, the Netherlands: Springer, 2007, pp. 221–45.

At the same time the empire was also losing patience with the colonial indigo plantations in Bihar, which continued to be a hotbed of recurrent peasant disturbances. Gandhi returned from South Africa to India in 1915 and immediately became the face of all-India nationalism. Gandhi decided to launch his movement on the indigo tracts of Champaran in Bihar in 1917, mobilizing the indigo peasantry to demand removal of the contracted obligation to grow indigo. Gandhi led movement in Champaran demanded a lowering of the size of land to be put to indigo under the terms of the contract and the planters were eventually forced to oblige. This was a major triumph for Gandhi and institutional nationalism in their fight against the colonial regime. But the symbolisms of Champaran were even graver for the empire. Champaran indigo tracts had become the symbol of colonial exploitation and a stage for the enactment by Gandhi of erasure of a political and moral wrongdoing. Imperial Britain saw the writing on the wall. The indigo tracts were also a huge political cost in terms of international image to Britain and of relatively meager economic advantage to an exceedingly small group of Britons. It simply made no sense for the empire to do anything exceptional to protect the plantation indigo industry.

Conclusion

Bengal Indigo and World History

What claims can we make about indigo knowledge on the basis of this history that unfolded on the Indian subcontinent but had built substantial connections with developments in other parts of the world? Was it the local version of a global knowledge system? Was it a body of knowledge that colonial planters appropriated to their own ends of profit making? The answer to both the questions must be in the affirmative. The two perspectives draw from a common claim of globality of indigo knowledge in the ultimate analysis but lead us to distinct understandings that are rich. The first line of reasoning allows us to see the cosmopolitan processes and "multiple logics" that went into the making of indigo knowledge on the stage of the world. The second leads to viewing the colonial appropriation of an aggregate of knowledge over which neither the colonial nor the imperial system could claim complete monopoly. Diverse constitutive forces went into the making of the knowledge of indigo culture in Bengal. And this book makes multiple genealogies of this knowledge system and its openness to a multitude of influences in South Asia a point of departure in telling the history of Bengal indigo. In his cultural readings of imperialism, that masterful analyst Edward Said reminded us as to "how oddly hybrid historical and cultural experiences are," emphasizing that "cultures actually assume more 'foreign' elements, alterities, differences, than they consciously exclude."[1] Said's observations about the eclectic nature of imperialism are equally valid for indigo knowledge

[1] Edward Said, *Culture and Imperialism*, New York: Alfred A. Knopf, 1993, p. 15.

on the Indian subcontinent. If one steps away from limiting frameworks, the knowledge surrounding indigo appears as a mosaic, and a mirror reflection of the plural historical experience of indigo and of intermixing of cultures along its odyssey across multiple continents.

Cosmopolitan and Colonial

The use of movement as a centerpiece of analysis offers an opportunity simultaneously to consider global flows and coloniality as tropes in the history of indigo knowledge in colonial South Asia. The context for fashioning of indigo knowledge in a protean form in the colony seems quite evident when mobility is viewed as a condition of its possibility.[2] The knowledge around indigo was made possible by the movement of indigo, of savants, and of planters across continents and of institutions and ideologies of production. This process of knowledge formation was distinctively cosmopolitan because of an all-around openness to multiple influences. The process of continual absorption was not cut short as the body of knowledge continued to evolve during the blue dye's rebirth as a colonial commodity in South Asia. Colonial power and hegemony became particularly relevant to the process in this phase even as the evolving knowledge of indigo on the Indian subcontinent remained simultaneously connected with dynamics beyond the colony and wider British imperial networks. But overall the history of indigo conveys a mode of knowledge formation that was remarkably more cosmopolitan in the early phase. The onset of developments associated with the arrival of synthetic dyes marked a transition to a more colonial form of knowledge generation.

A focus on diasporic and colonial planters as a primary category of actors in the history of Bengal indigo affords an opportunity to spotlight the changing dynamics of this knowledge formation. The European planters on the Indian subcontinent in the eighteenth and nineteenth centuries managed to remain remarkably connected with international networks of knowledge exchanges. A focus on this aspect of planters' relationship with knowledge challenges current historiographical understandings of colonial planters as a set of conservative people whose focus was primarily centered on agrarian economy and who indulged in exploitative

---

[2] The historiography of social knowledge in colonial South Asia has fluctuated between emphasis on syncretism and on difference in framing the relative lineages of the external and the indigenous.

management of labor they needed to obtain the raw material cheaply. The planters no doubt tapped into cheap labor of Indian peasants and laborers. But as historical agents they did much more than that. Indeed their success in the world economy derived also from their intense interest in "science" and best practices relating to indigo culture. Their ability to draw on this knowledge enabled them to produce indigo of an unmatched quality, which swiftly drove Spanish and French indigo out of the world market and established an era of dominance for Bengal indigo. Their dabbling into the world of knowledge also enabled them to produce indigo at a low cost that was tremendously profitable as a commodity in the international market dominated by them.

The developments associated with the arrival of synthetic indigo in many ways interrupted planters' relationship with science and transformed the nature of indigo knowledge in the colony from distinctly cosmopolitan to overwhelmingly colonial. As the crisis of synthetic's competition became more intense, the planters realized that their individual dabbling in science was no longer sufficient. The situation demanded the assistance of the emerging class of professional scientists in a much more dramatic way, which made them turn to state-sponsored science. It was at their request that the colonial state mobilized a distinctively colonial science in defense of plant indigo. The planters themselves interestingly no longer seemed to be at the cutting edge of science as their predecessors before 1890 had been. Rather, they increasingly appeared to be retrograde figures whose views of science were out of date. For the most part, European science had little to offer to them since its logic increasingly favored the synthetic product. Whereas their prior involvement in global circuits of knowledge required a certain cosmopolitanism until the last quarter of the nineteenth century, with the turn to the colonial state they were now dependent on a science that was typically proximate to colonial institutions in the late nineteenth and early twentieth centuries. Planters found themselves increasingly ensconced within the logic of this local colonial science, which was focused on a narrow set of goals of landed interests and expatriate Europeans in the colony. This colonial science was pushed to the background only during the final, brief phase of the First World War when temporary wartime contingencies alerted the home government to encourage plant indigo production in the colony as British dyers and printers faced shortages in the supply of German synthetic indigo. This phase, in any case, turned out to be inconsequential to the goals of reviving natural indigo and, in the end, only marked a brief phase of heightened imperial influence in the long history of indigo.

The Fundamental Diversity of Indigo Systems
on the Indian Subcontinent

There are in all some 300 species of Indigofera, distributed throughout the tropical regions of all the globe, with Africa as their head-quarters. India possesses some 40 species.[3]

The important account on indigo *Pamphlet on Indigo,* published from Calcutta in 1890, acknowledged the typological and historical diversity of the indigo plant in the world that was also mirrored on the Indian subcontinent. It gestured towards the Indian history of indigo as part of a global history, even implying, semantically at least, that the epicenter of the plant's history lay in Africa and a good part of its spread in the global South. Its writer was George Watt, reporter on economic products with the colonial government. He began his narration by asserting the diversity of indigo as a plant type and then went on to describe the different methods of indigo cultures existing in various regions of India. His testimony was not limited to Bengal and Bihar alone, which were admittedly by far the major areas of indigo production in the colony. It provided an account of indigo in different regions replete with a historical analysis of where in India it had been dominant before and where it had arisen recently. It focused on a comprehensive description of indigo in Lower Bengal; in the separate districts in north and south Bihar; in the North Western Provinces and Oudh; in Sind, Rajputana, Central India, and the central provinces; in Burma and Assam; and in Bombay and Madras, the last one having emerged as a producer of some note in recent times.

This diversity of indigo on the subcontinent was revealed in the different courses adopted by prior and coexisting indigo systems in different regions, each of which was unique and distinct. This aspect of the shared history of Bengal indigo with other indigo systems was evidently apparent even to George Watt, who was working within the colonial logic, assigned as he was with the responsibility of overseeing colonial commerce in commodities at the highest level. A detailed history of the subcontinent's indigo systems is still waiting to be written. But even a brief allusion to these precolonial and other temporally parallel colonial indigo systems helps mitigate the incorrect impression that may otherwise remain that the knowledge of Bengal plantations was independent in any sort of way, spontaneous in origin, or, say, attached to the single logic of colonial rule. In that sense, the singling out and identification of

---

[3] George Watt, *Pamphlet on Indigo,* Calcutta: 1890, p. 7.

other indigo systems bring to light new aspects of the structure of indigo knowledge on the subcontinent. They point to the possibility of analyzing multiple colonial and extracolonial logics even if all of the latter's interconnections to Bengal indigo cannot be fully established.

That the indigo system had a deep and flourishing history on the subcontinent is proven by the account of Abul Fazl, the Mughal chronicler who confirmed that indigo was produced in India for commerce in the sixteenth century at Bayana near Agra and at Sarkhej on the west coast in the province of Gujarat. The indigo in both of these regions was valued for its export market in Persia and in the Ottoman Empire.[4] Armenian and Indian merchants carried Indian indigo to these distant markets through well-established sea routes that linked Gujarat with ports in the western Indian Ocean. The entry of Indian indigo into the Ottoman Empire further enabled its passage via the Mediterranean to Italian ports and hence to European markets. The arrival of the Portuguese in the Indian Ocean in the sixteenth century and subsequently of the English and Dutch East India Companies in the early decades of the seventeenth century added a new dimension to the trade in Indian indigo. The Europeans offered their ships for moving indigo in the intra-Asian trade and emerged as significant traders of indigo themselves. Their shipment of indigo dye to Europe began to soar. As a result, first Lisbon and subsequently ports in Holland and England became major centers for the distribution of Indian indigo in the West.[5]

A peasant-based, variegated system of indigo culture evidently thrived in northern and western India in the sixteenth and seventeenth centuries, as the era of European trading companies dawned on the Indian subcontinent.[6] The two most prominent centers, Bayana and Sarkhej in the

---

[4] W. H. Moreland, *India at the Death of Akbar: An Economic Study*, Delhi: Atma Ram and Sons, 1962, pp. 96–9.

[5] Om Prakash, *The New Cambridge History of India: European Commercial Enterprise in Pre-Colonial India*, Cambridge: Cambridge University Press, 1998.

[6] For a description of seventeenth-century indigo trade and manufacture, see Ghulam Nadri, "Indigo Industry and Trade in Gujarat in the Seventeenth Century," unpublished M.Phil dissertation, Aligarh Muslim University, 1996. The cultivation and production of indigo for the most part remained in the hands of Indian peasants. Europeans generally bought indigo from Indian merchant middlemen or directly from the peasants. It was common for the European buyers to advance capital to Indian peasants. Nadri has referred to a "partial shift" from peasant controlled indigo manufacturing to merchant supervised processing after the 1630–1 famine in Gujarat when Indian capitalists and European factors took advantage of famine and poverty among peasants to insert themselves into indigo manufacturing. But this subversion of peasant agriculture did not happen everywhere. Cf. Ghulam Nadri, "Indigo Industry and Trade in Gujarat in the Seventeenth Century," pp. 11–13; W. H. Moreland, *India at the Death of Akbar*, pp. 104, 118 (note).

north and west respectively, which supplied indigo to Europeans, also showcased two distinct regional systems of indigo culture. The Bayana indigo was mostly produced through a "green leaf process" involving drenching of fresh leaves in vats immediately after they were harvested. Sarkhej, on the contrary, practiced a "dry leaf" process that involved cutting and drying of leaves before putting them into vats full of water, where they were then left submerged for four to five days. Sources have attested to the use of different numbers of vats in Sarkhej ranging from one to six. In contrast, a single vat was in use in Bayana. The shapes of vats in Sarkhej and Bayan were also distinct; the former region preferred rectangular vats and the latter round and shallow vats. In another instance of variation, flat indigo was turned out at Sarkhej while a round indigo was manufactured at Bayana. Despite the different stable traditions that can be generally associated with the two major centers, sources have also indicated the presence of Bayana features of manufacturing in Sarkhej, and vice versa.[7]

The system of indigo culture that flourished on the Coromandel Coast and its immediate hinterland roughly between the Godavari and Cauvery Rivers in peninsular India was also peasant based. This southern indigo was valued both for local use and for long-distance commerce. The indigo of Coromandel had traditionally met the dyeing requirements of local textiles manufactured by artisans in the region. But, in addition, it was exported to foreign countries along well-established trade routes going out of southern India and the west coast in Gujarat. In the early decades of the seventeenth century, the Dutch East India Company began exporting indigo from Coromandel to Holland. The Coromandel indigo was not of a uniform nature. Rather a number of varieties were recognized for their different constitutions and were variously named after their respective areas of origin such as Masulipatam, Pulicat, Tierepopelier, and Tegenapatam. They were also priced differently in the international markets.[8]

The world of Indian indigo was transformed over the course of several decades after the mid-seventeenth century. But many of the details of this transformation are lacking. Historians of Euro-Asian and intra-Asian trade have confirmed that the export of indigo from India receded in the latter half of the seventeenth century. This reflected a lower procurement

---

[7] The precolonial processes of culture are described in Ghulam Nadri, "Indigo Industry and Trade in Gujarat in the Seventeenth Century," pp. 8–10.

[8] Tapan Raychaudhuri, *Jan Company in Coromandel, 1605–1690*, The Hague: Martinus Nijhoff, 1962, pp. 162–4.

by European trading companies just as the Caribbean islands began to manufacture and supply indigo to Europe.[9] The existing histories of the Atlantic World have not quantified the total scale of indigo production in the Greater Antilles in the seventeenth century, which presumably replaced exports from India. The agricultural history of indigo on the Indian subcontinent through this transitory period also remains sketchy. So we do not definitively know the pace of substitution of Indian indigo by indigo from other parts of the world in the seventeenth and early eighteenth centuries.[10] But there are grounds to assume a longer history of indigo production in India after the mid-seventeenth century against the picture of a sharp decline in output that emerges from a singular focus on exports to Europe. The scale of output probably diminished more slowly than indicated. The remnants of these longer-surviving manufacturing traditions were quite evident in local pockets on the subcontinent long after large-scale export to Europe ceased.

There are indications, at least for some localities on the subcontinent, that the Indian traditions of manufacturing were still practiced as European colonial expansion started in the last quarter of the eighteenth century. The French planter from Mauritius De Cossigny de Palma experimented with dried leaves of indigo to produce the dye. He was an advocate of the West Indian system that involved processing freshly harvested leaves. But he had learned from the French Council of Pondicherry in southern India of this discrepant practice of manufacturing indigo used by the natives of India on the Coromandel Coast. On learning details of this new system he experimented to test the process and found it to be a

---

[9] K. N. Chaudhuri, *The Trading World of Asia and the English East India Company, 1660–1760*, Cambridge: Cambridge University Press, 1978; Om Prakash, *The New Cambridge History of India: European Commercial Enterprise in Pre-Colonial India*, Cambridge: Cambridge University Press, 1998, pp. 169–71, 192–3, 205, 226–7, 241.

[10] The peak period of production of indigo from Spanish America and Carolina lay ahead in the second half of the eighteenth century. For indigo production in the Spanish controlled territories in South and Central America, see, Troy S. Floyd, "Salvadorean Indigo and the Guatemalan Merchants: A Study in Central American Socio-Economic History, 1750–1800," unpublished Ph.D. dissertation, University of California, Berkeley, 1959; Jose Antonio Fernandez Molina, "Colouring the World in Blue: The Indigo Boom and the Central American Market, 1750–1810," Ph.D. dissertation, University of Texas, 1992; for indigo in the American colonies and early America, see, Virginia Gail Jelatis, "Tangled Up in Blue: Indigo Culture and Economy in South Carolina, 1747–1800," unpublished Ph.D. dissertation, University of Minnesota, 1999. The latter half of the eighteenth century was unquestionably the best era for Spanish and Carolina indigo, as argued by Jose Molina and Jelatis. Surprisingly there is very little direct reference to French indigo in the Caribbean in the English language references.

viable one even though he continued to believe in the superiority of his own system. The demonstration of the workability of the Indian system proved to him, as he wrote, that indigo could be produced by different methods. He also wanted further trials to be conducted to test whether the Indian method could be further improved and made better than the current system in use by European planters.[11] As the first batch of colonial entrepreneurs started their indigo operations in northern India they too came face to face with durable Indian traditions. Claude Martin, a Frenchman, who owned indigo factories in Oudh,[12] was deeply impressed with the Indian system of indigo manufacturing, which he first witnessed at Ambore, a region to the southwest of the Coromandel Coast. He indicated that Indians used a different species of indigo plant and that their system of manufacturing involved boiling of leaves in small earthenware. It can be inferred on the basis of the small scale of operation described by Martin that dye manufacturing was entirely in the hands of small peasants. Martin commended this system to fellow European planters particularly because it dispensed with the need to invest in the construction of large vats. So convinced had he become of the merit of this Indian system that he wished for Bengal planters to adopt the "Ambore system."[13] But it was William Roxburgh, the Scottish naturalist and East India Company official, who became the most prominent advocate of the Indian system of indigo manufacturing discovered during his appointment in Madras Province. He was put in charge of the East India Company's Garden at Samulcottah in the northern Circars, an administrative division overlapping with the Coromandel Coast. He reported the widespread practice of indigo manufacturing in the province as a unique system of manufacturing by natives using hot water. Referring to their typical method, he underlined the prevalence of indigo culture by Indian peasants "throughout the Northern provinces, or [northern] Circars" and "in many parts of the Carnatic."[14]

[11] De Cossigny de Palma, *Memoir Containing an Abridged Treatise of the Cultivation and Manufacture of Indigo*, Calcutta: 1789, first French edition published in 1779, pp. 130–6.

[12] Rosie Llewellyn-Jones, *A Very Ingenious Man: Claude Martin in Early Colonial India*, Delhi: Oxford University Press, 1999; *A Man of the Enlightenment in 18th Century India: The Letters of Claude Martin 1766–1800*, New Delhi: Permanent Black, 2003, pp. 268–70.

[13] "On the Manufacture of Indigo at Ambore by Lieutenant Colonel Claude Martin. (with) An Extract of a Treatise on the Manufacture of Indigo, by Mr. De Cossigny," London: Asiatic Researches, 1807, pp. 475–7, according to an old note of April 1791.

[14] "Process of Making Indigo on the Coast near Ingeram, Communicated by Mr. William Roxburgh," *Memoir Containing an Abridged Treatise on the Cultivation and Manufacture*

These Indian traditions proved resilient and durable as indigo production on the Indian subcontinent was reinvented by colonialists in the late eighteenth and nineteenth centuries. The peasant traditions survived in many localities, as so vividly chronicled by George Watt. But the durability of Indian traditions was not limited to local systems oriented toward small-scale consumption. The elements of admixture between the preexisting and the new system of indigo culture appeared through the long march of the nineteenth century in colonial India. The two major systems of indigo culture with sizable contribution to colonial export alongside Bengal and Tirhut, those in the North Western Provinces and Oudh and Madras specifically, showed significant remnants of peasant-based systems of manufacturing. The conflation of European and native participation in indigo manufacturing was indicated by Watt in his description of export of indigo from Bengal. He cited an 1889 report from Thomas and Company, significant traders of indigo based in Calcutta, to the effect that natives dominated by far the production of indigo in Lower Bengal and Doab. The latter included the prior seventeenth-century zone of peasant-dominated indigo culture in Bayana. The Indians also more or less matched Europeans in output in the Benares region while Europeans dominated production in Tirhut.[15] Indeed after the mid-nineteenth century, Watt also indicated, a major aspect of expansion of indigo in the North Western Provinces and Oudh was its dominance by Indians. The latter took the lead in the expansion of indigo cultivation in the area of canal irrigation, established factories under their own control, and produced indigo using their own methods.[16]

But it was the rise of the Madras system after the mid-nineteenth century that most spectacularly represented the preponderance of Indian methods of indigo culture in the nineteenth century. There is little doubt that the nineteenth-century knowledge of indigo production in Madras represented an amalgamation of varied Indian traditions of indigo production that had survived on the Coromandel Coast with a

of Indigo by M. De Cossigny De Palma, p. 157; William Roxburgh, "A Brief Account of the Result of Various Experiments Made with a View to Throw Some Additional Light on the Theory of This Artificial Production," *Transactions of the Society Instituted at London, for the Encouragement of Arts, Manufactures, and Commerce*, XXVIII (1811): 288. The later publication reflects Roxburgh's experiments in Madras between 1790 and 1793 before he was appointed the superintendent of the Botanical Garden in Calcutta.

[15] Bengal indigo in the nineteenth century typically included the aggregate of indigo produced in the multiple regions of Lower Bengal; Tirhut/Bihar, the dominant area of production; Doab; and Benares.

[16] George Watt, *Pamphlet on Indigo*, pp. 17, 40.

remarkable longevity. The indigo from Madras was commonly referred to as the *kurpah* indigo, a designation that represented a variation on the name of its most dominant place of origin, the Cuddapah district in Madras. Nellore was the other important district for production of *kurpah* indigo, while many other districts including South Arcot, Guntur, Kurnool, Chingleput, Krisna, Chittor, and Vizagapatam also participated in the production of dye in the South. *Kurpah* became established as the generic name for the indigo from Madras in the international markets. It was considered to be of a low quality and was priced lower than the higher-quality Bengal or Tirhut indigo. But nonetheless its importance should not be minimized. The Madras indigo established a niche for itself in the export market. Such was the scope of its expansion in the last few decades of the nineteenth century that George Watt took this additional supply to be the main contributing factor to a depression in the prices of indigo in the world markets. By the closing decade of the nineteenth century Madras indigo was a major industry of the Madras Presidency. Speaking in relative terms, in the year 1887–8, while Bengal, the dominant producer on the subcontinent, exported 9.7 million pounds of indigo, Madras exported 4.7 million pounds.[17]

The modern Madras system itself did not represent a single, pure tradition, but was rather a coalescence of numerous traditions itself, all of which were distinct from the dominant, normative system in Bengal. The Madras indigo was solely produced by Indian peasants in a small manufacturing arrangement just as in the past. So small was the scale of these Indian manufacturing establishments that government figures listed the number of Madras manufactories of *kurpah* indigo in terms of a count of vats, not factories. In addition, the distinct systems of indigo culture in Madras usually involved the "dry leaf process," incorporating a long delay between harvesting of indigo and its processing. Some of the

---

[17] See "Table No. III – *Foreign Exports of Indigo from India during the Past Ten Years,*" George Watt, *Pamphlet on Indigo,* p. 85. In a forthcoming article, the business historian Alexander Engel has suggested giving due attention to low-cost dyes as against "premium dyes" in assessing the changing nature of market during the long-drawn-out transition from natural to synthetic dyes. Engel argued that the dyes of mass consumption had a significant share of the market in quantitative terms and were primarily responsible for the longer sustenance of natural dyes even as the more expensive high-quality natural dyes swiftly lost out on price competition to the cheaper synthetic dyes. Such a framework allows us to resituate the importance of dyes like the *kurpah* indigo in the market. *Cf.* Alexander Engel, "Colouring Markets: The Industrial Transformation of the Dyestuff Business Revisited," forthcoming, *Business History.* I am grateful to Alexander Engel for sharing this yet-to-be published material with me.

peasants turned in dried leaves through a short stage of fermentation. Others avoided fermentation altogether. In some regions of Cuddapah fermentation was deliberately avoided even through fresh leaf was used in manufacturing. In still other regions peasants used the even more radically different system of boiling fresh leaves instead of fermenting them.[18] The entire range of practices in the Madras system bore similarity to the manifold methods earlier used on the Coromandel Coast. In that sense the system of indigo culture in Madras best represented the tenacity and survival of an indigenous, alternate system of indigo culture in nineteenth-century colonial India. These miscellaneous systems of manufacture had survived through to the world war period. A report from the time noted the prevalence of the wet leaf, dry leaf, and boiling processes in Madras in different regions, mostly under the control of peasants.[19] In short, the thesis of a hegemonic "West Indian" system overrunning the entire subcontinent seems manifestly untenable for describing the evolving relationship and dynamics among the multiple indigo systems on the subcontinent. The colonial logic failed to curb the basic heterogeneity of multiple indigo systems until the very end.

---

[18] George Watt, *Pamphlet on Indigo*, pp. 52–5.

[19] "Some remarks upon the manufacture in Madras and Behar," "Appendix A: Manufacture of Indigo by a Boiling Process in the Godavari District," "Appendix B: Manufacture of Indigo from the Dry Leaf," pp. 13–16, 22–6, enclosures, letter from Director of Agriculture, Madras, to Secretary, Revenue, Government of Madras, dated February 13, 1918, Government of India, Proceedings, Department of Revenue and Agriculture for the year 1919, India Office Records, z/p/1982.

Bibliography

Primary Sources

1. Archival Documents

National Archives of India, New Delhi

Government of India, Revenue and Agriculture, Civil Veterinary Administration, 1899, A

Government of India, Revenue and Agriculture, Agriculture, 1900–3, Part B

Bihar State Archives, Patna (India)

Government of Bengal, Proceedings of the Revenue Department
Agriculture Branch, 1896–1911

Government of Bihar and Orissa, Proceedings of the Revenue Department
Agriculture Branch, 1912–22

Oriental and India Office Collections, The British Library, London

Despatches Addressed to the Several Governments in India by the Secretary of State in Council, 1890–1919

European Manuscripts

H. T. Prinsep, "Four Generations in India," C 97

Curzon: F111/181; F11/182/F111/183

Government of Bengal, Proceedings of the Judicial Department, 1861–2

Government of Bengal, Proceedings of the Political Department
Police Branch, 1900–10

Government of Bengal, Proceedings of the Revenue Department
Land Revenue Branch, 1896–1911

Government of Bengal, Proceedings of the Revenue Department
Agriculture Branch, 1896–1911

Government of Bengal, Proceedings of the Home Department
Education Branch, 1890–1900
Government of India, Proceedings of the Revenue and Agriculture Department,
     1896–1922
Government of India, Proceedings of the Department of Commerce and Industry,
     1907, 1918
Papers Relating to Indigo Cultivation in Bengal
Report of the Indigo Commission, 1862
Report on the Administration of Bengal, 1892–3
Scientific Report of the ARI, Pusa, 1914–22

Public Record Office, Kew

Ministry of Overseas Development and predecessors: Tropical Products Institute
     and predecessors: Registered Files, 1895–1905
Board of Trade: Companies Registration Office: Files of Dissolved Companies,
     1890–1910

Royal Society, London

Proceedings of the Indian Government Advisory Committee of the Royal Society,
     1903–20

University of Leeds, Leeds

Papers of Clothworkers Laboratory

Manchester Archives and Local Studies, Manchester

Papers of Calico Printers' Association
Papers of British Cotton and Wool Dyers' Association

Gloucestershire Archives, Gloucestershire

Papers of William Playne's Company

National Agricultural Library, Beltsville, Maryland

Indian Planters' Gazette and Sporting News (1902–24)

2. *Personal Memoir*

Cyril Berkeley, *My Autobiography*

3. *Journals and Newspapers*

Agricultural Journal of India

Annals of Botany
Asiatic Researches
Calcutta Gazette
The Chemical Trade Journal
Journal of the Agricultural and Horticultural Society of India
Journal of the Asiatic Society of Bengal
Journal of the Chemical Society
Journal of the East India Association
Journal of the Society of Arts
Journal of the Society of Chemical Industry
Journal of the Society of Dyers and Colourists
Native Newspaper Reports, Bengal, 1890–1905
Textile Colourist
The Textile Manufacturer

4. *Published Primary Texts and Reports*

Anonymous. *A Short Description of the Province of South Carolina, With an Account of the Air, Weather, and Diseases, at Charlestown, 1763.* London: privately published, 1770.

Edwards, Bryan. *The History Civil and Commercial of the British Colonies in the West Indies, to which is added An Historical Survey of the French Colony in the Island of St. Domingue.* London: B. Crosby, 1798.

Foster, William. *The English Factories in India.* Oxford: Clarendon Press, 13 volumes, 1906–23.

Lambert, Claude Francois. *A Collection of Curious Observations on the Manners, Customs, Usages, Different Languages, Government, Mythology, Chronology, Ancient and Modern Geography, Ceremonies, Religion, Mechanics, Astronomy, Medicine, Physics, Natural History, Commerce, Arts, and Sciences, of the Several Nations of Asia, Africa, and America.* Translated by John Dunn, London, 1750.

Long, Edward. *The History of Jamaica, or a General Survey of the Antient and Modern State of the Island with Reflections on its Situation, Settlements, Inhabitants, Climate, Products, Commerce, Laws, and Government, Vol. III.* London: T Lowndes, 1774.

Pelsaert, Francisco., W. H. Moreland, and P. Geyl (trans. and ed.). *Jahangir's India: The Remonstrantie of Francisco Pelsaert.* Delhi: Idarah-i-Adabiyat-i-Delli, 2009.

Proceedings of the Meetings of the British Indian Association of Oudh, 1861–1865. Calcutta: Thacker, Spink, & Co., 1865. *Publications of the British Indian Association, 1863.*

Report of the Indian Famine Commission, Part II, Measures for Protection and Prevention. London: George Edward Eyre and William Spottiswoode, 1880.

Report of the Proceedings at a General Meeting of the Inhabitants of Calcutta, on the 15th of December, 1829. Extracted from the Bengal Hurkaru, etc. London: T Bretell, 1830.

Roy, Dipak. *A Hundred and Twenty-Five Years: The Story of J. Thomas & Company.* Calcutta: J Thomas & Company, not dated.

Schrottky, Eugene C. *The Principles of Rational Agriculture Applied to India and Its Staple Products.* Bombay: Times of India Office, 1876.

Tavernier, Jean Baptiste. *Travels in India by Jean-Baptiste Tavernier, Baron of Aubonne.* translated from the original French edition of 1676 by V. Ball, 2nd edition, edited by William Crooke, 2 volumes. London: Oxford University Press, 1925.

Voelcker, John Augustus. *Report on the Improvement of Indian Agriculture.* Delhi: Agricole Publishers, 1986, first published in 1893.

5. Tracts on Indigo

Anonymous. *Methods for Improving the Manufacture of Indigo: Originally Submitted to the Consideration of the Carolina Planters; and Now Published for the Benefit of all the British Colonies, Whose Situation is Favorable to the Culture of Indigo. To Which Are Added Several Public and Private Letters, Relating to the Same Subject, by an Experienced Dyer.* Devizes: T. Burrough, 1776.

Crokatt, James. *Further Observations Intended for Improving the Culture and Curing of Indigo in South Carolina.* London, 1747.

De Beauvais-Raseau. *L'Art de L'Indigotier*, Paris: L. F. Delatour, 1770, translated by Richard Nowland, *A Treatise on Indigo.* Calcutta: James White, 1794.

De Cossigny de Palma. *Memoir Containing an Abridged Treatise of the Cultivation and Manufacture of Indigo.* Calcutta, 1789, first French edition published in 1779.

Du Pratz, M. Le Page. *The History of Louisiana or of the Western Parts of Virginia and Carolina*, translated from French. London: T Becket, 1774.

Inglis, James. *Sport and Work on the Nepaul Frontier or Twenty Years Sporting Reminiscences of an Indigo Planter.* London: Macmillan & Co., 1878.

Lee, J. Bridges. *Indigo Manufacture.* Lahore, January 1892.

Miller, Philip. *The Gardener's Dictionary in Two Volumes.* London: printed by the author, 1743.

Minden, Wilson. *History of Behar Indigo Factories.* Calcutta: Calcutta General Printing, 2nd edition, 1908.

   Reminiscences of Behar. Calcutta: Calcutta General Printing, 2nd edition, 1908.

   Tirhoot and Its Inhabitants of the Past. Calcutta: Calcutta General Printing, 2nd edition, 1908.

Monnereau, Elias. *The Complete Indigo Maker. Containing an Account of the Indigo Plant; its Description, Culture, Preparation, and Manufacture, to Which is Added a Treatise on the Culture of Coffee*, translated from the French of Elias Monnereau, a planter in Saint Domingue. London: P. Elmsly, 1769.

Paul-Darrac, Pierre and Willem van Schendel. *Global Blue: Indigo and Espionage in Colonial Bengal.* Dhaka: University Press Limited, 2006.

Phipps, John. *A Series of Treatises on the Principal Products of Bengal: No. 1, Indigo.* Calcutta: Baptist Mission Press, 1832.

Reid, W. M. *The Culture and Manufacture of Indigo with Description of a Planter's Life and Resources.* Calcutta: Thacker, Spink and Co., 1887.

Watt, George. *Pamphlet on Indigo.* Calcutta: 1890.

6. Scientific Reports

Bloxam, W. Popplewell and H. M. Leake. with the assistance of R. S. Finlow. *An Account of the Research Work in Indigo, Carried Out at the Dalsingh Serai Research Station from 1903 to March 1904.* Calcutta: Bengal Secretariat Book Depot, 1905.

Bloxam, William P. *Report to the Government of India Containing an Account of the Research Work on Indigo Performed in the University of Leeds, 1905–1907.* London: His Majesty's Secretary of State, 1908.

Davis, W. A. *A Study of the Indigo Soils of Bihar: The Urgent Necessity of Immediate Phosphate Manuring If Crops Are to Be Maintained.* Calcutta: Agricultural Research Institute, 1918.

Howard, Albert and Gabrielle L. C. Howard. *First Report on the Improvement of Indigo in Bihar*, Bulletin No. 51. Calcutta: Agricultural Research Institute, Pusa, 1915.

Second Report on the Improvement of Indigo in Bihar, Bulletin No. 54. Calcutta: Agricultural Research Institute, Pusa, 1915.

Third Report on the Improvement of Indigo in Bihar, Bulletin No. 67. Calcutta: Agricultural Research Institute, Pusa, 1916.

Rawson, Christopher. *Report on the Cultivation and Manufacture of Indigo in Bengal.* Bradford: William Byles and Sons, 1899.

Report of the Indigo Research Station, Sirsiah, for the Year 1905–1906. British Library, ST 1882.

For the Year 1906–1907. British Library, ST 1882.

For the Year 1907–1908. British Library, ST 1882.

For the Year 1908–1909. British Library, ST 1882.

For the Year 1910–1911. British Library, ST 1882.

For the Year 1911–1912. British Library, ST 1882.

For the Year 1912–1913. British Library, ST 1882.

Report of the Indigo Research Station, Sirsiah, for the Year 1909–1910. Calcutta: Bihar Planters' Association, 1910. By C. Bergtheil, British Library, 07076.f.71.

Secondary Sources

Abelshauser, Werner et al. (ed.), *German History and Global Enterprise: BASF, the History of a Company.* Cambridge: Cambridge University Press, 2004.

Adal, Kristin. "The Problematic Nature of Nature: The Post-Constructivist Challenge to Environmental History," *History and Theory, Theme Issue* 42 (December 2003): 60–74.

Adas, Michael. *The Burma Delta: Economic Development and Social Change on an Asian Rice Frontier, 1852–1941.* Madison: University of Wisconsin Press, 1974.

Machines as the Measure of Men: Science, Technology, and Ideologies of Western Dominance. Ithaca, N.Y.: Cornell University Press, 1989.

Islamic and European Expansion: The Forging of a Global Order. Philadelphia: Temple University Press, 1993.

"Imperialism and Colonialism in Comparative Perspective," *International History Review* **XX** No. 2 (June 1998): 371–88.

Dominance by Design: Technological Imperatives and America's Civilizing Mission. Cambridge, Mass.: Belknap, 2006.

(ed.). *Essays on Twentieth-Century History*. Philadelphia: Temple University Press, 2010.

"AHR Conversation: On Transnational History," participation by C. A. Bayly, Sven Beckert, Matthew Connelly, Isabel Hofmeyr, Wendy Kozol, and Patricia Seed, *American Historical Review* **III** No. 5 (December 2006): 1140–64.

Alam, Ishrat. "New Light on Indigo Production Technology during the Sixteenth and Seventeenth Centuries," in A. J. Qaisar and S. P. Verma (eds.), *Art and Culture*. Jaipur: Publications Scheme, 1993.

Alden, Dauril. "The Growth and Decline of Indigo Production in Colonial Brazil: A Study in Comparative Economic History," *Journal of Economic History* **25** No. 1 (March 1965): 35–60, especially, 39–40.

Alexander, N. "Cultivation of Indigo, Read 13th August, 1829," *Transactions of the Agricultural and Horticultural Society of India* **II** (1836): 31–41.

Arnold, David. *Colonizing the Body: State Medicine and Epidemic Disease in Nineteenth-Century India*. Berkeley: University of California Press, 1993.

The Tropics and the Traveling Gaze. Delhi: Permanent Black, 2005.

Bagchi, Amiya K. *Private Investment in India, 1900–1939*. Cambridge: Cambridge University Press, 1972.

Bailey, Liberty H. *Plant Breeding: Being Five Lectures upon the Amelioration of Domestic Plants*. New York: Macmillan Company, 1904, 3rd edition, first published in 1895.

Balfour-Paul, Jenny. *Indigo in the Arab World*. London: Routledge Curzon, 1996.

Indigo. London: British Museum Press, 1998.

Ballantyne, Tony. *Orientalism and Race: Aryanism in the British Empire*. New York: Palgrave, 2002.

Ballard, George. "On the Culture of Indigo in Bengal, Read 10th June, 1829," *Transactions of the Agricultural and Horticultural Society of India* **II** (1836): 14–24.

Bancroft, Edward. *Experimental Researches containing the Philosophy of Permanent Colours and the Best Means of Producing them, by Dyeing, Calico Printing & c.*, Vol. 1. Philadelphia: Thomas Dobson, 1814.

Basalla, George. "The Spread of Western Science," *Science*, May 5 1967: 611–22.

Batie, Robert C. "Why Sugar? Economic Cycles and the Changing of Staples on the English and French Antilles, 1624–54," *Journal of Caribbean History* **8–9** (1976): 1–42.

Bayly, Christopher A. *New Cambridge History of India: Indian Society and the Making of the British Empire*. Cambridge: Cambridge University Press, 1988.

Imperial Meridian: The British Empire and the World, 1780–1830. New York and London: Longman, 1989.

Empire and Information: Intelligence Gathering and Social Communication in India, 1780–1870. Cambridge: Cambridge University Press, 1996.

The Birth of the Modern World, 1780–1914: Global Connections and Comparisons. Malden: Blackwell, 2004.

Bayly, Susan. *The New Cambridge History of India: Caste, Society and Politics in India from the Eighteenth Century to the Modern Age*. Cambridge: Cambridge University Press, 1999.

Beckles, Hilary M. *White Servitude and Black Slavery in Barbados, 1627–1715*, Knoxville: University of Tennessee Press, 1989.

Beer, John J. *The Emergence of the German Dye Industry*. Urbana: University of Illinois Press, 1959.

Beeson Kenneth H. Jr. *Fromajadas and Indigo: The Minorcan Colony in Florida*. Charleston, S.C.: History Press, 2006.

Bhabha, Homi (ed.). *Nation and Narration*. London: Routledge, 1990.

The Location of Culture. London: Routledge, 1994.

Bloor, David. *Knowledge and Social Imagery*. Chicago: University of Chicago Press, 1991, first published in 1976.

Boegner, Peggie P. and Richard Gachot. *Halcyon Days: An American Family through Three Generations*. New York: Old Westbury Gardens and Harry N. Abrams, 1986.

Bonneuil, Christophe. "Development as Experiment: Science and State Building in Late Colonial and Postcolonial Africa, 1930–1970," in Roy MacLeod (ed.), *Nature and Empire: Science and the Colonial Enterprise*. Chicago: University of Chicago Press, 2001.

Bose, Sugata. *The New Cambridge History of India, Peasant Labour and Colonial Capital: Rural Bengal since 1770*. Cambridge: Cambridge University Press, 1993.

A Hundred Horizons: The Indian Ocean in the Age of Global Empire. Cambridge, Mass.: Harvard University Press, 2006.

Bose, Sugata, and Kris Manjapara (eds.). *Cosmopolitan Thought Zones: South Asia and the Global Circulation of Ideas*. Basingstoke: Palgrave Macmillan, 2010.

Bowen, Huw V. *The Business of Empire: The East India Company and Imperial Britain, 1756–1833*. Cambridge: Cambridge University Press, 2006.

Bracey, Robert. "Jean Baptiste Labat," *New Blackfriars* 5 No. 151 (June 1924): 136–43.

Bray, Francesca. *The Rice Economies: Technology and Development in Asian Societies*. Berkeley: University of California Press, 1994.

Brock, William H. *Justus von Liebig: The Chemical Gatekeeper*. Cambridge: Cambridge University Press, 1997.

Browning, David. *El Salvador: Landscape and Society*. Oxford: Clarendon Press, 1971.

Buckland, C. E. *Bengal under the Lieutenant-Governors*. Calcutta: Deep Publications, 2nd edition, 1902.

Burton, Antoinette. *After the Imperial Turn: Thinking with and through the Nation*. London and Durham, N.C.: Duke University Press, 2003.

The Postcolonial Careers of Santha Rama Rau. London and Durham, N.C.: Duke University Press, 2007.

Chakrabarty, Dipesh. *Provincializing Europe: Postcolonial Thought and Historical Difference.* Princeton, N.J.: Princeton University Press, 2007.

Chandra, Bipan. *The Rise and Growth of Economic Nationalism in India.* New Delhi: People's Publishing, 1982, first published in 1966.

Chaplin, Joyce. *An Anxious Pursuit: Agricultural Innovation and Modernity in the Lower South, 1730–1815.* Chapel Hill and London: University of North Carolina Press, 1993.

Subject Matter: Technology, the Body, and Science on the Anglo-American Frontier, 1500–1676. Cambridge, Mass.: Harvard University Press, 2003.

Chapman, Stanley D. "The Agency Houses: British Mercantile Enterprise in the Far East, c. 1780–1920," *Textile History* 19 No. 2 (1988): 239–54.

Merchant Enterprise in Britain: From the Industrial Revolution to World War I. Cambridge: Cambridge University Press, 1992.

Charnley, Berris and Gregory Radick, "Plant Breeding and Intellectual Property Before and After the Rise of Mendelism: The Case of Britain," http://www.ipbio.org/berris.htm, accessed on June 30, 2010.

Chatterjee, Partha. *Nationalist Thought and the Colonial World: A Derivative Discourse.* London: Zed Books, 1986.

Chaudhuri, Kirti N. *The Trading World of Asia and the English East India Company, 1660–1760.* Cambridge: Cambridge University Press, 1978.

Chowdhury, Benoy B. *Growth of Commercial Agriculture in Bengal (1757–1900).* Calcutta: R. K. Maitra, 1964.

"Growth of Commercial Agriculture in Bengal, 1859–1885," *Indian Economic and Social History Review* 7 (1970): 25–60.

Clarke, John G. *La Rochelle and the Atlantic Economy during the Eighteenth Century.* Baltimore: Johns Hopkins University Press, 1981.

Cohn, Bernard. *An Anthropologist among the Historians and Other Essays.* Delhi: Oxford University Press, 1990.

Colonialism and Its Forms of Knowledge: The British in India. Princeton, N.J.: Princeton University Press, 1996.

Collins, Harry. "The TEA Set: Tacit Knowledge and Scientific Networks," *Science Studies* 4 No. 2 (April 1974): 165–86.

Changing Order: Replication and Induction in Scientific Practice. Beverly Hills, Calif.: Sage, 1991.

Coke, Thomas. *A History of the West Indies.* London: Frank Cass, 1971.

Connell, Raewyn. *Southern Theory.* Cambridge: Polity Press, 2007.

Connelly, Matthew. *Fatal Misconception: The Struggle to Control World Population.* Cambridge, Mass.: Belknap, 2008.

Coon, David L. "Eliza Lucas Pinckney and the Reintroduction of Indigo Culture in South Carolina," *Journal of Southern History* 42 No. 1 (February 1976): 61–76.

Costa, Albert B. *Michel Eugene Chevreul: Pioneer of Organic Chemistry.* Madison: State Historical Society of Wisconsin/Department of History, University of Wisconsin, 1962.

Courtenay, P. P. *Plantation Agriculture.* New York: Praeger, 1969.

Coventry, Bernard. "Rhea Experiments in India," *Agricultural Journal of India* 2 (1906): 1–14.

Cox, Jeffrey. *Imperial Fault Lines: Christianity and Colonial Power in India.* Stanford, Calif.: Stanford University Press, 2002.

Cronon, William. *Nature's Metropolis: Chicago and the Great West.* New York: W. W. Norton, 1991.

Daston, Lorraine. "Introduction: The Coming into Being of Scientific Objects," in Lorraine Daston (ed.), *Biographies of Scientific Objects.* Chicago: University of Chicago Press, 2000, pp. 1–14.

Dionne, Russell Jude. "Government Directed Agricultural Innovation in India: The British Experience." Unpublished Ph.D. dissertation, Department of History, Duke University, 1973.

Drayton, Richard. *Nature's Government: Science, Imperial Britain, and the "Improvement" of the World.* New Haven, Conn., and London: Yale University Press, 2000.

Duara, Prasenjit. *Rescuing History from the Nation: Questioning Narratives of Modern China.* Chicago and London: University of Chicago Press, 1995.

Dunn, Richard S. *Sugar and Slaves: The Rise of the Planter Class in the English West Indies, 1624–1713.* New York: W. W. Norton, 1972.

Edney, Matthew. *Mapping an Empire: The Geographical Construction of British India, 1765–1843.* Chicago and London: University of Chicago Press, 1997.

Elshakry, Marwa S. "Knowledge in Motion: The Cultural Politics of Modern Science, Translations in Arabic," *Isis* 99 (2008):701–30.

"When Science Became Western: Historiographical Reflections," *Isis* 101 (2010): 98–109.

Farrar, W. V. "Edward Schunck FRS: A Pioneer of Natural Products Chemistry," *Notes and Records of the Royal Society of London* 31 No. 2 (January 1977): 273–96.

Fergusson, Niall. *Empire: The Rise and Demise of the British World Order and the Lessons for Global Power.* New York: Basic Books, 2002.

Filgate, T. R. "The Bihar Planters' Association, Ltd.," in Arnold Wright (ed.), compiled by Somerset Playne, *Bengal and Assam Behar and Orissa: Their History, People, Commerce, and Industrial Resources.* London: Foreign and Colonial Compiling and Publishing Company, 1917, pp. 268–351.

Finlay, Mark. "The Rehabilitation of an Agricultural Chemist: Justus von Liebig and the Seventh Edition," *Ambix* 38 (1991): 155–67.

Fischer, Colin M. "Indigo Plantations and Agrarian Society in North Bihar in the Nineteenth and Early Twentieth Centuries," Unpublished Ph.D. dissertation, University of Cambridge, 1976.

Fisher, G. J. "Correspondence and Selections, Further Particulars Regarding the Nerium Indigo," Extract of a Letter from G. J. Fischer, dated Salem, 8th January 1845, to Dr. Robert Wight, of Coimbatore, *Journal of the Agricultural and Horticultural Society of India* IV Part 1 (January–December 1845): 129–31.

Fisher, Michael H. (&Sake Deen Mahomet) *The First Indian Author in English: Dean Mahomed (1759–1851) in India, Ireland, and England.* Delhi: Oxford University Press, 1996.

Fitzgerald, Deborah. *The Business of Breeding: Hybrid Corn in Illinois, 1890–1940*. Ithaca, N.Y., and London: Cornell University Press, 1990.

Floyd, Troy S. "Salvadorean Indigo and the Guatemalan Merchants: A Study in Central American Socio-Economic History, 1750–1800," Unpublished Ph.D. dissertation, University of California, Berkeley, 1959.

Foucault, Michel. *The Archaeology of Knowledge and the Discourse on Language*. Translated by A. M. Sheridan Smith, New York: Pantheon Books, 1972, French edition, 1969.

Frank, Andre Gunder. "India in the World Economy, 1400–1750," *Economic and Political Weekly*, July 27 1996, pp. 50–64.

Friedel, Robert. *Pioneer Plastic: The Making and Selling of Celluloid*. Madison: University of Wisconsin Press, 1983.

A Culture of Improvement: Technology and the Western Millennium. Cambridge, Mass.: MIT Press, 2007.

Furber, Holden. *John Company at Work: A Study of European Expansion in India in the Late Eighteenth Century*. Cambridge, Mass.: Harvard University Press, 1951.

Galison, Peter. *How Experiments End*. Chicago: University of Chicago Press, 1987.

Garfield, Simon. *Mauve: How One Man Invented a Color That Changed the World*. New York: W. W. Norton, 2001.

Garrigus, John. "Blue and Brown: Contraband Indigo and the Rise of a Free Colored Planter Class in French Saint-Domingue," *Americas* L2 (October 1993): 233–63.

Before Haiti: Race and Citizenship in French Saint Domingue. New York: Palgrave, 2006.

Gentleman's Magazine (May 1755): 256–8.

Gerber, Frederick H. *Indigo and the Antiquity of Dyeing*. Ormond Beach, Fla.: Gerber, 1977.

Goddard, Nicholas. *Harvests of Change: The Royal Agricultural Society of England, 1838–1988*. London: Quiller Press, 1988.

Golinski, Jan. *Making Natural Knowledge: Constructivism and the History of Science*. New York: Cambridge University Press, 1998.

Grove, Richard. "The East India Company, the Australians and the El Nino: Colonial Scientists and Ideas about Global Climatic Change and Teleconnections between 1770 and 1930," in Richard Grove, *Ecology, Climate and Empire: Colonialism and Global Environmental History, 1400–1940*. Cambridge: White Horse Press, 1997, pp. 124–46.

Guha, Ranajit. *A Rule of Property for Bengal: An Essay on the Idea of Permanent Settlement*. Durham, N.C., and London: Duke University Press, 1996.

Dominance Without Hegemony: History and Power in Colonial India. Cambridge, Mass.: Harvard University Press, 1998.

Elementary Aspects of Peasant Insurgency in Colonial India. Durham, N.C., and London: Duke University Press, 1999.

Haber, Ludwig F. *Poisonous Cloud: Chemical Warfare in the First World War*. Oxford and New York: Oxford University Press, 1986.

Habib, Irfan. *The Agrarian System of Mughal India, 1556–1707.* Bombay: Asia Publishing House, 1963.

Haraway, Donna. *Simians, Cyborgs and Women.* New York: Routledge, 1991. *Modest Witness @ Second Millennium.* New York: Routledge, 1997.

Headrick, Daniel R. *The Tools of Empire: Technology and European Imperialism in the Nineteenth Century.* New York: Oxford University Press, 1981.

The Tentacles of Progress: Technology Transfer in the Age of Imperialism, 1850–1940. New York: Oxford University Press, 1988.

"Botany, Chemistry, and Tropical Development." *Journal of World History* 7 No. 1 (1996): 1–20.

Technology: A World History. New York: Oxford University Press, 2009.

Power over Peoples: Technology, Environments, and Western Imperialism, 1400 to the Present. Princeton, N.J.: Princeton University Press, 2010.

Hearn, Lafcadio. *Two Years in the French West Indies.* New York: Harper and Brothers, 1890.

Homburg, Ernst. "The Role of Demand on the Emergence of the Dye Industry: The Roles of Chemists and Colourists," *Journal of the Society of Dyers and Colourists* 99 (November 1983): 325–32.

Hopkins, A. G. (ed.). *Global History: Interactions Between the Universal and the Local.* London: Palgrave Macmillan, 2006.

Howard, Albert and Gabrielle L. C. Howard. *Wheat in India: Its Production, Varieties and Improvement.* Calcutta: Thacker, Spink & Co., 1909.

Inden, Ronald. *Imagining India.* Bloomington: Indiana University Press, 2001, first published in 1990.

Jelatis, Virginia Gail. "Tangled Up in Blue: Indigo Culture and Economy in South Carolina, 1747–1800," Unpublished Ph.D. dissertation, University of Minnesota, 1999.

Johnson, Jeffrey. *The Kaiser's Chemists: Science and Modernization in Imperial Germany.* Chapel Hill: University of North Carolina Press, 1990.

Johnson, Jeffrey and Roy MacLeod. "The War the Victors Lost: The Dilemmas of Chemical Disarmament, 1919–1926," in Roy MacLeod and Jeffrey Johnson (eds.), *Frontline and Factory: Comparative Perspectives on the Chemical Industry at War, 1914–1924,* Dordrecht, the Netherlands: Springer, 2007, pp. 221–45.

Jones, Stephanie. *Merchants of the Raj: British Managing Agency Houses in Calcutta: Yesterday and Today.* Basingstoke: Macmillan, 1992.

Kaiwar, Vasant and Sucheta Mazumdar (eds.). *Antinomies of Modernity: Essays on Race, Orient, Nation.* Durham, N.C., and London: Duke University Press, 2003.

Kaminski, Arnold P. *The India Office, 1880–1910.* Westport, Conn.: Greenwood Press, 1986.

Karl, Rebecca E. *Staging the World: Chinese Nationalism at the Turn of the Twentieth Century.* Durham, N.C., and London: Duke University Press, 2002.

Kerr, Ian J. "Colonialism and Technological Choice: The Case of the Railways in India," *Itinerario* XIX 2 (1995): 91–111.

Khan, Iqtidar Alam. "Pre-Modern Indigo Vats of Bayana," *Journal of Islamic Environmental Design Research Center* (1989): 92–8.

Klein, Herbert S. *African Slavery in Latin America and the Caribbean*. New York: Oxford University Press, 1990.

Kling, Blair B. *The Blue Mutiny: The Indigo Disturbances in Bengal, 1859–1862*. Philadelphia: University of Pennsylvania Press, 1966.

Partners in Empire: Dwarkanath Tagore and the Age of Enterprise in Eastern India. Berkeley: University of California Press, 1976.

Knecht, Edmund, Christopher Rawson, and Richard Loewenthal. *A Manual of Dyeing: For the Use of Practical Dyers, Manufacturers, Students, and All Interested in the Art of Dyeing*, Vol 11, London: Charles Griffin and Company, 1893.

Kraft, Alison. "Pragmatism, Patronage and Politics in English Biology: The Rise and Fall of Economic Biology, 1904–1920," *Journal of the History of Biology* 37 No. 2 (2004): 213–58.

Kuhn, Thomas S. *The Structure of Scientific Revolution*. Chicago: University of Chicago Press, 1996, first published in 1962.

Kumar, Deepak. *Science and the Raj, 1857–1905*. Delhi: Oxford University Press, 1995.

"Science in Agriculture: A Study in Victorian India," *Asian Agri-History* 1 No. 2 (1997): 87–92

Kumar, Prakash. "Scientific Experiments in British India: Indigo Planters, Scientists, and the State, 1890–1930," *Indian Economic and Social History Review* 38 No. 3 (June–September, 2001): 249–70.

"Plantation Science: Improving Natural Indigo in Colonial India, 1860–1913," *British Journal for the History of Science* 40 No. 4 (December, 2007): 537–65.

"Transnational Knowledge and Colonial Indigo Plantations in South Asia," (In press, *Modern Asian Studies*).

Lake, Marilyn and Henry Reynolds. *Drawing the Global Colour Line: White Men's Countries and the International Challenge of Racial Equality*. Cambridge: Cambridge University Press, 2008.

Lakwete, Angela. *Inventing the Cotton Gin: Machine and Myth in Antebellum America*. Baltimore and London: Johns Hopkins University Press, 2005.

Lambert, David and Alan Lester (eds.). *Colonial Lives Across the British Empire: Imperial Careering in the Long Nineteenth Century*. Cambridge: Cambridge University Press, 2006.

Lane, Kris. *Pillaging the Empire: Piracy in the Americas, 1500–1750*. Armonk, N.Y.: M. E. Sharpe, 1998.

Latour, Bruno. "Give Me a Laboratory and I Will Raise the World," in Karin Knorr-Cetina and Michael Mulkay (eds.), *Science Observed: Perspectives on the Social Study of Science* Thousand Oaks, Calif.: Sage, 1983.

The Pasteurization of France (translated by Alan Sheridan and John Law), Cambridge, Mass.: Harvard University Press, 1998.

Latour, Bruno and Steve Woolgar. *Laboratory Life: The Construction of Scientific Facts*. Princeton, N.J.: Princeton University Press, 1986.

Lawes, J. B. "On Agricultural Chemistry," *Journal of the Royal Agricultural Society of England* 8 (1847): 226–60.

Levinstein, Herbert. "The Future of the Indigo Industry, with a Description of the Manufacture of Indigo from Naphthalene," *Journal of the Society of Dyers and Colourists* (June 1901): 138–42.

Liebig, Justus von. *Chemistry in Its Applications to Agriculture and Physiology*, Cambridge: John Owen, 1842, 3rd American ed.

Llewellyn-Jones, Rosie. *A Very Ingenious Man: Claude Martin in Early Colonial India*. Delhi: Oxford University Press, 1999.

A Man of the Enlightenment in 18th Century India: The Letters of Claude Martin 1766–1800. New Delhi: Permanent Black, 2003.

Lourdusamy, J. *Science and National Consciousness in Bengal*. Hyderabad: Orient Longman, 2004.

Ludden, David. *The New Cambridge History of India:An Agrarian History of South Asia*. Cambridge: Cambridge University Press, 1999.

Maat, Harro. *Science Cultivating Practice: A History of Agricultural Science in the Netherlands and Its Colonies, 1863–1986*. Boston: Kluwer Academic Publisher, 2001.

Martin, Claude. "On the Manufacture of Indigo at Ambore by Lieutenant Colonel Claude Martin. (with) an Extract of a Treatise on the Manufacture of Indigo, by Mr. De Cossigny," London: Asiatic Researches, 1807, pp. 475–77, according to an old note of April 1791.

MacKenzie, Donald and Graham Spinardi. "Tacit Knowledge, Weapons Design and the Uninvention of Nuclear Weapons," *American Journal of Sociology* 101 No. 1 (July 1995): 44–99.

MacLeod, Murdo. *Spanish Central America: A Socioeconomic History, 1520–1720*. Berkeley: University of California Press, 1973.

MacLeod, Roy. "Scientific Advice for British India: Imperial Perceptions and Administrative Goals, 1898–1923," *Modern Asian Studies* 9 No. 3 (1975): 343–84.

Macpherson, David. *The History of the European Commerce with India*. London: Longman et al., 1812.

Manela, Erez. *The Wilsonian Moment: Self-Determination and the International Origins of Anticolonial Nationalism*. Oxford and New York: Oxford University Press, 2007.

Manning, Patrick. *The African Diaspora: A History Through Culture*. New York: Columbia University Press, 2009.

Markovits, Claude, Jacques Pouchepadass, and Sanjay Subrahmanyam (eds.) *Society and Circulation: Mobile Peoples and Itinerant Cultures in South Asia, 1750–1950*. Delhi: Permanent Black, 2003.

Mazumdar, Sucheta. *Sugar and Society in China: Peasants, Technology, and the World Market*. Cambridge, Mass.: Harvard University Asia Center, 1998.

McClellan, James. *Colonialism and Science: Saint Domingue in the Old Regime*. Baltimore: Johns Hopkins University Press, 1992.

McCook, Stuart. *States of Nature: Science, Agriculture, and Environment in the Spanish Caribbean, 1760–1940*. Austin: University of Texas Press, 2002.

Meldola, Raphael. "The Synthesis of Indigo," *Journal of the Society of Arts* (April 19, 1901): 397–412.

Merton, Robert K. *The Sociology of Science: Theoretical and Empirical Investigations*. Edited by Norman W. Storer, Chicago and London: University of Chicago Press, 1973.

Metcalf, Thomas. *The New Cambridge History of India: Ideologies of the Raj*. Cambridge: Cambridge University Press, 1995.

Mintz, Sidney W. *Sweetness and Power: The Place of Sugar in Modern History*. New York: Penguin Books, 1986.

Molina, Jose Antonio Fernandez. "Colouring the World in Blue: The Indigo Boom and the Central American Market, 1750–1810," Ph.D. Thesis, University of Texas at Austin, 1992.

Moon Suzanne. *Technology and Ethical Idealism: A History of Development in the Netherlands East Indies*. Leiden: CNWS Publications, 2007.

Moreland, W. H. *India at the Death of Akbar, an Economic Study*. Delhi: Atma Ram and Sons, 1962.

Mukherjee, Rila. "Calcutta in the Eighteenth Century: Vignettes from Contemporary French and Scottish Travel Accounts," *Bengal Past and Present* 110 No. 210–11 (1991): 75–91.

Nadri, Ghulam. "Indigo Industry and Trade in Gujarat in the Seventeenth Century," Unpublished M.Phil. dissertation, Aligarh Muslim University, 1996.

  Eighteenth-Century Gujarat: The Dynamics of Its Political Economy, 1750–1800. Brill: Leiden and Boston, 2009.

Nandy, Ashis. *Alternative Sciences: Creativity and Authenticity in Two Indian Scientists*. Delhi: Oxford University Press, 1995.

Naoroji, Dadabhai. "Adjourned Meeting of the Bombay Branch of the East India Association, for Discussion of the Papers on 'the Poverty of India' Read by Mr. Dadabhai Naoroji," *Journal of the East India Association*, Vol. 10, pp. 83–96; 133–60.

Nieto-Galan, Augusti. *Colouring Textiles: A History of Natural Dyestuffs in Industrial Europe*. Norwell: Kluwer Academic Publishers, 2001.

Nordstrom, Carolyn. *Global Outlaws: Crime, Money, and Power in the Contemporary World*. Berkeley and Los Angeles: University of California Press, 2007.

Ogborn, Miles. *Global Lives: Britain and the World, 1550–1800*. New York: Cambridge University Press, 1998.

Palit, Chittabrata. *Tensions in Bengal Rural Society: Landlords, Planters, and Colonial Rule*. Hyderabad: Orient Longman, 1998, first published in 1975.

Palladino, Paolo and Michael Worboys. "Science and Imperialism," *Isis* 84 No. 1 (March 1993): 91–102.

Pande, Ishita. *Medicine, Race and Liberalism in British Bengal: Symptoms of Empire*. London and New York: Routledge, 2010.

Pares, Richard. *War and Trade in the West Indies, 1739–63*. London: Routledge, 1963.

Parry, Benita. "Problems in Current Theories of Colonial Discourse," *Oxford Literary Review* 9 Issue 1–2 (1987): 27–58.

Pati, Biswamoy and Mark Harrison. *Health, Medicine and Empire: Perspectives on Colonial India*. New Delhi: Orient Longman, 2001.

Perkin, A. G. "William Popplewell Bloxam," *Journal of the Chemical Society* 105 (1914): 1195–1200.

Perkin, A. G. and A. E. Everest. *The Natural Organic Colouring Matters.* New York: Longmans Green and Company, 1918.

Perkin, F. M. "The Present Condition of the Indigo Industry," *Nature* (November 1, 1900): 7–9.

"The Present Condition of the Indigo Industry," *Nature* (January 24, 1901): 302–3.

"Indigo and Sugar," *Nature* (May 2, 1901): 10–11.

Philip, Kavita. *Civilizing Natures: Race, Resources, and Modernity in Colonial South India.* New Brunswick, N.J.: Rutgers University Press, 2004.

Piddington, Henry. "On the Manufacture of Indigo. By Henry Piddington, Read 10th June, 1829" *Transactions of the Agricultural and Horticultural Society of India* II (1838): 30.

Pinch, Trevor J. and Wiebe E. Bijker. "The Social Construction of Facts and Artifacts: Or How the Sociology of Science and the Sociology of Technology Might Benefit Each Other," in Wiebe E. Bijker, Thomas P. Hughes, and Trevor J. Pinch (eds.), *The Social Construction of Technological Systems: New Directions in the Sociology and History of Technology.* Cambridge: MIT Press, 1995.

Polanyi, Michael. *Personal Knowledge.* London: Routledge & Kegan Paul, 1958.

Pouchepadass, Jacques. *Champaran and Gandhi: Planters, Peasants, and Gandhian Politics.* Delhi: Oxford University Press, 1999.

Land, Power and Market: A Bihar District under Colonial Rule 1860–1947. New Delhi: Sage, 2000.

Prakash, Gyan. "Science Gone 'Native' in Colonial India," *Representations* No. 40 (Autumn 1992): 154–78.

Another Reason: Science and the Imagination of Modern India. Princeton, N.J.: Princeton University Press, 1999.

Prakash, Om. *New Cambridge History of India: European Commercial Enterprise in Pre-Colonial India.* New York: Cambridge University Press, 1998.

Pratt, Mary Louise. *Imperial Eyes: Travel Writing and Transculturation.* London and New York: Routledge, 2008, 2nd edition.

Raj, Kapil. *Relocating Modern Science: Circulation and the Construction of Knowledge in South Asia and Europe, 1650–1900.* Basingstoke: Palgrave, 2007.

Rajan, Ravi S. *Modernizing Nature: Forestry and Imperial Eco-Development, 1800–1950.* Oxford and New York: Oxford University Press, 2006.

Ramanna, Mridula. *Western Medicine and Public Health in Colonial Bombay 1845–1895.* Chennai: Orient Longman India, 2002.

Rao, Amiya and B. G. Rao. *The Blue Devil: Indigo and Colonial Bengal.* Delhi: Oxford University Press, 1992.

Ray, Indrajit. *Bengal Industries and the British Industrial Revolution (1757–1857).* London and New York: Routledge, 2010.

Raychaudhuri, Tapan. *Jan Company in Coromandel, 1605–1690.* The Hague: Martinus Nijhoff, 1962.

Reed, Peter. "The British Chemical Industry," *British Journal for the History of Science* 25 (1992): 113–25.

Rehling, H. "Result of Trials Given to Various Seeds at Chandamaree Factory, Rungpore, Communicated by H. Rehling," *Journal of the Agricultural and Horticultural Society of India* IV Part 1 (January–December 1845): 27–30.

Reid, D. J. "Ten Years Practical Experience of Java Indigo in Bihar," *Agricultural Journal of India* XII (1917): 1–26.

Reinhardt, Carsten and Anthony Travis. *Heinrich Caro and the Creation of Modern Chemical Industry.* Dordrecht: Kluwer, 2000.

Richards, John F. "Early Modern India and World History," *Journal of World History* 8 No. 2 (1997): 197–209.

Robb, Peter. "British Rule and Indian Improvement," *Economic History Review* 34 No. 4 (November 1981): 507–23.

   Rural India: Land, Power and Society under British Rule. London: Curzon Press, 1983.

   "Bihar, the Colonial State and Agricultural Development in India, 1880–1920," *Indian Economic and Social History Review* 25 No. 2 (1988): 205–35.

Rocke, Alan. *The Quiet Revolution: Hermann Kolbe and the Science of Organic Chemistry.* Berkeley: University of California Press, 1993.

Rosenberg, Nathan. *Inside the Black Box: Technology and Economics.* New York: Cambridge University Press, 1982.

Rossiter, Margaret W. *The Emergence of Agricultural Science: Justus Liebig and the Americans, 1840–1880.* New Haven, Conn.: Yale University Press, 1975.

Roy, Tirthankar. *Traditional Industry in the Economy of Colonial India.* New York: Cambridge University Press, 1999.

Russell, E. J. *A History of Agricultural Science in Great Britain, 1620–1954.* London: George Allen & Unwin, 1946.

Said, Edward. *Culture and Imperialism.* New York: Alfred A. Knopf, 1993.

Sandberg, Gosta. *Indigo Textiles: Technique and History.* Asheville, N.C.: Lark Books, 1989.

Sarkar, Sumit. *The Swadeshi Movement in Bengal, 1903–1908.* New Delhi: People's Publishing House, 1994.

   "Orientalism Revisited: Saidian Frameworks in the Writings of Modern Indian History," *Oxford Literary Review* 16 (1994): 205–24.

   Writing Social History. New York and Delhi: Oxford University Press, 1997.

Sassen, Saskia. "Globalization or Denationalization," *Review of International Political Economy* 10 No. 1 (2003): 1–22.

   Territory, Authority, Rights: From Medieval to Global Assemblages. Princeton, N.J., and Oxford: Princeton University Press, 2006.

   "Introduction: Deciphering the Global," in Saskia Sassen (ed.), *Deciphering the Global: Its Scales, Spaces, and Subjects.* New York: Routledge, 2007.

   Sociology of Globalization. New York: W. W. Norton, 2007.

Schendel, Willem van. *The Bengal Borderland: Beyond State and Nation in South Asia.* London: Anthem Press, 2005.

Schiebinger, Londa. *Plants and Empire: Colonial Bioprospecting in the Atlantic World,* Cambridge, Mass.: Harvard University Press, 2004.

Scott, James C. *The Moral Economy of the Peasant: Rebellion and Resistance in Southeast Asia*. New Haven, Conn.: Yale University Press, 1977.

The Art of Not Being Governed: An Anarchist History of Upland Southeast Asia. New Haven, Conn.: Yale University Press, 2010.

Shapin, Steven and Simon Schaffer. *Leviathan and the Air Pump: Hobbes, Boyle, and the Experimental Life*. Princeton, N.J.: Princeton University Press, 1985.

A Social History of Truth: Civility and Science in Seventeenth Century England. Chicago: University of Chicago Press, 1994.

Sharma, Jayeeta. *Empire's Garden: Assam and the Making of India*. Durham, N.C., and London: Duke University Press, 2011.

Shukla, Prabhat Kumar. *Indigo and the Raj: Peasant Protests in Bihar, 1780–1917*. Delhi: Pragati Publication, 1993.

Singh, S. B. *European Agency Houses in Bengal, 1783–1833*. Calcutta: Firma K. L. Mukhopadhyay, 1966.

Sinha, Mrinalini. *Specters of Mother India: The Global Restructuring of an Empire*. London and Durham, N.C.: Duke University Press, 2006.

Sinha, N. K. *Economic History of Bengal from Plassey to the Permanent Settlement, Vol. I*. Calcutta: K. L. Mukhopadhyay, 1956.

Sivaramakrishnan, K. *Modern Forests: Statemaking and Environmental Change in Colonial Eastern India*. Stanford, Calif.: Stanford University Press, 1999.

Sivasundaram, Sujit. "Sciences and the Global: On Methods, Questions, and Theory," *Isis* 101 (2010):146–58.

Stoler, Ann L. "Rethinking Colonial Categories: European Communities and the Boundaries of Rule," *Comparative Studies in Society and History* 31 (1989): 134–61.

Stoler Ann L. and Frederick Cooper. "Between Metropole and Colony: Rethinking a Research Agenda," in Frederick Cooper and Ann L. Stoler (eds.), *Tensions of Empire: Colonial Cultures in a Bourgeois World*. Berkeley: University of California Press, 1997, pp. 1–56.

Stone, Ian. *Canal Irrigation in British India: Perspectives on Technological Change in a Peasant Economy*. Cambridge: Cambridge University Press, 2002, first published in 1984.

Subrahmanyam, Sanjay. "Connected Histories: Notes Towards a Reconfiguration of Early Modern Eurasia," *Modern Asian Studies* 31 No. 3 (1997): 735–62.

Sur Abha. *Dispersed Radiance: Caste, Gender, and Modern Science in India*. New Delhi: Navyana, 2011.

Taylor, C. B. "Further Particulars Regarding the Manufacture of Indigo from *Nerium tinctorum*, Communicated by C. B. Taylor," *Journal of the Agricultural and Horticultural Society of India* V Part 1 (January–December 1846): 77–8.

Tilley, Helen. *Africa as a Living Laboratory: Empire, Development, and the Problem of Scientific Knowledge, 1870–1950*. Chicago and London: University of Chicago Press, 2011.

Tivedi, K. K. "Innovation and Change in Indigo Production in Bayana, Eastern Rajasthan," *Studies in History* 10 No. 1 n.s. (1994): 53–79.

Travis, Anthony S. *The Rainbow Makers: The Origins of the Synthetic Dyestuffs Industry in Western Europe*. Bethlehem, Pa.: Lehigh University Press, 1993.

"Heinrich Caro and Ivan Levinstein: Uniting the Colours of Ludwigshafen and Lancashire," in Ernst Homburg, Anthony S. Travis, and Harm G. Schröter (eds.), *The Chemical Industry in Europe, 1850–1914: Industrial Growth, Pollution, and Professionalization*. Boston: Kluwer Academic, 1998.

Travis, Anthony S. and Harm G. Schröter (eds.). *The Chemical Industry in Europe, 1850–1914: Industrial Growth, Pollution, and Professionalization*. Boston: Kluwer Academic, 1998.

Tripathi, Amales. *Trade and Finance in the Bengal Presidency, 1793–1833*. Calcutta: Oxford University Press, 1976, 2nd edition.

Tsing, Anna L. *In the Realm of the Diamond Queen*. Princeton, N.J.: Princeton University Press, 1993.

Friction: An Ethnography of Global Connection. Princeton, N.J., and Oxford: Princeton University Press, 2005.

Vernon, Keith. "Science for the Farmer? Agricultural Research in England, 1909–23," *Twentieth Century British History* 8 No. 3 (1997): 310–33.

Vileisis, Ann. "Are Tomatoes Natural?" in Martin Reuss and Stephen H. Cutcliffe (eds.), *The Illusory Boundary: Environment and Technology in History*. Charlottesville: University of Virginia Press, 2010, pp. 211–48.

Worboys, Michael. "Science and Imperialism in the Development of the Colonial Empire, 1895–1940," Unpublished D. Phil., University of Sussex, 1979.

Worster, Donald. *Nature's Economy: A History of Ecological Ideas*. Cambridge: Cambridge University Press, 1977.

Yafa, Stephen. *Cotton: The Biography of a Revolutionary Fiber*. New York: Penguin Books, 2005.

Yang, Anand A. *Bazaar India: Markets, Society, and the Colonial State in Gangetic Bihar*. Berkeley: University of California Press, 1998.

Index

Printed in the United States
By Bookmasters